THE PROPERTIES AND APPLICATIONS OF ZEOLITES

Special Publication No 33

The Properties and Applications of Zeolites

The Proceedings of a Conference organised jointly by
the Inorganic Chemicals Group of The Chemical Society
and The Society of Chemical Industry

The City University, London, April 18th—20th, 1979

Edited by
R. P. Townsend, The City University, London

The Chemical Society
Burlington House, London, W1V 0BN

British Library Cataloguing in Publication Data
The Properties and applications of zeolites.—
  (Chemical Society. Special Publications; No. 33 ISSN 0557-618X)
  1. Zeolites – Congresses
  I. Townsend, R. P. II. Chemical Society. *Inorganic Chemicals Group*
  III. Society of Chemical Industry
  549'.68 QE 391.Z5

ISBN 0-85186-670-0

Printed in Great Britain by
Whitstable Litho Ltd., Whitstable, Kent

FOREWORD

The papers included in this Special Publication of The Chemical Society were presented and discussed at a Symposium at City University from 18th to 20th April 1979. The meeting was held under the joint auspices of the Society's Inorganic Chemicals Group, a subject group of its Industrial Division, and the Colloid and Surface Chemistry Group of the Society of Chemical Industry.

In planning the Symposium the aims of the Organising Committee were to bring together zeolite chemists from all parts of the world, both by asking well-known authorities to provide the plenary lectures and by inviting contributions from research workers active in the field. The achieved result was an informative meeting and lively discussion, and this publication which is intended to provide an up-to-date account of developments rather than a comprehensive review of zeolite chemistry.

Particular thanks are due to Dr. R.P. Townsend (City University) for the editorial effort necessary in compiling this publication, and to Mr. K.J. Matterson (Honorary Secretary and Treasurer, Inorganic Chemicals Group) through whose organising activities much of the success of the Symposium itself was due. Attendance by more than two hundred participants, from over twenty countries, led to a small financial surplus which the Organising Committee agreed shall be donated for the purpose of encouraging young research workers in the zeolites field.

K.S.W. Sing,   R. Thompson,
Colloid and Surface Chemistry Group,   Inorganic Chemicals Group,
Society of Chemical Industry.   The Chemical Society.

Organising Committee

Professor K.S.W. Sing (Chairman), K.J. Matterson (Secretary), S.A. Mitchell, Dr. L.V.C. Rees, C.V. Roberts, Dr. R.P. Townsend, Dr. T.V. Whittam, Dr. A.L. Smith (Chairman of Colloid and Surface Chemistry Group SCI), Professor R. Thompson (Chairman of Inorganic Chemicals Group CS), A.G. Cubitt (Editor, CS Special Publications), H.L. Bennister (Industrial Division, CS).

CONTRIBUTORS

R. AIELLO. University of Calabria, Italy

J.W. ARMOND. BOC TechSep, London, U.K.

R.M. BARRER. Imperial College, London, U.K.

W.D. BASLER. University of Hamburg, West Germany

A. BEZUS. Heyrovský-Institute of Physical Chemistry and
  Electrochemistry, Prague, Czechoslovakia

F.O. BRAVO. UMIST, Manchester, U.K.

D.W. BRECK. Union Carbide Corporation, Tarrytown, New York, U.S.A.

M. BÜLOW. Central Institute for Physical Chemistry, Berlin,
  East Germany

J.W. CARTER. University of Birmingham, U.K.

C. COLELLA. University of Calabria, Italy

A. CREMERS. Catholic University of Leuven, Belgium

W.O. DALY. University of Bradford, U.K.

J. DWYER. UMIST, Manchester, U.K.

A. DYER. University of Salford, U.K.

P.C. ELMES. Llandough Hospital, Penarth, U.K.

H. ENAMY. University of Salford, U.K.

P. FLETCHER. The City University, London, U.K.

I.M. GALABOVA. Higher Chemical Technological Institute, Sofia,
  Bulgaria

W.H. GRANVILLE. University of Bradford, U.K.

G.A. HARALAMPIEV. Higher Chemical Technological Institute,
  Sofia, Bulgaria

R. HAUL. University of Hanover, West Germany

W. HEINTZ. University of Hanover, West Germany

R.V. HERCIGONJA. University of Belgrade, Yugoslavia

vi

H. KACIREK. University of Hamburg, West Germany

J. KÄRGER. Karl Marx University, Leipzig, East Germany

J.R. KIOVSKY. Norton Company, Akron, Ohio, U.S.A.

M. KOČIŘÍK. Heyrovský-Institute of Physical Chemistry and
Electrochemistry, Prague, Czechoslovakia

G.T. KOKOTAILO. Mobil Research and Development Corporation,
Paulsboro, New Jersey, U.S.A.

P.B. KORADIA. Norton Company, Akron, Ohio, U.S.A.

H. LECHERT. University of Hamburg, West Germany

C.T. LIM. Norton Company, Akron, Ohio, U.S.A.

U. LOHSE. Central Institute for Physical Chemistry, Berlin,
East Germany

P.R. LOWER. The City University, London, U.K.

F. MAHONEY. University of Nottingham, U.K.

A. MAES. Catholic University of Leuven, Belgium

W.M. MEIER. ETH, Zurich, Switzerland

A. NASTRO. University of Calabria, Italy

H. PFEIFER. Karl Marx University, Leipzig, East Germany

B.B. RADAK. Boris Kidrič Institute of Nuclear Sciences,
Belgrade, Yugoslavia

V.M. RADAK. Boris Kidrič Institute of Nuclear Sciences,
Belgrade, Yugoslavia

L.V.C. REES. Imperial College, London, U.K.

C.W. ROBERTS, Laporte Industries Ltd., Widnes, U.K.

R. RUDHAM. University of Nottingham, U.K.

D.M. RUTHVEN. University of New Brunswick, Canada

W. SCHIRMER. Central Institute for Physical Chemistry, Berlin,
East Germany

W. SCHWEITZER. University of Hamburg, West Germany

M.J. SCHWUGER. Henkel KGaA, Dusseldorf, West Germany

K.S.W. SING. Brunel University, Uxbridge, U.K.

H.G. SMOLKA. Henkel KGaA, Dusseldorf, West Germany

M.S. SPENCER. ICI Ltd., Billingham, U.K.

H. STACH. Central Institute for Physical Chemistry, Berlin,
    East Germany

A. STOCKWELL. University of Nottingham, U.K.

H. STREMMING. University of Hanover, West Germany

H. THAMM. Central Institute for Physical Chemistry, Berlin
    East Germany

R. THOMPSON. Borax Holdings Ltd., Chessington, U.K.

R.P. TOWNSEND. The City University, London, U.K.

D.E.W. VAUGHAN. W.R. Grace & Co., Columbia, Maryland, U.S.A.

J. VERLINDEN. Catholic University of Leuven, Belgium

A.A. VLČEK. Heyrovský-Institute of Physical Chemistry and
    Electrochemistry, Prague, Czechoslovakia

T.V. WHITTAM. ICI Ltd., Billingham, U.K.

K.-P. WITTERN. University of Hamburg, West Germany

D. ZAMBOULIS. UMIST, Manchester, U.K.

S.P. ZHDANOV. Institute of Silicate Chemistry, Leningrad,
    U.S.S.R.

A. ZIKÁNOVÁ. Heyrovský-Institute of Physical Chemistry and
    Electrochemistry, Prague, Czechoslovakia

CONTENTS

Introduction   1

Session 1: Diffusion Processes

P1. Plenary Lecture: Sorption Kinetics and
     Diffusivities in Porous Crystals   3
     By R.M. Barrer

P2. NMR and Sorption Kinetic Studies of Diffusion
     in Porous Silica   27
     By R. Haul, W. Heintz, and H. Stremming

P3. Diffusion in Zeolites: A Review of Recent
     Progress   43
     By D.M. Ruthven

P4. Diffusion of C_6 Hydrocarbons in NaX and MgA
     Zeolites   58
     By A. Zikánová, M. Kočiřík, A. Bezus, A.A. Vlček,
     M. Bülow, W. Schirmer, J. Kärger, H. Pfeifer, and
     S.P. Zhdanov

     Discussion on Session 1   71

Session 2: Separation Processes

P5. Plenary Lecture: The Adsorption Separation
     Process   76
     By J.W. Carter

P6. The Practical Application of Pressure Swing
     Adsorption to Air and Gas Separation   92
     By J.W. Armond

P7. Molecular Sieves for Industrial Separation
     and Adsorption Applications   103
     By C.W. Roberts

P8. Oxygen Enrichment of Air on Alkaline Forms of
     Clinoptilolite   121
     By I.M. Galabova and G.A. Haralampiev

ix

Discussion on Session 2   130

Session 3: Sorption Processes

P9A. Pentasil Family of High Silica Crystalline
     Materials   133
     By G.T. Kokotailo and W.M. Meier

P9B. Evaluation of a New Zeolitic Catalyst for
     Selective Catalytic Reduction of NO_x from
     Stationary Sources   140
     By J.R. Kiovsky, P.B. Koradia, and C.T. Lim

P10. NMR Investigations of Mobility Mechanisms of
     Benzene Molecules in the Cavities of Faujasite
     Type Zeolites and Connections with Catalytic
     and Sorption Properties of These Substances   164
     By H. Kacirek, H. Lechert, W. Schweitzer, and K.-P. Wittern

P11. NMR study of the Equilibrium Exchange Rate of
     Water between and of the Intermolecular Proton
     Exchanges in the Large and Small Pores of Type
     Y Zeolites   174
     By W.D. Basler

P12. Gas Adsorption Rate Data Using Zeolitic Molecular
     Sieves   184
     By W.O. Daly and W.H. Granville

P13. Heats of Immersion of Co-Substituted X and Y
     Zeolites   198
     By R.V. Hercigonja, B.B. Radak, and V.M. Radak

P14. The Influence of Dealumination of Synthetic Y
     Zeolites on the Equilibrium of Adsorption of
     C_6 - Hydrocarbons   204
     By W. Schirmer, H. Thamm, H. Stach, and U. Lohse

     Discussion on Session 3   214

Session 4: Ion Exchange and Modification Processes

P15. Plenary Lecture: Binary and Ternary Ion
     Exchange in Zeolite A   218
     By L.V.C. Rees

P16. Sodium Aluminium Silicates in the Washing
   Process. Part VII: Counterion Effects   244
   By M.J. Schwuger and H.G. Smolka

P17. Natural Chabazite for Iron and Manganese Removal
   from Water   258
   By R. Aiello, C. Colella, and A. Nastro

P18. Influence of Outgoing Cation in Ion Exchange.
   Transition Metal Ion Exchange in K-Y Zeolite   269
   By A. Maes, J. Verlinden, and A. Cremers

P19. The Mobility of Cations in Synthetic Zeolite A   279
   By A. Dyer and H. Enamy

   Discussion on Session 4   288

Session 5: Catalysis

P20. Plenary Lecture: Industrial Uses of Zeolite
   Catalysts   294
   By D.E.W. Vaughan

P21. Catalysis on X and Y Zeolites Containing
   Cupric Ions   329
   By F. Mahoney, R. Rudham, and A. Stockwell

P22. Catalysis by Highly Siliceous Zeolites   342
   By M.S. Spencer and T.V. Whittam

P23. The Catalytic Activity of Transition Metal
   Exchanged Mordenites Towards Methane Oxidation   353
   By P. Fletcher, P.R. Lower, and R.P. Townsend

P24. Catalysis on Transition-Metal Ion-Exchanged
   Zeolites   369
   By F. Bravo, J. Dwyer, and D. Zamboulis

   Discussion on Session 5   385

Session 6: Zeolites in the Future

P25. Plenary Lecture: Potential Uses of Natural and
   Synthetic Zeolites in Industry   391
   By D.W. Breck

Invited Lecture: Hazards of Fibrous Mineral Dusts   423
   including Zeolites
   By P.C. Elmes

Introduction

The great interest in zeolites, and their widespread industrial application, which has its origin partly in the pioneering work of the late 1940's and early 1950's, but also especially in the discovery of their catalytic properties later in the latter decade, shows no sign of abating twenty years later. In fact the reverse is true; the intensive research that continues in both academic institutions and industry seems to be accelerating with every year that passes. Such are the number of papers that are now being published on zeolites that only this year the United Kingdom Chemical Information Service (UKCIS) have started producing, in conjunction with the Chemical Abstracts Service of the American Chemical Society, a separate CA Select on this subject alone, with between fifty and sixty abstracts being selected by computer every fortnight.

It is not surprising therefore that there has been no shortage of conferences on the subject. Of particular note have been the International Conferences on Molecular Sieves organised by the International Zeolite Association, which were held in 1967, 1970, 1973 and 1977. The Fifth International Conference in this series is due to be held in 1980 in Naples, Italy, and will bring together work on both natural and synthetic zeolites.

Thus a pertinent question might be, why hold yet another conference? Two considerations were foremost in the minds of the organisers of this Conference, which was jointly sponsored by the Society for Chemistry and Industry and the Chemical Society. These were to consider together both the properties and applications of zeolites, and the relationships between these two aspects. Considering properties first of all, the crystalline stereoregular structures of zeolites make them particularly suitable for fundamental studies on both equilibrium and kinetic aspects of sorption. Both these topics were discussed in this Conference; the new light that discrepancies between NMR and sorptiometric measurements have thrown on the mechanism of diffusion in these structures is just one example of the new and deeper understanding which has developed in the last few years regarding zeolite properties. Secondly, applications of zeolites are increasing at an almost bewildering rate, and an emphasis on this was felt to be appropriate. Indeed, deeper studies into the fundamental properties of zeolites and into their application are complementary to one another, as, for example, in the relatively new idea of making use of the ion exchange properties of zeolites to sequester calcium and magnesium in detergent solutions, a subject which was covered extensively in this Conference.

The order in which papers are presented in this volume follows the Conference Programme. Thus, Session 1, which was concerned with the problems of inter- and intracrystalline diffusion in zeolites was opened with a plenary lecture on the subject by Professor R. M. Barrer. Session 2, which dealt mainly with applications of zeolites in separation processes, was opened with a plenary lecture on chemical engineering aspects of separation, given by Dr. J.W. Carter. In contrast the emphasis in Session 3 was more on fundamental properties, with an opening lecture by Professor W. H. Meier and Dr. G. T. Kokotailo on the new "Pentasil" family of zeolites. Session 4 was concerned with ion exchange, and here the connection between an understanding of fundamental properties and their practical application was emphasised in the plenary lecture by Dr. L.V.C. Rees. Emphasis on catalysis was not strong in this Conference, because of the recent conference on zeolite catalysis sponsored by the Hungarian Chemical Society and held in September 1978 in Szeged, Hungary. However, Session 5 included a plenary lecture by Dr. D.E.W. Vaughan, who provided an over-view of catalysis, and some new work in this area was also presented. The Conference finished with Session 6, in which Dr. D.W. Breck presented a plenary lecture on both new and old applications of zeolites, and Professor Elmes discussed the evidence for or against there being a possible health hazard associated with certain fibrous minerals, including erionite.

Discussion of papers was strongly encouraged, and the resulting comments are found at the end of each relevant section in this volume.

Finally, thanks are due to the Organising Committee for their work in preparing the programme, and in particular to Professor K.S.W. Sing, who was Chairman of both the Committee and the Conference.

The City University,   Rodney P. Townsend
London.
October 16, 1979.

Sorption Kinetics and Diffusivities in Porous Crystals

by

R.M. Barrer

Physical Chemistry Laboratories, The Chemistry Department

Imperial College, London SW7 2AY

Summary

Attention has been drawn to some experimental problems in sorption
kinetics, such as the basic requirement to find which of several
possible processes is rate controlling. Ways to determine this step
have been considered. Because intracrystalline diffusion is not
necessarily rate controlling some prior measurements of diffusivities
may need to be repeated or reassessed. Where on the basis of appropriate criteria and experimental methods true intracrystalline diffusivities
have been obtained attention has been drawn to the relationship between
differential intrinsic and self-diffusivities, as well as to the useful
rate coefficient, k, based upon the $\sqrt{t}$ law, $M_t/M_\infty = k\sqrt{t}$. This expression is often valid over the early stages whether diffusivities are or
are not concentration dependent, and k/A (A = external area of crystals)
is independent of crystal size or shape distributions. Some properties
of self- and intrinsic diffusivities have been considered for n-paraffins
in Na-X where sieve effects are absent; for n-paraffins in (Na,Ca)-A
where such effects are present; for the small polar water molecule in
a range of crystal sieves of different mesh size; and for groups of
sorbates in crystals where the sieve mesh differentiates strongly between
the diffusing species on the basis of molecule size and shape. The
dominant role the exchange cation can play has been discussed.

Introduction

Rates of sorption in zeolites and derived rate coefficients such as
the diffusivities, D, attract increasing attention as the technical and
scientific interest in molecular sieves grows. In what follows some
problems involved in studying and interpreting sorption kinetics will be
outlined and some properties of diffusivities discussed. These
diffusivities may be self-diffusion coefficients and differential and
integral intrinsic diffusion coefficients.

Rate-determining Steps

In large crystals there is no doubt that one measures intracrystalline diffusivities. Examples are the work of Tiselius on diffusion of
water in single crystals of heulandite[1] and of ammonia in analcime;[2]
and that of Barrer and Fender on self- and intrinsic diffusivities of

water in heulandite, gmelinite and chabazite. [3]   However in beds of
fine synthetic zeolite powders or powder compacts, in which the
individual crystals are normally in the size range 0.1 to 10 μm, the
situation is more complex.   The possible rate controlling steps which
can arise or have been suggested are:

1. Intracrystalline diffusion
2. Intercrystalline diffusion
3. Control by slow transmission through surface skins on crystal
   surfaces
4. Heating on sorption, or cooling on desorption, with consequent
   time dependent heat dissipation and shifts of sorption equilibria.

These steps will be considered in reverse order.

Heat Effects

Sorption is exothermal, desorption is endothermal.   For permanent
gases the heats of sorption are small, but for dipolar and quadrupolar
molecules (e.g. H_2O, NH_3, CO_2) and also for non-polar bigger molecules
(benzene, paraffins, cycloparaffins) they are large.   If the sorption
occurs only slowly there is time for the heat to be dissipated and
quasi-isothermal conditions to prevail, but the greater the heat and
the more rapid the sorption the greater the heating of the bed of sorbent,
the temperature being a function of positional co-ordinates and time.
Then, for an activated diffusion, where $D = D_o \exp -E/RT$, D also becomes
$D(x,y,z,t)$.   Heat effects in these circumstances can be reduced by
using shallow layers of sorbent.

Another influence of non-isothermal conditions (heating on sorption,
cooling on desorption) is to tend to make sorption governed by intracrystalline activated diffusion faster than desorption.   In studies of
sorption kinetics heat effects have normally been neglected, although
some recent efforts have been made to allow for them.[4]

Surface Skins

The existence of rate-controlling surface skins in sorption kinetics
in zeolites was inferred from the large discrepancies sometimes found
between intracrystalline self-diffusivities measured from pulsed field
gradient NMR and these diffusivities measured from sorption kinetics
(Table 1[5]).   These two diffusivities can differ by many orders of magnitude and it seemed plausible to explain the difference as the result of
a surface resistance which greatly slows down the rate of sorption or
desorption but of course does not influence the values of D* obtained by
pulsed field gradient NMR.[5]

For total control by transmission through a surface skin one has for sorption at constant external pressure into spheres all of the same radius, r_o, and of total external surface area A,[13]

$$\frac{Q_t - Q_o}{Q_\infty - Q_o} = 1-\exp\left(-\frac{3\alpha t}{r_o}\right) = 1 - \exp\left(-\frac{\alpha A t}{V}\right) \tag{1}$$

In this expression α is the transmission coefficient through the skin, defined by

$$\alpha = j_s / (C_\infty - C) \tag{2}$$

where j_s denotes the sorbate flux density, C is the concentration inside the skin and C_∞ is its value at equilibrium. Q_t, Q_o and Q_∞ are amounts sorbed at t = t, 0 and ∞ respectively. V is the total volume of the crystals.

Kinetic control according to eqn 1 may be contrasted with control by intracrystalline diffusion, which, over the early stages of the sorption at constant pressure, is normally governed by

$$\frac{Q_t - Q_o}{Q_\infty - Q_o} = \frac{M_t}{M_\infty} = \frac{2A}{V}\sqrt{\frac{D_i t}{\pi}} \tag{3}$$

where D_i is the intracrystalline diffusivity, assumed independent of concentration. Kinetically, therefore, the two kinds of control are easily differentiated. In practice for fast sorptions it becomes difficult to test the $\sqrt{t}$ equation (eqn 3), but where sorption is slow enough one has nearly always found the $\sqrt{t}$ equation to be obeyed,[3,14,15] so that control by transmission through surface skins can be ruled out in such cases. Figs. 1 and 2 show examples of the initial validity of the $\sqrt{t}$ equation for gases diffusing in levynite and mordenite[14] and for water in chabazite, gmelinite and heulandite.[3]

Table 1. Comparison of self-diffusivities, D* ($cm^2 s^{-1}$) from pulsed field gradient NMR and from sorption[a] kinetics[5]

System	Temp. (°C)	Loading in NMR (molecules per cavity)	D* (NMR)	D* (sorption rates)
CH_4/Ca-rich A	23	5	2×10^{-5} [6]	5×10^{-10} [7]
C_2H_6/Ca-rich A	23	3	2×10^{-6} [8]	10^{-10} [7]; 6×10^{-12} [9]
C_3H_8/Ca-rich A	23	4	5×10^{-8} [6]	3×10^{-11} [7]
n-C_7H_{16}/Na-X	164	0.6	5×10^{-5} [10]	3×10^{-9} [11]
cyclo-C_6H_{12}/Na-X	164	1.3	4.5×10^{-5} [12]	4×10^{-9} [11]
C_6H_6/Na-X	164	1	2×10^{-6} [6]	10^{-10} [11]

(a) D* from sorption kinetics have involved extrapolation to $\theta \to 0$, so that

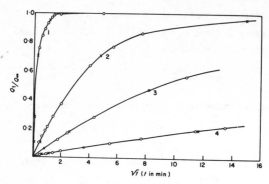

Fig.1.   Rates of diffusion of gases in levynite and mordenite at constant
         volume, variable pressure, when sorption is near the Henry law range[14]

Curve 1.   Ne in Li-mordenite at -185°C
Curve 2.   Ne in Ca-mordenite at -185°C
Curve 3.   Kr in Ba-mordenite at 24°C
Curve 4.   Kr in levynite at 0°C

⊙ = experimental points, X = points calculated from the appropriate
solution [14] of the diffusion equation.   (Ref.16, p.276)

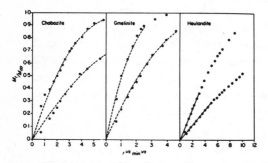

Fig.2.   Kinetics of sorption of water at constant pressure into chabazite,
         gmelinite and heulandite. [3]   The dashed lines are calculated curves.
         The experimental points are :

Chabazite, ⊙ , 75.4°C, ●, 30.8°C
Gmelinite, ⊙ , 62.5°C, ●, 31.7°C
Heulandite, ⊙ , 77.8°C, ●, 37.4°C (Ref.16, p.276)

The larger the crystals for a given sorbate-sorbent system the less important should surface skins become and the more nearly should pulsed field gradient NMR and sorption rates give the same diffusivity. Fig.3 shows a test of this for CH_4 diffusing in chabazite[5] which appears to confirm the above argument. However further consideration is required, as indicated below.

Intra- and Inter-crystalline Diffusion

One may obtain the half-life of the sorption process for D_i independent of amount sorbed from the relation

$$t_{\frac{1}{2}} = 2.1 \times 10^{-2} \, r_o^2/D_i \qquad (4)$$

where $t_{\frac{1}{2}}$ is in seconds, r_o in cm and D_i in $cm^2 \, s^{-1}$. $t_{\frac{1}{2}}$ is the time when $M_t/M_\infty = 1/2$. Table 2 gives values of $t_{\frac{1}{2}}$ for different values of crystal radius, r_o, and of D_i [16]. The dashed line divides approximately those systems above this line where the values of r_o and D_i are such that one could not hope to measure D_i from sorption kinetics, and those systems below the line where D_i could be found from sorption kinetics. The self-diffusivities in Table 1 found from pulsed field gradient NMR are all so large that for $r_o \sim 1 \, \mu m$ the half-lives are so minute that D_i could never be measured from sorption kinetics. It is thus an unavoidable conclusion assuming the correctness of the NMR values of D_i^* that the sorption rate measurements must be controlled by a process other than intracrystalline diffusion. This process could be intercrystalline flow, even in thin beds, and the only explanation of Fig. 3 need not be that there is control by transmission through a surface skin.

Table 2. $t_{\frac{1}{2}}$ in seconds for [16] :

r_o (cm) D_i (cm²s⁻¹) =	10^{-6}	10^{-8}	10^{-10}	10^{-12}	10^{-14}
10^{-5}	2.1×10^{-6}	2.1×10^{-4}	2.1×10^{-2}	2.1	2.1×10^{2}
10^{-4} (= 1 μm)	2.1×10^{-4}	2.1×10^{-2}	2.1	2.1×10^{2}	2.1×10^{4}
10^{-3}	2.1×10^{-2}	2.1	2.1×10^{2}	2.1×10^{4}	2.1×10^{6}
10^{-2}	2.1	2.1×10^{2}	2.1×10^{4}	2.1×10^{6}	2.1×10^{8}
10^{-1}	2.1×10^{2}	2.1×10^{4}	2.1×10^{6}	2.1×10^{8}	2.1×10^{10}
1	2.1×10^{4}	2.1×10^{6}	2.1×10^{8}	2.1×10^{12}	2.1×10^{12}

Table 2 also shows that the sorption kinetics procedure comes into its own in synthetic zeolite powders with $r_o \sim 1 \, \mu m$ for $D_i < 10^{-12} cm^2 s^{-1}$, where now the pulsed field gradient NMR method would not be applicable; and that for large enough crystals ($r_o \geq 0.1$ cm) the sorption rate measurements will also give values of D_i as big as $10^{-6} cm^2 s^{-1}$. Problems arise in sorption kinetics, as Fig.3 and Table 1 illustrate, when the D_i or D_i^* are large and crystal size small. It is then that the NMR procedure becomes the only practicable one. Table 2 gives guidance as to the range of usefulness of each method.

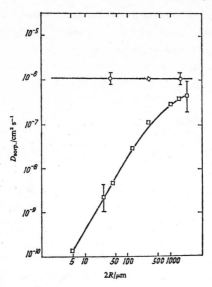

Fig.3.   Intracrystalline self-diffusivity ($\Diamond$) from pulsed field gradient
NMR and apparent intracrystalline diffusivity ($\square$) from sorption rates
of CH_4 in chabazites at $0^{\circ}C$ as functions of crystallite radii.[5]

We now need means of distinguishing sorption rates controlled by
intercrystalline or by intracrystalline flow of sorbate.   Consider
linear diffusion in a cylindrical column of zeolite powder bounded top
and bottom by planes $x = 0$ and $x = \ell$.   Sorption occurs only after flow
through $x = 0$, all other surfaces being impermeable due to the container
walls.   The bed consists of crystallites all of radius r_o.   For control,
at constant pressure of sorbate gas at $x = 0$, by intercrystalline Knudsen
flow the early time $\sqrt{t}$ equation is

$$\frac{M_t}{M_\infty} = \frac{2}{\ell} \sqrt{\frac{D_{eff} \, t}{\pi}} \tag{5}$$

where D_{eff} is the effective diffusivity.   On the other hand if intracrystalline diffusion is rate controlling, the pressure now being the same
throughout the bed,

$$\frac{M_t}{M_\infty} = \frac{6}{r_o} \sqrt{\frac{D_i t}{\pi}} \tag{6}$$

If these eqns are each used in turn to interpret the same kinetic run one
must find

$$\frac{9 D_i}{r_o^2} = \frac{D_{eff}}{\ell^2} \tag{7}$$

If $\ell = 1$ cm and $r_o = 3 \times 10^{-4}$ cm then $D_i = D_{eff} \times 10^{-8}$, but how can we find which
interpretation is correct?

One simple criterion for isothermal sorption conditions is based on the temperature coefficient of the apparent diffusivity. D_i is normally exponentially temperature dependent ($D_i = D_o \exp -E/RT$) whereas a Knudsen diffusivity, D_g, depends on $\sqrt{T}$.

A second criterion uses the method of moments. The first moment is defined as

$$E_{w_1} = \int_0^t t \, \frac{d(M_t/M_\infty)}{dt} \, dt \qquad (8)$$

and it has been shown [17] that

$$E_{w_1} = \frac{\ell^2}{pD_g \, \nu(\nu+2)} + E_o \qquad (9)$$

where p is the fraction of the molecules in the bed which is in the gas phase, $\nu = 1$ for linear diffusion, 2 for cylinder diffusion and 3 for diffusion in a sphere. ℓ is a characteristic dimension, the depth of the bed for eqn 5, or the radius of the cylinder or sphere. E_o is the part of E_{w_1} which depends on intracrystalline diffusion. If beds of different depths, ℓ, are used having the same porosities, and keeping p constant (readily achieved in self-diffusion; approximated to in interval runs), plots of E_{w_1} against ℓ^2 will serve to give $(3p \, D_g)^{-1}$ as slope and E_o as

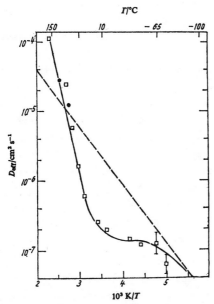

Fig. 4. D_{eff} for cyclohexane in Na-X [5] with a sorbate concentration of $\sim$ 2 molecules per large cavity (□), and comparison with D_i^* (dashed line), and corresponding values from sorption rates, obtained using the method of moments (●).

intercept. Thence it is easy to find which of the two terms on the
r.h.s. of eqn 9 is dominant, and so is the rate-controlling step.
When this method was applied to sorbates such as cyclohexane in compacts
of Na-X from about 0.15 up to 0.5 cm thick with intercrystalline
porosities between 0.40 and 0.23 cm^3 per cm^3 of compact it was found that
sorption was in all cases controlled kinetically by intercrystalline
diffusion.[18]

Pulsed field gradient NMR provides a third method of differenti-
ating control by inter- or intra-crystalline self-diffusion. The spin
echo amplitude, χ, is given by[19,20]

$$\ell n\ \chi(\sigma g) = \ell n\ \chi(0) - \gamma^2 \sigma^2 g^2 \Delta\ D_{eff} \tag{10}$$

where γ is the gyromagnetic ratio and σ, g and Δ are respectively the
width and amplitude of, and the time interval between the gradient pulses.
D_{eff} is the effective diffusivity, given by

$$D_{eff} = \left\{ D_i^* + \frac{p\ D_g^*}{\gamma^2 \sigma^2 g^2 \tau\ pD_g^* + 1} \right\} \simeq \frac{\langle w^2 \rangle}{6\Delta} \tag{11}$$

In eqn 11 $\langle w^2 \rangle$ is the mean square displacement of a molecule over the
interval Δ and τ is the mean lifetime of sorbate molecules within the
intracrystalline space. Two extremes occur:
1. $\tau \gg \Delta$ and so $\langle w^2 \rangle^{\frac{1}{2}} \ll$ crystal radius. Eqn 11 then reduces to
   $D_{eff} \simeq D_i^*$.
2. $\tau \ll \Delta$ so that $\langle w^2 \rangle^{\frac{1}{2}} \gg$ crystal radius. Eqn 11 gives in this
   case $D_{eff} \simeq p\ D_g^*$.
Fig. 4 illustrates these two extreme situations for the self-diffusion
of cyclohexane in Na-X,[5] as the plot of $\ell n\ D_{eff}$ vs $1/T$. Below $-65°C$
$\langle w^2 \rangle^{\frac{1}{2}} \ll$ crystal radius and so D_i^* is obtained. Above $10°C$ $\langle w^2 \rangle^{\frac{1}{2}} \gg$
crystal radius and so the curve gives $p\ D_g^*$, the temperature coefficient
of which is largely that of p. Between -65 and $+10°C$ there is an
intermediate region in which translational diffusion is restricted to the
intracrystalline space of each crystal because the thermal energy is
inadequate for the majority of molecules to evaporate from the crystals.
Accordingly D_{eff} for a given Δ is nearly independent of T. This result
serves to estimate crystal size since for restricted diffusion in small
spheres of radius r_o it has been shown[21] that

$$D_{eff} \simeq r_o^2/5\Delta \tag{12}$$

With $\Delta = 6$ ms and in the restricted diffusion range Fig.4 gives $D_{eff} \simeq$
$1.5 \times 10^{-7}\ cm^2 s^{-1}$ so that $r_o \simeq 0.7\ \mu m$, as compared with $\sim 1\ \mu m$ found by
photomicrography.

Relations between Intracrystalline Diffusivities

For diffusion of a species A admixed with tracer A (A*), and with
concentrations C_A and C_A^* of A and A* respectively, exact relations have

been developed[22,23] between the differential intrinsic diffusivity, $D_i(C)$ and the differential self-diffusivity, $D_i^*(C)$. One form of the relationship is[24]

$$D_i(C) = D_i^*(C) \frac{d\ln p}{d\ln C} \left[1 + \frac{\beta(C)C}{\alpha(C)} \right] \tag{13}$$

where β and α are connected with the straight (L_{AA}) and the cross-coefficient (L_{AA}^*) of the irreversible thermodynamic formulation of intracrystalline flow of A and A* by the equations:

$$L_{AA}/C = \alpha C_A + \beta C_A^2$$
$$L_{AA}^*/C = \beta C_A C_A^* \tag{14}$$

$C = C_A + C_A^*$ is constant throughout the crystal in any one experiment and p is the equilibrium pressure of sorbate when C is its intracrystalline concentration.

The term $\left[1 + \frac{\beta(C)C}{\alpha(C)} \right]$ for water diffusing near saturation of the zeolite in chabazite, heulandite and gmelinite can be estimated approximately from the measurements of D_i, D_i^* and $d\ln p/d\ln C$ of Barrer and Fender[3]. The results in Table 3 suggest that $\left[1 + \frac{\beta(C)C}{\alpha(C)} \right]$ is small compared with $d\ln p/d\ln C$ for these particular systems. In a further analysis of the relation between D_i and D_i^* using a simple microkinetic model Kärger[24] found the term $[1 - \gamma(C)p]^{-1}$ replacing $[1 + \beta(C)C/\alpha(C)]$ in eqn 13, where $\gamma(C)$ is a coefficient which should not vary much with C. According to this analysis as C and hence p decrease the above correction term should approach unity. For Type I isotherms in Brunauer's classification[25] the term $d\ln p/d\ln C$ also decreases monotonically with C, becoming unity in the Henry's law range (C = kp).

If the term $[1 + \beta(C)C/\alpha(C)]$ is omitted from eqn 13 Darken's[26] approximate relation is obtained:

$$D_i(C) = D_i^*(C) \frac{d\ln p}{d\ln C} \tag{15}$$

while in the Henry's law range, from what has been said in the previous paragraph, one finds

$$D_i = D_i^* \tag{16}$$

where both diffusivities cease to be functions of concentration. Thus, sorption kinetics in the Henry's law range also serve to measure self-diffusivities but outside this range D_i^* is obtained only approximately by the use of Darken's relationship (eqn 15). Where D_i is a function of concentration and the isotherm is not very rectangular one may add sorbate in small increments, with each addition strictly at constant pressure,[27] to obtain an average D_i over the small concentration step. In this way $D_i(C)$ can be found.

In the general case where the concentration step becomes large one may also consider an integral diffusion coefficient $\tilde{D}_i(C)$ which is related

Table 3   Relation between D_i^* and D_i for water in several zeolites
near saturation of the zeolite [3]

Zeolite	$T(^{\circ}C)$	$D_i^* \times 10^8$ $(cm^2 s^{-1})$	$\frac{d\ln p}{d\ln C}$ ⊙	$D_i^* \frac{d\ln p}{d\ln C} \times 10^6$ $(cm^2 s^{-1})$	$D_i(expt) \times 10^6$ $(cm^2 s^{-1})$	$[1 + \frac{\beta(C)}{\alpha(C)} \cdot C]$
Chabazite B	75	46.2	23.0 [+]	10.6_5	11.7	1.10
	65	31.8	24.0 [+]	7.6_5	8.9	1.17
	55	21.4	25.5 [+]	5.4_5	6.6	1.21
	45	14.1	27.0 [+]	3.8_0	4.8	1.26
	35	9.0	28.0 [+]	2.50	3.4	1.36
Heulandite	75	9.77	30.0	3.00	4.9	1.63
	65	6.07	32.0	1.95	3.5	1.79
	55	3.70	34.0	1.26	2.4	1.90
	45	2.19	35.5	0.78_0	1.6	2.05
	35	1.24	37.0	0.46_5	1.05	2.26
Gmelinite	55	7.33	26.5	1.95	2.3	1.18
	45	4.99	28.0	1.40	1.5	1.07
	35	3.31	29.5	0.97	1.0	1.03

⊙ Found from nearly linear flat top of isotherm

[+] From isotherm in chabazite sample A.

to the differential value $D_i(C)$ by [15]

$$\widetilde{D}_i(C) = \frac{1}{C} \int_0^C D_i(C) \, dC \tag{17}$$

or, for an interval C_1 to C_2

$$\widetilde{D}_i(C_1, C_2) = \frac{1}{C_2 - C_1} \int_{C_1}^{C_2} D_i(C) \, dC \tag{18}$$

Alternatively, these integral values may be related to $D_i(C)$ by

$$D_i(C) = \frac{d}{dC}\left[\widetilde{D}_i(C)C\right] = \widetilde{D}_i(C) + C\frac{d\widetilde{D}_i(C)}{dC} \tag{19}$$

or

$$D_i(C_1, C_2) = \frac{\partial}{\partial C_1}\left[\widetilde{D}_i(C_1, C_2)(C_1 - C_2)\right]_{C_2} = \widetilde{D}_i(C_1, C_2) + (C_1 - C_2)\left[\frac{\partial\widetilde{D}_i(C_1, C_2)}{\partial C_1}\right]_{C_2} \tag{20}$$

These integral diffusivities are involved particularly where the isotherm is very rectangular so that step-wise addition of sorbate, with each addition at constant pressure, is no longer practicable. [15] In the Henry's law range where D_i does not depend on C eqns 17, 18 and 19 show that

$$\widetilde{D}_i(C) = \widetilde{D}_i(C_1, C_2) \tag{21}$$

and, with eqn 15, that

$$\widetilde{D}_i = D_i = D_i^* \tag{22}$$

Some Self-diffusivities in Zeolites

From the foregoing discussion the differential self-diffusion coefficient is seen to be a basic rate coefficient influenced by temperature, concentration of diffusant, its molecular shape and size, and the nature of the zeolite. The intrinsic diffusivity is influenced in addition by the term $d\ln p/d\ln C$, and by a term involving the straight and cross-coefficients.

<u>n-Paraffins</u>. Pulsed field gradient NMR has been used to study the self-diffusion of alkanes n-C_4 to n-C_{18} in zeolite X in the range -100 to + $200^{\circ}C$.[28] This study of self-diffusion has demonstrated the following properties:

1.   For comparable degrees of filling of the intracrystalline pore space D_i^* decreased with increasing carbon number.

2.   At constant temperature D_i^* decreased monotonically for each paraffin with increasing amount sorbed.

3.   The values of D_i^* at a given temperature and for a given paraffin were within an order of magnitude of the self-diffusivities in the corresponding liquid paraffin.

4. D_i^* for each paraffin depended exponentially upon temperature for a constant amount sorbed.

5.   Whereas the heat of sorption for a given degree of filling of intracrystalline pore space increased continuously with increasing carbon number, the activation energies for self-diffusion approached an asymptotic limit and were never more than a rather small fraction of the heat of sorption (Fig.5).[28]

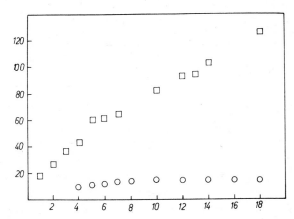

Fig.5.   The heats of sorption, and energies of activation for self-diffusivities, of n-alkanes in Na-X zeolite,[28] as functions of carbon number.   Upper curve, heat of sorption, lower curve activation energy. Ordinate is in kJ mol^{-1}; abscissa is carbon number.

It has been suggested[29] that the above behaviour of the energy of activation for D_i^* means that as the paraffin chains increase in length each unit diffusion process tends more and more to involve segmental rotation of part only of the molecule around a C-C bond.   This results in changes in position of the centre of mass of the molecule, but not necessarily in translation of all parts of the molecule simultaneously. The longer the chains the more probable the segmental mechanism becomes, and if this mechanism tends for larger chains to involve on average segments of similar size, eventually nearly independent of carbon number, then, as observed, the activation energy would become almost constant. Unit diffusion steps by segmental rotation can also account for the decrease in D_i^* with increasing carbon number, because the number of such steps required for translation of an entire molecule from one 26-hedral cavity to the next will increase with chain length.

Zeolite X is extremely open in structure and no appreciable molecule sieving effects based on constriction by the 12-ring windows are expected for n-paraffins, since the windows have free diameters of ~ 7.4Å compared with the critical cross-sectional dimension of an extended n-paraffin chain of ~ 4.9Å. In zeolite Ca-A, however, the 8-ring windows leading to and from the 26-hedra of Type I[30] have free diameters of only 4.2Å and accordingly act as energy barriers. The diffusivities D_i^* decrease and the activation energies become considerable, according to Quig and Rees[31] for C_5, C_6 and C_7 n-paraffins in (Na,Ca)-A with 32.8% exchange of Na^+ by Ca^{2+} (Table 4). The radio-tracer method rather than pulsed field gradient NMR was used, so that if surface skins were present they could also influence the results.

Table 4. Self-diffusivities of some n-paraffins in (Na,Ca)-A (32.8% exchange of Na^+ by Ca^{2+}), near saturation of the zeolite [31]

n-alkane	D_i^* ($cm^2 s^{-1}$) at		E^*
	$25°C$	$100°C$	$kJ mol^{-1}$
$C_5 H_{12}$	8.5×10^{-16}	4.2×10^{-13}	76.5 (18.3[†])
$C_6 H_{14}$	3.5×10^{-17}	1.2×10^{-13}	99.0 (23.7[†])
$C_7 H_{16}$	1.15×10^{-17}	1.2×10^{-13}	115.0 (27.5[†])

† Values in kcal mol^{-1}

Water. As a contrast with non-polar and relatively flexible n-alkanes one may consider the small very polar water molecule (equilibrium diameter ~ 2.8Å). Some values of D_i^* are given in Table 5. In the very restricted intracrystalline environment of analcime even water encounters a high energy barrier in the average unit diffusion step, but, as the zeolites become more open, the energy barriers decrease and D_i^* increases rapidly, until in zeolite X D_i^* is close to the value for bulk water. The energy barriers remain somewhat larger than in liquid, however, and indicate an enhanced "stickiness" of intrazeolitic as compared with liquid water. This, independently of true molecule sieving, is an expected result of the very polar and ionic nature of the zeolites, and the rigidity of the confining anionic framework. Nevertheless diffusivities in open zeolites such as Na-X, extrapolated to $-2°C$, are much larger than in ice.

Table 5. D_i^* for water in several zeolites near saturation of the zeolite, and E in $D_i^* = D_o \exp -E/RT$ [32]

Diffusion medium	Free dimensions of windows in Å	D_i^* (cm^2 s^{-1}) (at T$^\circ$C)	E (kcal mol^{-1})	Method
Analcime [33]	2.2 - 2.4	$1.9_7 \times 10^{-13}$ (46°)	17.0± 0.3	Tracer, HTO
Natrolite [34]	2.6 x 3.9	-	15.0	NMR
Heulandite [3]	2.4 x 6.1 3.2 x 7.8 3.8 x 4.5	$2.0_7 \times 10^{-8}$ (45°)	11.0± 0.3	Tracer, D_2O
Chabazite [3]	3.7 x 4.2	$1.2_6 \times 10^{-7}$ (45°)	8.7± 0.3	Tracer, D_2O
Gmelinite [3]	6.9 3.4 x 4.1	5.8×10^{-8} (45°)	8.1± 0.3	Tracer, D_2O
Na-X [35]	$\simeq 7.4$	$2.1_1 \times 10^{-5}$ (40°)	6.9	NMR
Ca-X [35]	$\simeq 7.4$	$2.4_1 \times 10^{-5}$ (40°)	6.8	NMR
Ca-Y [35]	$\simeq 7.4$	-	5.6	NMR
Ice [36,37]	-	1×10^{-10} (-2°)	13.5±1.1	Tracer H_2O^{18}; Dielectric relaxation
Water [35,38]	-	$3.8_7 \times 10^{-5}$ (45°)	4.6	Tracer H_2O^{18}; Dielectric relaxation

The Rate Constant of the $\sqrt{t}$ Law

For concentration dependent intrinsic diffusivities in systems where the isotherms are very rectangular there are problems in interpreting sorption kinetics, because the required <u>constant pressure</u> additions of small increments of sorbate are virtually impossible to make, as noted earlier. This is because for most of the isotherm the equilibrium pressures are so small that excessively large doser volumes of gas at the minute pressures would be needed to avoid rapid changes in pressure with time.

This difficulty led Barrer and Clarke [15] to use an alternative procedure for C_4, C_6 and C_9 n-alkanes diffusing into each of a series of (Na,Ca)-forms of zeolite <u>A</u> with regulated Ca contents. The zeolite was first allowed to take up known amounts, Q_o, of sorbate which were varied in different runs, and was then exposed to the same constant and substantial pressure of sorbate for each run, so that for a given hydrocarbon the final uptake, Q_∞, was the same in all runs. The kinetics followed the $\sqrt{t}$ law over the early stages, as shown in Fig. 6,[15] so that

$$\frac{M_t}{M_\infty} = \frac{Q_t - Q_o}{Q_\infty - Q_o} = k\sqrt{t} \tag{23}$$

The rate constant, k, can then serve well to characterise the kinetics, because, if J(t) is the total flux through the surfaces of all crystals then

$$M_t = \int_0^t J(t) \, dt \tag{24}$$

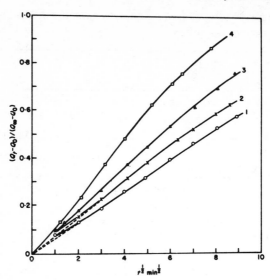

Fig.6.   Plots of $M_t/M_\infty = (Q_t-Q_0)/(Q_\infty-Q_0)$ against $\sqrt{t}$ for n-C_6H_{14} in 30.55%
Ca-exchanged zeolite (Ca,Na)-$\underline{A}$ at constant pressure and at 348 K. [15]

$Q_\infty = 120.05$ mg g^{-1}

Curve 1 : $Q_0 = 0$ Curve 3 : $Q_0 = 63.37$ mg g^{-1}
Curve 2 : $Q_0 = 48.08$ mg g^{-1} Curve 4 : $Q_0 = 96.99$ mg g^{-1} (Ref.16,p.281)

so that

$$\frac{d}{dt^{\frac{1}{2}}}[M_t/M_\infty] = \frac{2J(t)t^{\frac{1}{2}}}{Q_\infty - Q_0} = k \qquad (25)$$

Thus, the rate constant, k, is defined; and it is seen that $J(t)$ over the
region of validity of the $\sqrt{t}$ law varies as $(Q_\infty-Q_0)$ and as $t^{-\frac{1}{2}}$.   At a
reference time of unity (e.g. 1 minute) $t^{\frac{1}{2}} = 1$ and so

$$\frac{2J(1)}{Q_\infty-Q_0} = k \qquad (26)$$

If both sides of eqn 26 are divided by the external area of the crystallites,
A, then k/A is a characteristic of the system which is independent of the
state of subdivision of the crystals and of their shape or size distributions.
A particular interest of the $\sqrt{t}$ law is that for kinetics controlled by intracrystalline diffusion the law appears to hold initially when D_i is $D_i(C)$ as well
as when D_i is independent of concentration.   Examples of plots of k/A against
$(Q_\infty-Q_0)$ at each of several temperatures are shown for n-C_4, n-C_6 and n-C_9 in
Fig.7. [15]   k/A does not depend much on $(Q_\infty-Q_0)$ and can show small increases or
decreases as $(Q_\infty-Q_0)$ increases.   It has the expected considerable temperature
coefficient associated with activated diffusion.   The ease with which the coefficient k can be measured in appropriate conditions makes k/A very useful in
sorption kinetics.   It is a primary experimental quantity rather than a
secondary one, derived from the kinetics with the aid of assumptions.

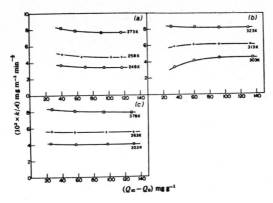

Fig.7.   Examples of plots of k/A from M_t/M_∞ = k/t as functions of $(Q_\infty-Q_0)$ in
34.10% Ca-exchanged (Ca,Na)-A.[15]

(a) n-C_4H_{10} ; (b) n-C_6H_{14} ; (c) n-C_9H_{20} .   (Ref.16, p.283)

Intrinsic Diffusivities

The further interpretation of k for constant pressure sorption experiments involving rectangular isotherms rests upon an assumption leading to
integral diffusivities $\tilde{D}_i$.   When D_i is independent of concentration

$$k = \frac{2A}{V} \left(\frac{D_i}{\pi} \right)^{\frac{1}{2}}$$ (27)

where V is the total volume of the crystals of zeolite in the sorbent bed.
The minimum assumption that can be made [15] when $D_i = D_i(C) = D_i(Q)$ is to replace D_i in eqn 27 by $\tilde{D}_i$, where

$$\tilde{D}_i (Q_\infty-Q_0) = \int_{Q_0}^{Q_\infty} D_i(Q)\,dQ$$ (28)

or

$$D_i(Q_0,Q) = - (Q_\infty-Q_0) \left(\frac{\partial \tilde{D}_i}{\partial Q_0} \right)_{Q_\infty} + \tilde{D}_i$$ (29)

Accordingly, with

$$\frac{k}{A} = \frac{2}{V} \left(\frac{\tilde{D}_i}{\pi} \right)^{\frac{1}{2}}$$ (30)

one may evaluate $\tilde{D}_i$ as a function of $(Q_\infty-Q_0)$ for fixed Q_∞.   Thence, from
plots of $\tilde{D}_i$ against $(Q_\infty-Q_0)$, eqn 29 was used to find D_i as a function of Q_0
for n-C_4 , n-C_6 and n-C_9 paraffins in 30.55, 32.54, 34.10 and 35.48% Ca-exchanged (Na,Ca)-A at several temperatures. [15]   As examples, the values
of D_i are given for n-hexane in Table 6.   The intrinsic diffusivities have
values which depend strongly upon temperature and are very sensitive to the
% of exchange of Na by Ca.   At the highest temperatures (373 K and 348 K)
the diffusivities decrease with increasing Q_0, but at the lower temperatures
(323 to 298 K) the diffusivity is not very dependent upon the amount sorbed.
The behaviour of D_i in relation to Q_0 can be considered in terms of

Table 6. D_i for n-hexane in 30.55, 32.54, 34.10 and 35.48% Ca-exchanged (Na,Ca)-A. [15]

% Ca	T/K	Q_∞/mg g^{-1}	Q_0/mg g^{-1}	$10^{13} \times D_i$/cm^2 s^{-1}
30.55	323	122.5	0	0.09
			5	0.09
			20	0.08
			40	0.07
	373	116.86	0	1.9_3
			5	1.5_8
			20	0.8_3
			40	0.35
			60	0.18
32.54	303	136.19	0	0.13_6
			20	0.13
			50	0.13
			70	0.13
			95	0.11
	348	130.58	0	2.6_2
			5	2.6_2
			20	2.2_2
			40	1.8_9
			45	1.6_5
			60	1.4_8
34.10	303	136.06	0	0.70
			5	0.70
			20	0.70
			40	0.76
			60	0.83
			80	0.89
			95	1.03
	323	130.65	0	2.3_0
			5	2.3_0
			20	2.2_0
			40	2.0_8
			60	2.0_8
			80	2.0_6
			90	2.0_5
35.48	298	136.9	0	1.9_2
			5	2.1_0
			20	2.2_0
			40	2.2_1
			60	2.4_8
			80	2.6_5
			90	2.8_5
	308	133.4	0	3.2_0
			5	3.2_0
			20	3.2_0
			40	3.2_0
			60	3.2_5
			80	3.3_5
			90	3.4_9

eqn 13. The lower the temperature the more rectangular are the isotherms, and so outside the Henry's law range $d\ell np/d\ell nC = d\ell np/d\ell nQ_o$ increases rapidly with Q_o. On the other hand, one expects $D_i^*(C)$ to decrease as Q_o increases because the molecules of diffusant are impeded more and more by other molecules. The jump probability of a molecule from the nth to the $(n+1)^{th}$ cavity next to it should inter alia be a function of $(1-\theta)_{(n+1)}$. It seems likely from the results in Table 6 that, for very rectangular isotherms as Q_o increases, the factor $d\ell np/d\ell nQ_o$ for n-hexane in the zeolites <u>A</u> increases D_i rather more rapidly with θ than the diminishing jump probability from one 26-hedral cavity to the next decreases it, while as the isotherms become less rectangular (at the higher temperatures) the reverse behaviour is found. Accordingly D_i can increase or decrease with Q_o or remain nearly independent of it. For an ideal interstitial diffusion where one molecule fills a cavity and the isotherm obeys Langmuir's equation ($K = \theta/p(1-\theta)$) we expect that D_i^* will be proportional to $(1-\theta)$, but from the Langmuir isotherm we also find $d\ell np/d\ell nC = d\ell np/d\ell n\theta = (1-\theta)^{-1}$. Thus the Darken product $D_i^* \, d\ell np/d\ell n\theta$ should be independent of θ. D_i (eqn 13) may then vary little with concentration.

The 32.8% Ca-exchanged (Na,Ca)-<u>A</u> of Quig and Rees [31] (Table 4) is very near in composition to the 32.54% Ca-exchanged form of Barrer and Clarke [15] (Table 6). For n-hexane near saturation of the zeolite D_i^* was $3.5 \times 10^{-17} cm^2 s^{-1}$ at $25^{\circ}C$ and from E^* in Table 4 is estimated as 6.8×10^{-17} at $30^{\circ}C$. This value is ~ 160 times less than D_i when $Q_o = 95 \ mg \ g^{-1}$ (Table 6). For rectangular isotherms near saturation of the zeolite one expects that $d\ell np/d\ell nC$ will be very large (as illustrated for water in Table 3) and that the further correction term in eqn 13 may be appreciable although much smaller (cf Table 3, columns 4 and 7). Thus, as the comparison of D_i^* and D_i illustrates, D_i should become much larger than D_i^* as saturation is approached, but of course the two become equal in the dilute Henry's law range. Whereas D_i could increase or decrease with Q_o (Table 6), D_i^* in an ideal energetically uniform sorbent should only decrease.

From eqn 13 one also finds the relation between E and E^*, the activation energies for D_i and D_i^* respectively:

$$E = E^* + RT^2 \left\{ \frac{\partial}{\partial T} \ell n \left[\frac{d\ell np}{d\ell nC} \left(1 + \frac{\beta(C)C}{\alpha(C)} \right) \right] \right\}_{Q_o} \tag{31}$$

Thus $E \neq E^*$ except in the Henry's law range when the second term on the r.h.s. of eqn 31 tends to zero. The values of E for D_i are given in Table 7, and for n-C_6H_{14} E for 32.54% Ca-exchanged (Na,Ca)-<u>A</u> ($E = 58.2 \ kJ \ mol^{-1}$) is considerably less than E^* for 32.8% Ca-exchanged <u>A</u> of Table 4 ($E^* = 99 \ kJ \ mol^{-1}$). This difference suggests a negative contribution from the second term on the r.h.s. of eqn 31 (cf Table 3 for the case of water).

With one irregularity E in Table 7 diminishes as the % exchange of Na^+ by Ca^{2+} increases. In addition, both E^* (Table 4) and E (Table 7) increase

Table 7.   Activation energies E in zeolite (Na,Ca)-$\underline{A}$ for D_i in

$D_i = D_o \exp (-E/RT)$

Hydrocarbon	E (kJ mol^{-1}) for % Ca exchanges:			
	30.55	32.54	34.10	35.48
n-C_4H_{10}	40.7	47.3	36.4	-
n-C_6H_{14}	61.1	58.2	48.5	38.9
n-C_9H_{20}	74.5	67.4	60.7	53.1

with chain length.   The range in chain lengths is insufficient to know whether eventually E* or E would approach an asymptotic limit, as suggested by the results for E* for n-paraffins up to C_{18} in Na-X.[28]   The two situations are, however, different, as pointed out earlier, in that the 12-ring windows in Na-X impose little or no restriction or energy barrier to n-alkane diffusion, but the 8-ring meshes of (Ca,Na)-$\underline{A}$ do.

Diffusivities and Molecular Dimensions

According to the free dimensions of the "mesh" or "window" through which diffusing molecules must migrate intracrystalline diffusivities will change very strongly with molecular dimensions.   The windows available for molecule sieving are 6-, 8-, 10- and 12-rings with various degrees of distortion and of obstruction by cations.   Crystals with restricted windows differentiate between groups of small molecules and exclude big ones altogether;   crystals with larger windows differentiate between groups of bigger molecules, and so on.

To illustrate this behaviour, Fig. 8 shows the very different sorption rates of O_2, N_2 and Ar in levynite at -184°C [39] and in Ca-mordenite at -78°C,[14] both of which behave as fine-mesh sieves.   Fig. 9 shows a similar behaviour when the large molecules 1,3,5-trimethyl-, 1,3,5-triethyl- and 1,3,5-triisopropyl-benzene were sorbed in the wide mesh sieve, Na-Y, at 30°C.[40]   Some diffusivities (D_i or D_i^*) and activation energies (E or E*) in K-$\underline{A}$ and K-mordenite are shown in Table 8 in relation to molecular dimensions. In the systems illustrated the isotherms are often near or in the Henry law range, in which $D_i = D_i^*$. D_o in column 5 is the pre-exponential constant in the expressions $D_i = D_o \exp -E/RT$ or $D_i^* = D_o \exp -E^*/RT$.   With the dimensional change from 3.2Å (Ne) to 3.94Å (Kr) the diffusivity in K-$\underline{A}$ at 25°C changes by 10^5-fold and the activation energy from 7.0 to 16.4 kcal mol^{-1}. The corresponding changes in K-mordenite at -78°C are equally large.

The part played by the exchangeable cations in moderating molecule sieving was already demonstrated in 1945,[43] and was examined more fully in 1949 [14].   Table 9 shows how the rate constant D_i/r_o^2 (min^{-1}) changed at -78°C with different cations in mordenite.   The NH_4-mordenite in the table,

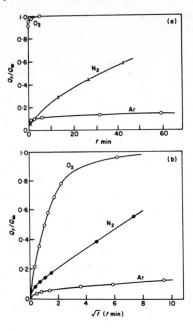

Fig.8. (a) Sorption rates of O_2, N_2 and Ar in levynite at $-184^\circ C$.[39] For O_2, $Q_\infty = 10.02$, for N_2 it is 9.77 and for Ar, 10.01, all in cm^3 at s.t.p. g^{-1}.

(b) Sorption kinetics of O_2, N_2 and Ar in Ca-mordenite at $-78^\circ C$.[14] (Ref.16, p.290)

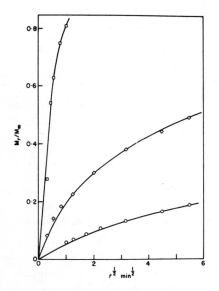

Fig.9. Sorption rates at $30^\circ C$ in Na-Y of liquid 1,3,5-trimethyl- (□), 1,3,5-triethyl- (◇) and 1,3,5-triiso-propyl-benzene (○).[40] (Ref.16, p.293)

Table 8. Relation between diffusivities and molecular dimensions [41]

Zeolite	Molecule	Equilibrium Dimensions (Å)	D_i or D_i^*/cm^2 s^{-1} (and T°C)	D_0 $(cm^2 s^{-1})$	E or E* kcal mol⁻¹
K-<u>A</u> [42]	Ne	3.2	2.9×10^{-13} (20)	4.8×10^{-8}	7.0
	H_2	2.4 × 3.1	2.0×10^{-12} (20)	4.5×10^{-6}	9.9
	Ar	3.8_3	1.6×10^{-17} (20)	3.7×10^{-8}	12.6
	N_2	3.0 × 4.1	9.8×10^{-18} (20)	1.5×10^{-5}	16.2
	Kr	3.9_4	1.8×10^{-18} (20)	6.7×10^{-7}	16.4
K-mordenite [14]	H_2	2.4 × 3.1	2.7×10^{-13}(-78)	1.64×10^{-10}	2.5
	O_2	2.8 × 3.9	2.0×10^{-15}(-78)	1.59×10^{-10}	4.4
	N_2	3.0 × 4.1	9.2×10^{-16}(-78)	2.0×10^{-10}	4.8
	Ar	3.8_3	2.4×10^{-16}(-78)	5.6×10^{-7}	8.4
	Kr	3.9_4	1.8×10^{-18}(-78)	$2.5_6 \times 10^{-7}$	10.0

Table 9. D_i/r_o^2 (min⁻¹) for argon in mordenites at -78°C [14]

Cationic form	D_i/r_o^2
Ca-mordenite	1.51×10^{-8}
K-mordenite	2.8×10^{-6}
Ba-mordenite	5.5×10^{-6}
Na-mordenite	3.8×10^{-6}
Li-mordenite	4.0×10^{-4}
NH_4(H)-mordenite	1.35×10^{-3}

having been outgassed above 300°C, was probably largely the H-form. Exchanging the cation has changed the rate constant D_i/r_o^2 for Ar in mordenite at -78°C by 10^5-fold. Other examples of the very large effects of cation exchanges on diffusivities are well known (e.g. Na- and Ca-<u>A</u>) and have major practical applications. The monovalent cations in mordenite gave D_i/r_o^2 in the sequence K < Na < Li < H, which is the inverse order of the ionic radii. Cations act by partially or wholly obstructing the channels along which diffusion occurs.

Two treatments have been developed to describe the obstructions imposed by cations. One of these, due to Hammersley,[44] is the percolation theory, appropriate for 3-dimensional channel systems such as arise in zeolite <u>A</u>. When the exchange $2Na^+ \rightleftarrows Ca^{2+}$ occurs the cations Na^+ which are blocking the six 8-ring windows which link a given 26-hedron of Type I with each of six other such 26-hedra are not replaced on those sites by Ca^{2+}, which go instead to other more preferred sites. Thus the windows are progressively cleared, and when about 33% exchange of Na^+ by Ca^{2+} has occurred the diffusivities for n-paraffins increase dramatically,[15] a result predicted by Hammersley's treatment.

Hammersley's theory is inappropriate for zeolites in which the channels are all parallel and are not inter-connected (1-dimensional channel systems, as found in mordenite, offretite, zeolite $\underline{L}$ or mazzite (zeolite Ω)). For this situation the obstructions (which may be cations, or impurity molecules like H_2O or NH_3, that are immobile under the conditions in which another diffusant is being investigated) form periodic barriers along each channel. To pass these barriers the diffusant requires a larger activation energy per unit diffusion step than in the intervals between each pair of obstructions. This situation was treated in a manner formally analogous to that involved in one-dimensional heat flow along a chain of parallel radiating slabs. [45,46,47]

Each highly conducting lamina or slab corresponds with the region between two adjacent barriers and each barrier with the space between slabs. Even when the activation energy for passing barriers is not very large their effect on the diffusivities can be very great. [46] If for example there was one such barrier for every 5 normal unit diffusion steps, and E for transmission past the barrier was 7.2 kcal mol^{-1} compared with 2.7 kcal mol^{-1} per normal diffusion step, it was estimated for certain conditions that rates of sorption would be reduced by about 3×10^5-fold. That very large reductions are indeed observed is shown by the diffusivities of several gases at -183°C in mordenite modified by pre-sorption of NH_3 (Fig. 10 [46]).

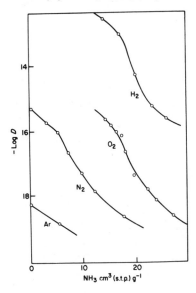

Fig.10. The diffusivities (cm^{2}s^{-1}) of H_2, O_2, N_2 and Ar in mordenite at -183°C, as functions of the amount of pre-sorbed NH_3. [46] (Ref.16 p.307)

A second example of obstruction, this time by cations, is provided by the one-dimensional channel system in a synthetic (K, TMA)-offretite (TMA denotes tetramethyl ammonium). When there was a significant amount of TMA in the wide channels the zeolite did not admit n-hexane, but as the TMA was progressively removed from these channels, by ion-exchange with smaller cations, n-hexane was taken up with increasing ease. [48] The large TMA cations in the wide channels are the barriers which, according to their number as modified by ion exchange, control the sorption kinetics of n-hexane. When the channels were freed of the barriers the crystals also readily imbibed 2,2-dimethyl butane.

Concluding Remark

It has been the purpose of this paper to draw attention to some problems in interpreting sorption kinetics and to methods for their solution, as well as to discuss and interpret the inter-relationships between, and some properties of, intrinsic differential and self-diffusivities of molecules in crystal sieves. The subject is making rapid progress and is of importance for various technical applications of zeolites. On the fundamental side a greater range of measurements of high accuracy must be made in systems where intracrystalline diffusion is rate-controlling. This will permit theoretical interpretations to be refined and extended.

References

[1] A. Tiselius, *Z. Phys.Chem.*, 1934, *A 169*, 425.

[2] A. Tiselius, *Z. Phys. Chem.*, 1935, *A 174*, 401.

[3] R.M. Barrer and B.E.F. Fender, *J. Phys. and Chem. Solids*, 1961, *21*, 12.

[4] e.g. J. Crank "The Mathematics of Diffusion" (Oxford) 1956, Chap. 13.

[5] J. Kärger and J. Caro, *J. Chem. Soc. Faraday I*, 1977, *73*, 1363.

[6] J. Caro and J. Kärger, unpublished results, see ref. 5.

[7] D.M. Ruthven and K.F. Loughlin, *Trans. Faraday Soc.*, 1971, *67*, 1661.

[8] J. Caro, J. Kärger, G. Finger, H. Pfeifer and R. Schöllner, *Z. Phys. Chem. (Leipzig)*, 1976, *257*, 903.

[9] W.W. Brandt and W. Rudloff, *J. Phys. Chem. Solids*, 1965, *26*, 741.

[10] J. Karger, S.P. Shdanov and A. Walter, *Z. Phys. Chem. (Leipzig)*, 1975, *256*, 319.

[11] D.M. Ruthven, 68th Ann. Meeting, Amer. Inst. Chem. Engineers, Los Angeles, 1975.

[12] J. Kärger, P. Lorenz, H. Pfeifer and M. Bülow, *Z. Phys. Chem. (Leipzig)*, 1976, *257*, 209.

[13] J. Kärger and P. Hermann, *Ann. Phys. (Leipzig)*, 1974, *31*, 277.

[14] R.M. Barrer, *Trans. Faraday Soc.*, 1949, *45*, 358.

[15] R.M. Barrer and D.J. Clarke, *J. Chem. Soc. Faraday I*, 1974, *70*, 535.

[16] R.M. Barrer "Zeolites and Clay Minerals as Sorbents and Molecular Sieves", Academic Press, 1978, p.318.

[17] M. Kocirik and A. Zikanova, *Ind. Eng. Chem. Fund.*, 1975, *13*, 347.

[18] J. Kärger, M. Bülow, P. Struve, M. Kocirik and A. Zikanova, *J. Chem. Soc. Faraday I*, 1978, *74*, 1210.

[19] D. Beckert and H. Pfeifer, *Ann. Phys. (Leipzig)*, 1965, *16*, 262.

[20] J. Kärger, *Z. Phys. Chem. (Leipzig)*, 1971, *248*, 27.

[21] J.E. Tanner and E.O. Stejskal, *J. Chem. Phys.*, 1968, *49*, 1768.

[22] R. Ash and R.M. Barrer, *Surface Sci.*, 1967, *8*, 461.

[23] J. Kärger, *Surface Sci.*, 1973, *36*, 797.

[24] J. Kärger, *Surface Sci.*, 1976, *57*, 749.

[25] S. Brunauer, "The Adsorption of Gases and Vapours", (Oxford) 1944, p.150.

[26] L.S. Darken, *Trans. Am. Inst. Mining Met. Engrs.*, 1948, *175*, 184.

[27] Ref. 16, p.280.

[28] J. Kärger, H. Pfeifer, M. Rauscher and A. Walter, *Z. Phys. Chem. (Leipzig)*, 1978, *259*, 784.

[29] R.M. Barrer, Symposium on the Characterisation of porous Solids, Neuchatel, 9-13 July 1978.

[30] Ref. 16, p.36 (Table 4) and p.38 (Fig.4).

[31] A. Quig and L.V.C. Rees, in Proc. 3rd Internat. Conference on Molecular Sieves, Sept. 3-7th, Zurich, 1973, p.277.

[32] Ref.16, p.314

[33] A. Dyer and A. Molyneux, J. Inorg. Nucl. Chem., 1968, 30, 829.

[34] S.P. Gabuda, Dokl. Akad. Nauk. SSSR, 1962, 146, 840.

[35] C. Parravano, J.D. Baldeschweiler and M. Boudart, Science, 1967, 155, 1535.

[36] W. Kuhn and M. Thurkauf, Helv. Chim. Acta, 1958, 41, 938.

[37] L. Onsager and L.K. Runnels, Proc. Nat. Acad. Sci., 1965, 50, 208.

[38] J.H. Wang, C.V. Robinson and I.S. Edelman, J. Amer. Chem. Soc., 1953, 75, 466.

[39] R.M. Barrer, Nature, 1947, 159, 508.

[40] C.N. Satterfield and C.S. Cheng, Am. Inst. Chem. Eng., 68th National Meeting, Houston 1971, Feb.28th - Mar.4th, Adsorption Pt I Paper 16f.

[41] From ref. 16, Table 8, p.292.

[42] P.L. Walker Jr, L.G. Austin and S.P. Nandi, in "Chemistry and Physics of Carbon", Vol.2, (Dekker) 1966, pp.257-371.

[43] R.M. Barrer, J. Soc. Chem. Ind., 1945, 64, 130.

[44] J.M. Hammersley in "Methods in computational Physics" Vol. I (Academic Press) 1963, p.281.

[45] H.S. Carslaw and J.C. Jaeger "Conduction of Heat in Solids" (Oxford) 1947, p.332.

[46] R.M. Barrer and L.V.C. Rees, Trans. Faraday Soc., 1954, 50, 852 and 989.

[47] Ref. 16, p.309 et seq.

[48] R.M. Barrer and D.A. Harding, Separation Sci., 1974, 9, 195.

NMR and Sorption Kinetic Studies
of Diffusion in Porous Silica

R. Haul, W. Heintz and H. Stremming

Institut für Physikalische Chemie und Elektrochemie,
Universität Hannover, Callinstr. 3A, D 3000 Hannover 1,
Fed. Republ. Germany

Abstract
Diffusion coefficients from rates of ad- and desorption
of benzene, cyclohexane and pyridine in spherical particles of
porous silica and from pulsed NMR measurements are compared over
a range of temperatures and surface coverages. The sorption
kinetics can be described by simultaneous diffusion and heat
conduction processes on the basis of the NMR self diffusion data.

1. Introduction
Material transport of sorbable gases in porous solids,
in particular in catalysts, has frequently been studied by a
variety of methods: Permeation techniques, kinetics of ad- and
desorption and tracer exchange experiments. More recently, NMR
pulse methods have been used to measure diffusivities by means
of the field gradient technique and the analysis of relaxation
times. It is therefore of interest to compare the results from
rates of sorption and NMR measurements in order to ascertain
whether the same elementary kinetic processes are revealed by
both methods.

In an earlier study on sorption kinetics of benzene in
porous plugs of compacted Aerosil it was found that the NMR
diffusion coefficients of the order of 10^{-5} cm^2/s at room temperature were larger by a factor of about two compared with values
from rates of adsorption[1]. Drastic differences were, however,
observed for cyclohexane in microporous thorium oxide for which
diffusion coefficients of 10^{-14} cm^2/s had been found from sorption
experiments[2] compared with NMR values which were larger by about
seven orders of magnitude[3]. In a number of studies on diffusion

of hydrocarbons in zeolites Kärger et al.[4] have found that self diffusivities are much larger than values from sorption measurements and that activation energies are lower. Interparticle transport and surface resistance effects were suggested to explain these discrepancies. A comparison between NMR and sorption data for diffusion in zeolites has been reviewed by Ruthven[5].

In a study on sorption and desorption of n-butane in zeolite NaX Doelle and Riekert[6] showed that apparent diffusion coefficients can be strongly influenced by interparticle transport effects and that the sorption process is affected by intermediate temperature changes due to the heat of adsorption.

The interpretation of these different findings is still controversial. In the present study, therefore a model system consisting of spherical porous silica particles of equal size has been investigated with respect to material transport under non-isothermal conditions as well as diffusion.

2. Experimental

2.1 Sorption measurements. Rates of ad- and desorption were measured with a vacuum microbalance (Cahn R.G., Paramount, USA; sensitivity $\pm$ 1 μg) at basic residual pressures $< 10^{-5}$ mbar[7]. The small V2A steel pan (240mg) with the sample was suspended in a 50 cm long double walled glass tube. The temperature was controlled ($\pm$ 0.03 $^{\circ}$C) by circulating thermostated water through the jacket as well as an inserted glass spiral. In a second glass tube a reference balance pan was suspended and in addition a pan with an identical sample was fixed in the same position as in the first tube. An iron-constantan thermocouple was brought in direct contact with the surface of adjoining particles. Furthermore, an NTC thermistor was attached with water glass to two silica particles suspended in the vapour. In this way temperature changes, typically 0.5 to 1 $^{\circ}$C, could be followed simultaneously with the rates of ad-or desorption.

Pressures in the sorption volume (7 1) were measured with a differential fused quartz Bourdon gauge (Texas Instruments, Houston, USA) with an accuracy of $\pm$ 10^{-3} mbar. The rates of sorption were measured at constant pressure in small differential steps corresponding to changes in amount adsorbed about 10 mg/g adsorbent. The weight changes were automatically recorded with an integrating digital voltmeter at one second intervals.

2.2 NMR measurements. A 60 MHz pulse spectrometer (B-KR-302,Bruker Physik AG, Karlsruhe) with a signal averaging device (Fabri Tek 1072, Nicolet Instr., Madison, USA) was used[7]. The following techniques were applied:

(i)   Rates of sorption were measured by following the signal intensity at intervals of 10 sec. Spin concentrations were calibrated with known amounts of sorbed cyclo-hexane.

(ii) Self diffusion coefficients were determined by means of the pulse field gradient spin echo method[8].

(iii) Self diffusion coefficients were derived from an analysis of the relaxation times[9].

2.3 Materials. The porous silica adsorbent consisted of regular spherical particles of uniform size (radius 0.03 cm).The samples were prepared by a special technique of hydrolytic polycondensa-tion of polyethoxysiloxane[10] in heterogeneous phase and were kind-ly supplied by Prof. Unger, University Mainz.The silica samples (porosity $\varepsilon = 0.53$, density of solid $\boldsymbol{g}^S = 2.2$ g/cm^3, density of porous adsorbent $\boldsymbol{g}^{\varepsilon} = 1.03$ g/cm^3) are characterized by nitrogen BET surface areas and relatively narrow pore size distributions, with no indications for micro pores. The following samples were used:

1) for NMR measurements: surface area 270 m^2/g, mean pore dia-meter 6 nm, weight 300 mg, height 20 mm.
2) for sorption measurements: 270 m^2/g, 6 nm, 50 mg, 2mm.
3) for sorption measurements: 327 m^2/g, 7.2 nm, 60 mg, mono-particle layer.

The absorptives cyclohexane, benzene and pyridine (grade "Uvasol", Merck, Darmstadt) were dried with Linde molecular sieve.

3. Evaluation of experiments

Diffusion from a well stirred fluid into a sphere of radius r_0 with an initial uniform concentration c_1 and a constant concentration c_0 at the surface is given by[11]

$$(1) \qquad \frac{M_t}{M_\infty} = 1 - \frac{6}{\pi^2} \sum_{n=1}^{\infty} \frac{1}{n^2} \exp\left(-n^2 \pi^2 \frac{D \cdot t}{r_0^2}\right)$$

M_t and M_∞ are the amounts of substance taken up at the time t and at equilibrium respectively, D is the diffusion coefficient considered as independent of concentration and temperature.

The simplified solution for large times

$$(2) \qquad \frac{M_\infty - M_t}{M_\infty} = \frac{6}{\pi^2} \exp\left(-\pi^2 \frac{D \cdot t}{r_o^2}\right)$$

is frequently used for the evaluation of sorption kinetic measurements.

For small times the solution is

$$(3) \qquad \frac{M_t}{M_\infty} = \frac{6\sqrt{Dt}}{r_o}\left(\pi^{-\frac{1}{2}} + 2\sum_{n=1}^{\infty} ierfc\frac{n\,r_o}{\sqrt{Dt}}\right) - \frac{3Dt}{r_o^2}$$

By approximation diffusion coefficients are often obtained from a plot of M_t versus $\sqrt{t}$.

Since heat is exchanged on ad- and desorption, the diffusion of sorbable gases in porous adsorbents does not occur under iso-thermal conditions. Thus material transport is determined by an interplay of diffusion and heat conduction processes.

In an early paper Wicke[12] has pointed out that heat conduction is a determining factor for the rate of adsorption and has mathematically treated this influence on the later stages of the sorption process. For diffusion of water vapour into a package of textile fibres, a problem also of practical significance, Henry[13] has given solutions for the coupled differential equations of diffusion and heat conduction for various boundary conditions. However, no heat transfer to the surroundings is considered. Crank[11] has given mathematical solutions for some cases in which the change of surface temperature can be evaluated on the assumption that the uptake of adsorptive is proportional to $\sqrt{t}$.

For the present case of sorption in a spherical porous adsorbent suspended in a gas of constant pressure of the adsorptive and dissipation of the heat of adsorption to the ambient atmosphere, an analytical solution of the system of coupled inhomogeneous partial differential equations for material and heat conservation becomes considerably involved.

As a step to this solution we have considered the following much simpler case. It is assumed that the diffusion coefficient is independent of temperature and concentration during the differential ad- or desorption change. Further the heat conductivity of the porous adsorbent is supposed to be so large that the sphere is at a uniform temperature. Heat is exchanged at the surface at a rate which for small temperature differences θ between the adsorbent surface and ambient atmosphere is given by $K \cdot \theta$, where k is a heat transfer coefficient.

From equation (1) dM/dt is obtained and inserted in the equation expressing conservation of heat at the surface of the sphere:

$$(4) \qquad q_{st} \varrho^\varepsilon \frac{\partial M}{\partial t} = c_p^s \, \varrho^\varepsilon \, \frac{\partial \theta}{\partial t} + \frac{3 k \theta}{r_o}$$

where q_{st} is the isosteric heat of adsorption, ϱ^ε the density of the porous adsorbent and c_p^s the specific heat of the solid adsorbent including the sorbate.
This leads to:

$$(5) \qquad \frac{\partial \theta}{\partial t} + \frac{3 k \theta}{r_o \, c_p^s \, \varrho^\varepsilon} = \frac{6 \, q_{st} \, M_\infty \, D}{c_p^s \, r_o^2} \sum_{n=1}^\infty \exp(- n^2 \pi^2 \frac{D \, t}{r_o^2})$$

The solution is:

$$(6) \qquad \theta = \sum_{n=1}^\infty \frac{6 \, q_{st} \, \varrho^\varepsilon \, M_\infty \, D}{3 k r_o - n^2 \pi^2 c_p^s \varrho^\varepsilon D} [\exp(-n^2 \pi^2 \frac{D t}{r_o^2}) - \exp(- \frac{3 k}{c_p^s \, r_o \varrho^\varepsilon} t)]$$

for $\theta = 0$ at t = 0 and t = ∞.

4. Results

4.1 Adsorption isotherms

The amount adsorbed at equilibrium in each gravimetric rate
experiment yields one point of the corresponding adsorption isotherm. While the isotherms of benzene and cyclohexane (fig.1)
belong to type II, the isotherm of pyridine (fig. 2) follows
type III and can thus not be evaluated by the BET method.

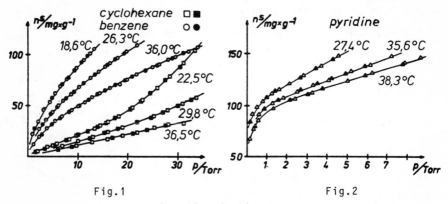

Fig.1   Fig.2

Sorption isotherms
adsorption: empty, desorption: full symbols

On the basis of the N_2 surface area the usual cross section
of 41 $Å^2$ is obtained for benzene (c = 6), while a value of
25 $Å^2$ would result for pyridine (c = 104). This small cross
section may suggest an upright orientation of the adsorbed
pyridine molecule. The large c-value indicates strong interaction
which is attributed to hydrogen bonding as deduced from IR
studies [14,15]. For cyclohexane an estimated cross section of
38 $Å^2$ was used to indicate surface coverage [16].

From the adsorption isotherms the following isosteric heats of
adsorption were evaluated, weight adsorbed 100 mg/g: Cyclohexane
10.0, benzene 10.5 and pyridine 14.3 kcal/mol.

4.2 Diffusion coefficients from sorption measurements. A typical
example for graphical evaluation of effective diffusion coefficients on the basis of equation (2) for large times is shown in
fig.3a (cyclohexane). The extrapolated straight line leads to
the expected intercept. A plot of M_t/M_∞ versus $\sqrt{t}$ according to
the first term of equation (3) does not result in a straight

line for small times. All sorption data were treated with a
computer program according to equations (2) and (3) respectively.
Comparison of the experimental uptake curve with the calculated
curves for D_e $(t \to \infty)$ = $8 \cdot 10^{-8}$ and D_e $(t \to o)$ = $7 \cdot 10^{-6}$ cm^2/s
clearly demonstrates the deficiency of the isothermal treatment
of the sorption process.

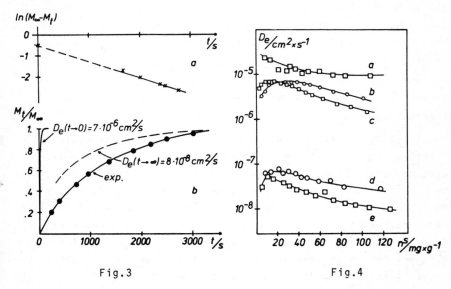

Fig.3　　　　　　　　　　　　　　　Fig.4

Fig.3 a) Evaluation of $D_e(t \to \infty)$, equat. (2),b) Uptake curve for
　　　cyclohexane, T= 22.5°C, M_∞ = 18 mg/g. Calculated (equat.1)
　　　for D_e $(t \to \infty)$= $8 \cdot 10^{-8}$ and D_e $(t \to o)$= $7 \cdot 10^{-6}$ cm^2/s.

Fig.4　　Comparison of diffusion coefficient for cyclohexane
　　　a) NMR, fieldgradient, self diffusion D_e^* (△), sample 1, b)
　　　sorpt. kinetics, self diffus. D_e^* $(t \to o)$ sample 2,c)
　　　dito, D_e $(t \to o)$,d) sorpt. kinetics, NMR, signal inten-
　　　sity, D_e (NMR), sample 1,e) sorpt. kinetics, D_e $(t \to \infty)$,
　　　sample 2.

　　　D_e values from sorption kinetics are plotted versus surface
coverage for cyclohexane in fig.4 . Remarkably, the diffusion
coefficients are larger by two orders of magnitude for the
initial (D_e $(t \to o)$,curve c) than for the final(D_e $(t \to \infty)$,curve e)
region of the sorption process. This is true for all three
systems studied. At low coverage ($\theta \approx 0.1$) the diffusion coefficients show a maximum as had been observed and discussed already
earlier[17].

4.3 Diffusion coefficients from NMR measurements. Since the
signal intensity is directly proportional to the number of spins
and thus to the concentration of adsorbate, the NMR method
enables also the measurements of diffusion coefficients under
non-equilibrium conditions corresponding to the gravimetric
experiments. Comparing the results (curve d and e in fig.4) it
has to be considered that the amount and the packing of the
samples were different. The order of magnitude and the trend of
the diffusion coefficients, however, is the same.

The self diffusion coefficients D_e^* ($\triangle$) from pulsed field
gradient measurements are shown in curve a fig.4. These are still
larger than the D_e ($t \rightarrow o$) and D_e^* ($t \rightarrow o$) values from sorption
experiments.

The temperature dependence of D_e^*($\triangle$) is shown for cyclohexane in fig.5 and for benzene and pyridine in fig.6. In all
three cases the slope of the curves decreases distinctly at
lower temperatures. In this range, below about 0°C, the diffusion
coefficients are dependent on the pulse distance $\triangle$. For experimental reasons the variation of $\triangle$ is limited. The influence of
the pulse distance indicates the effect of restricted diffusion
which has been discussed in detail for diffusion in zeolites by
Kärger[18]. The present paper will be mainly concerned with a
comparison between diffusion coefficients from NMR and sorption
data in the temperature range above about 10°C. In this case
the diffusion coefficients are independent of the pulse distance.

In addition to the results obtained by the field gradient
technique, diffusion coefficients in the lower temperature range
were also derived from an analysis of relaxation times for the
three systems[7]. On the basis of the treatment by Beckert[9] it is
possible to derive jump times $\mathcal{T}_{tr.}$ as well as jump distances λ for
two-dimensional translational diffusion from the T_1 minima of
the intermolecular relaxation contributions. The temperature
dependence of the surface self diffusion coefficients $D^{S*} = \lambda^2 / 4\mathcal{T}_{tr.}$
is indicated in fig. 5 and 6. The following parameters were
obtained; correlation time $\mathcal{T}_c$ at the T_1 minima $5.3 \cdot 10^{-9}$ sec.,
cyclohexane: T_1 minimum 103 K, jump distance λ 3.8 Å, activation
energy 1.04 kcal/mol; pyridine: 159 K, 3.1 Å, 1.4 kcal/mol;
benzene: 102 K, 3.6 Å, 1.1 kcal/mol. Further details of the
relaxation time measurements will be published elsewhere.

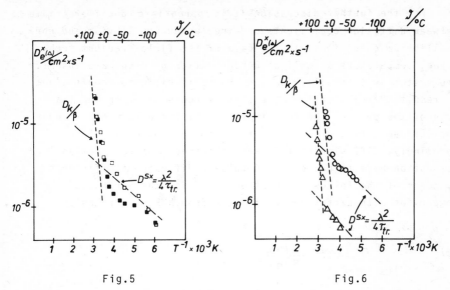

Fig.5   Fig.6

NMR self diffusion coefficients D_e^* ($\triangle$), surface coverage $\theta = 1$. cyclohexane, pulse distance $\triangle$ 4ms $\square$, 10ms $\blacksquare$, benzene 5ms $\circ$, pyridine, secondary echo 80ms $\triangle$.

The self diffusion coefficients obtained from relaxation times are in good agreement with the values from the field gradient method in the lower temperature range. They are evidence for the preponderance of surface transport at these temperatures.

5. Discussion

In principle, material transport in porous media can occur within the pore volume and/or along the pore walls. According to a theoretical treatment which goes back to Damköhler[19], for the case of molecular flow the following relation holds between the effective diffusion coefficient D_e, the pore volume Knudsen diffusion coefficient D_k and the surface diffusion coefficient D^S:

$$(7) \qquad D_e = D_k/\beta + D^S$$

$$\text{where} \quad \beta = \frac{\partial c^S}{\partial c^g} = \frac{1-\varepsilon}{\varepsilon} RT \varrho^S \frac{\partial n^S}{\partial P} \gg 1$$

is the slope of the adsorption isotherm.

For the further discussion it is appropriate to differentiate
between diffusion coeffcients obtained under equilibrium and non-
equilibrium conditions. In the case of the field gradient technique, temperature, as well as concentration gradients are absent.
The system is in equilibrium and thus self diffusion coefficients
D^* result. Marking of the molecules is achieved by the nuclear
spin of the protons. On the other hand diffusion can occur under
the influence of a gradient of the chemical potential. Corre-
spondingly, diffusion coefficients obtained from rates of ad- and
desorption have to be converted to self diffusion coefficients
by means of the following relation [17,20] for the purpose of
comparison with diffusion coeffcients from NMR measurements

$$(8) \qquad D^x = D \; \frac{d \ln c}{d \ln a}$$

where a is the activity and c the concentration of the adsorptive
per unit volume of adsorbent. For ideal behaviour of the vapour
a=p, where the term $d \ln c / d \ln p$ can be obtained from equlibrium
sorption measurements. Equation (8) is a special form of the
Darken equation for binary systems if one component, in this
case the silica framework, is immobile.

5.1 Self diffusion coefficients. From the temperature dependence
of the effective self diffusion coefficients (Fig. 5 and 6)
it can be seen that D_e^* (Δ) shows a steep increase above
temperatures of about $10^{\circ}C$ for benzene and cyclohexane and about
$30^{\circ}C$ for pyridine. The slopes result in activation energies
which are just below the heats of adsorption (section 4.1). This
indicates that desorption takes place, leading to increased
transport in the pore volume as the temperature is raised while
at lower temperatures surface diffusion prevails (section 4.3).
Over the whole temperature range, transport is given by a
weighted superposition of both mechanisms according to equation
(7).

The pore volume diffusion coefficient can be calculated
for Knudsen flow conditions if the pore texture is described
by equivalent parallel cylindrical capillaries.

$$(9) \qquad D_k = \frac{8}{3} \frac{\varepsilon}{A_v} \left(\frac{2RT}{\pi M} \right)^{1/2}$$

where M is the mole mass of the adsorptive and A_v the surface area
per unit volume of adsorbent. In this way the straight lines
denoted D_k/β were calculated in fig. 5 and 6. The deviations of
the experimental values at lower temperatures are due to surface
diffusion. The straight lines in this range were calculated on
the basis of relaxation times data (section 4.3). It can thus be
concluded that the proposed model correctly describes the molecular mobility of the adsorption in the pore system.

5.2 Non-equilibrium diffusion coeffcients. A comparison between
NMR self diffusion coefficients and diffusion coefficients from
rates of ad-and desorption (fig. 4, section 4.2 and 4.3) clearly
demonstrates the large differences and the fact that the sorption
process can not be described by a uniform diffusion coefficient
under isothermal conditions (fig.3b). As has already been
mentioned, similar discrepancies were found for the mobility of
hydrocarbons in zeolites by Kärger et al[4] and Ruthven et al[5]. On
the other hand Wicke[12] and more recently Doelle and Riekert[6]have
emphasized the effect of intermediate temperature changes due to
the heat of adsorption.

As a typical example for the findings in the present investigation an uptake curve for benzene is shown in fig.7 together
with the simultaneous temperature changes. It can be seen that
under the prevailing experimental conditions a temperature
maximum of 0.6°C has been recorded after about 20 sec., while the
completion of the sorption process requires more than 300 sec.
In other experiments the temperature maximum was even larger than
1°C depending on the amount adsorbed during a differential
sorption step. The course of the temperature change clearly illustrates the relatively rapid uptake by diffussion in the initial
range of sorption and the rather slow establishment of equilibrium due to heat exchange with the surroundings.

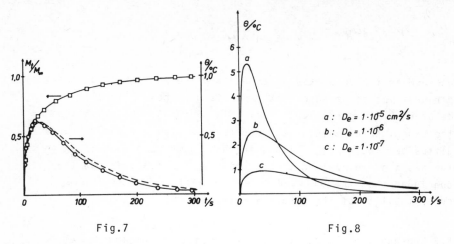

Fig.7   Fig.8

Fig.7 Uptake curve and temperature change: benzene,
25°C, n^S = 62.7 mg/g $\triangleq$ θ = 0.64, M_∞ = 11.1 mg/g adsorbent.
calculated temperature curve--- for D_e = 1.10^{-6}cm^2/s,
k = 5·10^{-5} cal/cm^2·s·K, r_0 = 0.03 cm.

Fig.8 Calculated temperature difference θ, variation of diffusion
coefficient D, k = 5.10^{-5} cal/cm^2·s·K, r_0 = 0.03cm,
M_∞ = 11.1 mg/g.

For comparison with the experimental results the intermediate
temperature changes θ were calculated with equation (6) for
variations of the diffusion coefficient D (fig.8), the heat transfer coefficient k (fig.9), the radius of the spherical adsorbent
r_0 (fig.10) and the amount adsorbed M_∞ during a sorption step. The
following constant parameters were chosen: specific heat
c_p^S = 0.21 cal/g·K, density of porous adsorbent ρ^ϵ = 1.03g/cm^3,
isosteric heat of adsorption q_{st} = 147 cal/g.

The following conclusions can be drawn. The temperature
maximum increases with rising diffusion coefficient and is shifted
to slightly smaller times (fig.8). On the other hand smaller heat
transfer coefficients result in larger temperature maxima at
larger times(fig.9). The drastic influence of the particle size
is shown in fig.10. For particles with radii less than 0.01 cm,
as had been used, for instance, in previous sorption studies in
zeolites [4d,6], the maximum would occur already at times less than
2 sec. As the particle size increases, the temperature effects
become more and more blurred. The maximal temperature change

θ_{max} is directly proportional to M_{∞}, the amount adsorbed during a sorption step, without altering the time at which the maximum occurs.

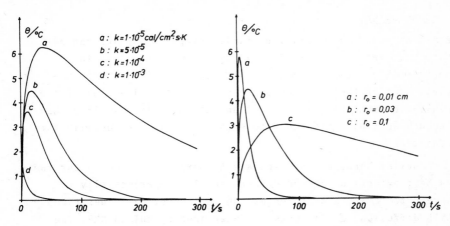

Fig.9   Fig.10

Calculated temperature difference θ

Fig. 9 Variation of heat transfer coefficient k

$\quad$ $D = 5 \cdot 10^{-6} cm^2/s$, $r_0 = 0.03 cm$, $M_{\infty} = 11.1 mg/g$.

Fig.10 Variation of particle radius r_0

$\quad$ $D = 5 \cdot 10^{-6} cm^2/s$, $k = 5 \cdot 10^{-5} cal/cm^2 \cdot s \cdot K$, $M_{\infty} = 11.1 mg/g$.

For comparison with the experimental results, a theoretical temperature curve is shown in fig.7. The calculation is based on the NMR diffusion coefficient D_e^* ($\triangle$) = $1 \cdot 10^{-5} cm^2/s$ and a heat transfer coefficient $k = 5 \cdot 10^{-5}$ $cal/cm^2 \cdot s \cdot K$, while the ratio $\theta_{max}(exp)/\theta_{max}$ (theoret.) = 0.3 has been adjusted. In principle, the course of the temperature change is correctly described. As has been stated, the present theoretical treatment is an approximation in which a constant diffusion coefficient is assumed as well as uniform temperature within the single adsorbent particle suspended in the vapour. In reality, the sample consisted of a number of particles placed on a metal pan. Although the packing of the rather large particles used in the present study has no influence on the material transport, it will affect the heat conduction properties.

Besides the heat effects treated in the present paper, it is conceivable that narrow passages in the pore texture of certain adsorbents could influence the sorption process. For the silica

particles studied here, there is no evidence for such effects
since the NMR self diffusion coefficients are independent of
the pulse distance in the temperature range discussed.

6. Conclusions

Rates of sorption in small particles of porous solids can not
simply be treated by Fickian diffusion. This can lead to
considerable inconsistencies: diffusion coefficients from the
final part of the sorption process would be smaller than those
evaluated from the initial part and these in turn smaller than
the NMR self diffusion coefficients measured under equilibrium
conditions.

During sorption intermediate temperature changes occur due
to heat effects of ad-and desorption. The experimentally observed temperature differences can be theoretically described
with diffusion coefficients as determined by pulsed NMR. The
treatment indicates the dependence of heat effects on the
diffusion coefficient, the heat transfer coefficient, the
particle size and the amount adsorbed in a differential sorption
step.

Acknowledgement

The authors wish to thank Prof. Dr. K. Unger, Universität
Mainz, who kindly supplied the well characterized silica samples.
Thanks are due to Prof. Dr. B. Boddenberg, Universität Dortmund,
for discussions in the early stages of the work. Financial
support from Deutsche Forschungsgemeinschaft und Fonds der Chemischen Industrie is gratefully acknowledged.

References

1. B. Boddenberg, R. Haul and G. Oppermann,
   Advan. Mol. Relax. Process, 1972, 3, 61.

2. H. Knözinger, H. Jeziorowski and W. Meye,
   J. Colloid Interface Sci., 1975, 50, 283.

3. E. Riensche, Dissertation, Techn. Univ. Hannover 1976,

4a) J. Kärger, Z. Phys. Chem (Leipzig), 1973, 252, 299.
 b) J. Kärger, P. Lorenz, H. Pfeiffer and M. Bülow,
    Z. Phys. Chem. (Leipzig), 1976, 257, 209.

 c) M. Bülow, J. Kärger, N. van Phat and W. Schirmer,
    ibid., 1976, 257, 1205.
 d) J. Kärger, M. Bülow and N. van Phat,
    ibid., 1976, 257, 1217.

5. D.M. Ruthven, in J.R. Katzer (editor), Molecular Sieves II,
   ACS Symposium Series, 1977, 40, 320.

6. H.J. Doelle and L. Riekert, in J.R. Katzer (editor),
   Molecular Sieves II, ACS Symposium Series, 1977, 40, 401.

7. W. Heintz, Dissertation, Techn. Univ. Hannover, 1977.
   H. Stremming, Dipl.-thesis, Techn. Univ. Hannover, 1978.

8. J. Kärger, Z. Phys. Chem. (Leipzig), 1971, 248, 27.

9. D. Beckert, Ann. Phys. (Leipzig), 1967, 20, 220.
   D. Beckert, ibid. , 1969, 23, 98.

10. K. Unger, J. Schick-Kalb and B. Straube,
    Colloid Polym. Sci., 1975, 253(8), 658.

11. J. Crank, "The Mathematics of Diffusion",
    Clarenden Press, Oxford, 1956, p. 86, p. 318.

12. E. Wicke, Kolloid-Zeitschr., 1939, 86, 167.

13. P.S.H. Henry, Proc. R.Soc. London, Ser. A,
    1939, 171, 215.
    P.S.H. Henry, Discussion of the Faraday Soc.,
    1948, 3, 243.

14. H. Knözinger, Surf. Sci., 1974, 41, 339.

15. G. Curthoys, V.Y. Davydov, A.V. Kiselev, S.A. Kiselev
    and B.V. Kuznetsov, J. Colloid Interfac Sci.,
    1974, 48, 58.

16. N. Smith, C. Pierce and H. Cordes, J. Am. Chem. Soc.,
    1950, 72, 5595; R. Sh. Mikhail and F.A. Shebl,
    J. Colloid Interface Sci. 1970, 32, 505.

17 B. Boddenberg, R. Haul and G. Oppermann,
    Surf. Sci., 1970, 22, 79.

18. J. Kärger, Z. Phys. Chem. (Leipzig), 1971, 248,27.

19. G. Damköhler, Z. Phys. Chem.,1935, A174, 222.

20. L.S. Darken, Trans. A.I.M.E., 1948, 1975, 184;
    R.M. Barrer and B.E.F. Fender, J. Phys. Chem. Solids,
    1964, 21, 12; R. Ash and R.M. Barrer,
    Surf. Sci.,1967, 8, 46; J. Kärger, Surf. Sci.,
    1973, 36, 797.

Diffusion in Zeolites: A Review of Recent Progress

by

Douglas M. Ruthven
Department of Chemical Engineering
University of New Brunswick
Fredericton, N.B., Canada

ABSTRACT

The results of recent experimental studies of sorption kinetics are reviewed. Topics covered include thermal effects, comparative studies with different zeolite samples and with different size fraction of the same crystals, a comparison between sorption and NMR diffusivities, chromatographic methods and diffusion in a binary adsorbed phase.

Introduction

The rational design and optimization of adsorption separation processes requires a detailed knowledge of both the kinetics and equilibrium of adsorption. This requirement and the academic advantage of studying diffusional behaviour in the regular lattice of a zeolite, rather than in the random pore structure of other microporous adsorbents, have stimulated extensive research. The majority of kinetic studies have employed classical sorption rate measurements but recently NMR methods have been applied. The results of these investigations show remarkable discrepancies between the diffusivities for apparently similar systems. Many of the discrepancies between the results of sorption rate measurements can be rationally accounted for by the intrusion of other mass transfer[1] or heat transfer resistances[2]. Diffusivities have been reported which are substantially smaller for desorption compared with adsorption, under similar conditions[3,4], but detailed analysis shows that such differences can generally be explained by a failure to account correctly for the concentration dependence of the diffusivity in a non-linear system[5]. However, the results of recent kinetic studies with various samples of 4A and 5A zeolite crystals reveal that the differences in diffusional properties between samples are very much greater than has been previously assumed. Therefore some of the apparent discrepancies between diffusivities measured in different laboratories probably reflect genuine differences in the zeolite samples.

There have been four recent reviews[6-9] which between them summarize most of the earlier work including detailed comparisons of the results of sorption and NMR studies. The present paper will therefore deal only with the more recent work, much of which has not yet been formally published.

Heat Transfer Effects in Sorption Rate Measurements

Diffusion coefficients for adsorbed gases have often been calculated from experimental transient uptake curves on the assumption that the system may be considered isothermal. Because of the heat of adsorption this approximation is valid only when diffusion is relatively slow. The potential significance of thermal effects has been demonstrated experimentally by Eagan and Anderson[10] and, more recently, by Dolle and Riekert[11]. However, apart from the early work of Armstrong and Stannett[12], which appears to have been largely ignored by subsequent investigators, there has, until recently, been little detailed analysis[2,13,14]. There are three potential thermal resistances which must be considered; external heat transfer from the surface of the pellet or crystal aggregate, heat transfer from the surface of an individual crystal/conduction

through the aggregate and heat conduction within an individual crystal.
Analysis of the time constants for these processes shows that, under all
physically reasonable conditions, the first of these processes will be by far
the slowest. It is therefore a good approximation to consider the temperature
to be uniform throughout the adsorbent sample but, as a result of the finite
external heat transfer resistance, there may be a significant temperature difference between the adsorbent and the ambient fluid.

There are two distinct ways in which the heat transfer rate may affect
the uptake rate. During the initial part of the curve the temperature dependence of the diffusivity is the main effect leading to somewhat faster adsorption
compared with an isothermal system[10] (and somewhat slower desorption) but in
the final stages it is the temperature dependence of the equilibrium position
which is dominant. The first of these effects can be eliminated by making
measurements differentially over a sufficiently small concentration step but
the second effect is independent of the step size and leads to the same in-
hibition of both adsorption and desorption rates.

Theoretical Model. To obtain the expression for the uptake curve under non-
isothermal conditions requires the simultaneous solution of the differential
heat and mass balance equations, subject to the time dependent boundary
condition arising from the temperature dependence of the equilibrium. Our
analysis, which is described in more detail elsewhere[14], is restricted to
systems in which the change in sorbate concentrations is sufficiently small for
the equilibrium relationship to be considered linear and the diffusivity constant. The zeolite sample is considered as an assemblage of uniform spherical
crystals in which the mass transfer rate is controlled entirely by intracrystalline diffusion. Since the dominant heat transfer resistance is external,
the temperature is considered uniform throughout the sample and the heat
transfer rate is represented by Newton's Law.

Subject to these approximations the differential heat and mass balance
equations may be solved simultaneously to obtain an expression for the non-
isothermal uptake curve, for a step change in ambient sorbate concentration at
time zero[14]:

$$m_t/m_\infty = 1 - \sum_{n=1}^{\infty} \frac{9\{(q_n\cot q_n-1)/q_n^2\}^2\exp(-q_n^2\tau)}{\frac{1}{\beta} + \frac{3}{2}\{q_n\cot q_n(\frac{q_n\cot q_n-1}{q_n^2}) + 1)\}} \tag{1}$$

where q_n is given by the roots of:

$$3\beta(q_n\cot q_n-1) = q_n^2 - \alpha \tag{2}$$

The uptake curve thus depends on two dimensionless parameters $\alpha \equiv (ha/\rho c_s)/$
(D/r^2) and $\beta \equiv (\Delta H/c_s)(\partial q^*/\partial T)$ in addition to the diffusional time constant
D/r^2. For either $\alpha \to \infty$ (rapid heat transfer) or $\beta \to 0$ (negligible heat of
sorption or temperature dependence of the equilibrium) the roots of eqn. 2
become simply $q_n = n\pi$ so that eqn. 1 reduces to the familiar expression for
isothermal diffusion[15]:

$$m_t/m_\infty = 1 - \frac{6}{\pi^2} \sum_{n=1}^{\infty} \frac{1}{n^2} e^{-n^2\pi^2\tau} \tag{3}$$

The limiting case of heat transfer control ($\alpha \to 0$) is obtained by replacing
$(q_n\cot q_n-1)$ in eqns. 1 and 2 by $-q_n^2/3$, the first term of the series expansion,
and considering only the first term of the summation:

$$m_t/m_\infty = 1 - \frac{\beta}{1+\beta} \exp\{-\alpha\tau/(1+\beta)\} \tag{4}$$

It is evident that, provided the differential step is small enough to validate the assumptions of the model (D and $\partial q^*/\partial T$ constant) the response curves for adsorption and desorption will be identical and independent of the step size. Varying the step size is therefore not a sensitive test for the intrusion of thermal effects, contrary to our previous assumption [16].

For all values of α and β the curves reduce in the long time region to a simple exponential decay. It is therefore not always possible to detect the intrusion of thermal effects simply from an examination of the later parts of the uptake curves. Comparison of the uptake curves calculated from eqns. 1 and 2 with the isothermal solution (eqn. 3) shows that the isothermal model is a valid approximation only if $\alpha/\beta > \sim 60$. Since the parameter α is directly proportional to the external area/volume for the sample, varying the depth and configuration of the adsorbent provides a convenient experimental test for the intrusion of heat transfer resistance.

Comparison of Theory and Experiment

Experimental uptake curves for three systems, selected to illustrate the main features of the model are analyzed below.

n-Butane-Linde 13X (crystal diameter 2.8µm). For this system diffusion is rapid and uptake rates are essentially controlled by heat transfer. The uptake curves, which show a very rapid initial uptake followed by a slow exponential approach to equilibrium, are well described by eqn. 4. The parameters obtained from the analysis of one set of such curves (at 88°C) are summarized in Table 1 and the form of the uptake curves is shown in figure 1. Values of βc_s were calculated from the equilibrium data and the values of c_s derived using the values of β from the kinetic uptake curves are essentially constant and close to the estimated total heat capacity. Values of the heat transfer coefficient are also essentially constant and close to the value estimated from the combined effects of radiation (Stefan's Law with emissivity 0.8) and conduction ($N_{Nu} = 2Rh/k = 2$).

Using the Langmuir expression as an approximation for the isotherm:

$$\Theta = q^*/q_m = bc/(1+bc); \quad b = b_o \exp(-\Delta H/RT) \tag{5}$$

it follows that the parameter β is given by:

$$\beta = \left(\frac{\Delta H^2}{RT_0}\right) \frac{Rq_m}{c_s} \Theta(1-\Theta) \tag{6}$$

It is evident that β will reach its maximum value at $\Theta = 1/2$ so that under conditions of heat transfer control the uptake rate will be rapid at both low and high concentrations, passing through a minimum in the intermediate range. This trend is shown by the diffusional time constants, arbitrarily calculated from the 80% uptake times according to the isothermal model. Similar trends have been reported[17] in the apparent diffusivities for several rapidly diffusing systems and, in view of the present analysis, such data should be re-examined.

n-Butane-Linde 5A (3.6µm diameter crystals)

For this system diffusion is relatively slow and thermal effects have only a minor effect on the uptake rate. Figure 1 shows uptake curves for three different bed depths, measured over the same pressure step. Also shown is the much faster uptake curve obtained with a sample of small (2.8µm) 13X crystals under otherwise similar conditions. Since the equilibria for 13X and 5A are similar the only significant difference between the 13X and 5A runs is in the crystal structure. The large difference in uptake rates therefore confirms the dominance of intracrystalline diffusional resistance. The curves for the two smaller bed depths are close to the theoretical curve for

Table 1

Analysis of Experimental Uptake Curves for n-Butane
in Linde 13X Crystals (2.8 μm diameter) at 88°C

Pressure Step (Torr)	Final Concentration (mmoles/gm)	From Equilib. βc_s	$(ha/\rho\, c_s)$	From Uptake Curves		Apparent D/r^2 (sec^{-1})
				β	c_s (cals/gm.deg.)	
78-60	1.945	.08	0.072	0.163	0.49	0.077
43-32	1.775	.15	0.074	0.37	0.41	0.02
32-24	1.67	.21	0.063	0.40	0.53	0.018
24-17	1.52	.20	0.07	0.46	0.44	0.012
16-13	1.4	.281	0.075	0.50	0.55	0.012
10-9	1.17	.284	0.078	0.56	0.51	0.0097
8.8-7	1.01	.32	0.08	0.61	0.52	0.0093
7.0-5.2	0.84	.27	0.073	0.49	0.55	0.0105
4.8-38	0.67	.21	0.08	0.45	0.47	0.015
3.0-2.8	0.52	.20	0.074	0.33	0.59	0.023
2.8-2.1	0.43	.16	0.081	0.28	0.56	0.036
2.0-1.5	0.34	.12	0.091	0.20	0.59	0.06
1.5-1.0	0.29	.075	0.071	0.15	0.52	0.10

Estimation of c_s

Weight of sample = 14 mg, heat capacity = 0.22 cals/gm.deg.
Weight of pan ≈ 20 mg, heat capacity = 0.21 cals/gm.deg.
Estimated total heat capacity = $0.22 + \dfrac{20 \times 0.21}{14} = 0.52$ cals/gm.deg.

Estimation of h

Radiation contribution: $h' = 4.4 \times 10^{-12} T^3 = 2.05 \times 10^{-4}$ cals/cm².sec.deg.
Conduction contribution $h'' = k/R = 5 \times 10^{-5}/0.16 = 3.1 \times 10^{-4}$ cals/cm².sec.deg.
Total estimated heat transfer coefficient = 5.2×10^{-4} cal/cm².sec.deg.

an isothermal system. With the largest sample there is a minor effect of thermal resistance evident above 70% uptake. The experimental curve agrees well with the theoretical curve calculated from eqn. 1 and 2 with the same value of D/r^2 and with values of β and $ha/\rho c_s$ estimated the equilibrium data and the experimental heat transfer coefficient.

For this system the non-isothermal analysis essentially confirms the validity of the isothermal approximation[16].

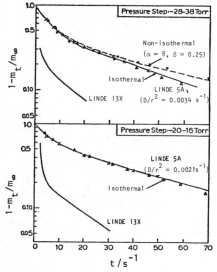

Fig. 1 Experimental and theoretical uptake curves for nC_4H_{10} at 323K in Linde 5A (x = 14 mg, ▲ = 22 mg, ● = 52 mg).

Fig. 2 Experimental and theoretical uptake curves for CO_2-5A (34μm) at 323K. Non-Isothermal curves are from eqns. 1 and 2.

CO_2-5A (34μm, lab synthesized crystals, 65% Ca^{++}, 323K). For this system diffusion is rapid and the heat of sorption is large (∿ 10 kcal/mole), so that uptake rates are controlled by the combined effects of heat transfer and intracrystalline diffusion (eqn. 1). Experimental uptake curves, measured over comparable pressure steps, with two different samples of the same zeolite crystals (10.3 mg on a wide pan and 37 mg on a smaller pan) are shown in figure 2. The large difference in uptake rates for the two different sample configurations can be quantitatively accounted for by heat transfer effects. Values of $ha/\rho c_s$ and c_s for the two samples were obtained directly from experiments at higher CO_2 pressures under conditions of complete heat transfer control (eqn. 4). Values of β were then calculated from the equilibrium data and the uptake curve for the smaller sample was matched to eqn. 1 to determine the diffusional time constant (D/r^2). The theoretical curves calculated from eqn. 1 with this value of D/r^2 and the appropriate values of α and β for the two sample configurations fit the experimental data well. Although the uptake curve for the smaller sample deviates significantly from the isothermal limit, it is still possible to derive a reasonably reliable value of D/r^2 from the non-isothermal model. However, in the case of the larger sample the uptake rate is almost entirely controlled by heat transfer and under these conditions no reliable information about the diffusional time constant can be obtained. Over the entire measurable range (t > ∿ 5 secs) the uptake curve for the larger sample can be very closely approximated by the isothermal model (eqn. 3) with $D/r^2 = 0.0015$ sec^{-1}. However, this is fortuitous since the uptake rate is in fact thermally limited. The value of D/r^2 derived assuming isothermal behaviour is thus more than an order to magnitude too small.

Effect of Crystal Size on Sorption Rates

To explain the large discrepancy between NMR self diffusivities and the corrected diffusivities derived from sorption rate measurements it has been suggested that uptake rates may be controlled by a surface barrier rather than by intracrystalline diffusion[7]. Evidence in favour of this hypothesis has been presented by Karger, Bulow and co-workers who, in a series of sorption rate studies for several sorbates in CaA and MgA zeolites, found that the apparent diffusional time constants (D/r^2), derived from the uptake curves, did not show the expected dependence on crystal radius, varying more nearly with r than with r^2[18,19]. However, the form of the uptake curves, both in our own experiments and in the experiments of Karger and Bulow, is generally more consistent with diffusion control (initial uptake proportional to $\sqrt{t}$ rather than t).

We carried out a similar series of studies in which we measured uptake rates for CH_4, C_2H_6, N_2 and CO_2 in different size fractions of 4A zeolite crystals and for CF_4 and nC_4H_{10} in the 5A form. To eliminate uncertainties arising from the use of different zeolite samples we prepared our own crystals by Charnell's method[20] and separated the crystals into different size fractions. Representative results are shown in figures 3-8.

For the slower diffusing systems the absence of significant thermal or bed diffusional resistance was confirmed directly by varying the sample configuration. Both the form of the uptake curves and the difference in sorption rates between different sized crystals were well represented by the simple isothermal diffusion model. This is illustrated, for the N_2-4A system, in figure 3. Diffusivities for several different sorbates in 4A, calculated from the time constants, are plotted against crystal size in figure 4 and it is evident that the diffusivities for the different size fractions are consistent, within the limits of experimental error.

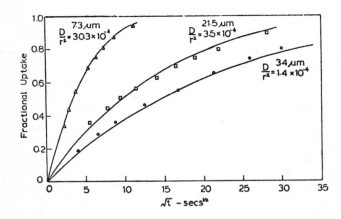

Fig. 3   Representative uptake curves for N_2 at 273K in 3 different size fractions of 4A zeolite crystals. Points are experimental. Curves are calculated for isothermal diffusion, with due allowance for crystal size distribution ($D = 4 \times 10^{-10}$ cm^2 sec^{-1} for all curves).

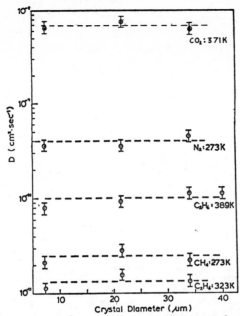

Fig. 4. Diffusivities calculated from uptake curves for different size fractions of 4A zeolite crystals. Error bars denote $\pm$ 15%. Data are for low concentration region in which D is not concentration dependent.

For some of the faster diffusing systems (nC_4H_{10} in 7.3μm 5A and CO_2 in 7.3 μm 4A) the effects of heat transfer resistance become significant, especially at higher sorbate concentrations. The uptake curves for such systems, when analyzed according to the non-isothermal model (eqn. 1), gave values of D/r^2 which were consistent with the data for the larger crystals in which diffusion was essentially isothermal. However, if the uptake curves are interpreted assuming isothermal behaviour, erroneously low values of D/r^2 are obtained (see figures 5 and 6). This is one way in which an apparent dependence of diffusivity on crystal size can arise. Such an effect can also arise from other time delays since any time delay will be proportionately more significant for the smaller crystals in which the value of D/r^2 is larger.

For all the systems we investigated our results show that, provided due allowance is made for heat transfer effects in the faster diffusing systems, the uptake curves are entirely consistent with the assumption that the dominant resistance to mass transfer is intracrystalline diffusion. We find no evidence of either a surface barrier or a systematic variation of diffusivity with crystal size. However, when different samples of zeolite crystals are compared we find large differences in diffusivity, as illustrated in figures 7 and 8. In general the diffusivities for the small commercial zeolite crystals are much smaller than the values for our own crystals but even between different commercial samples of similar crystal size there are also large differences in some cases. Despite the large differences in diffusivity the relative rates of diffusion of different sorbates in the various zeolite samples are similar. The differences in diffusivity appear to be associated with differences in the pre-exponential factor and the activation energies for the different samples are similar (Table 2). This suggests that the differences may arise from differences in the cation distribution since blocking a certain fraction of the windows will generally lead to a reduction in diffusivity with no significant change in activation energy[21]. These differences probably arise from differences in the initial dehydration procedure[22] which can affect the cation distribution. Such "pore closure" effects have been discussed by Breck[23].

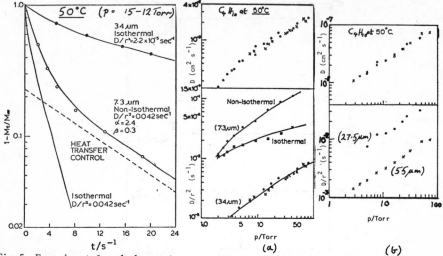

Fig 5. Experimental and theoretical uptake curves for n-butane in 5A (65% Ca^{++}). Diffusion is isothermal in 34μm crystals and non-isothermal in 7.3μm crystals but diffusivities for both curves are consistent (D $\simeq$ 6x10^{-9} cm^2.sec^{-1}).

Fig. 6 Diffusional time constants and diffusivities for butane in 5A zeolite. (a) 34μm and 7.3μm (65% Ca^{++}) at 50°C (b) 55μm and 27.5μm (95%Ca^{++}) at 50°C Diffusivities for both crystal size fractions are consistent.

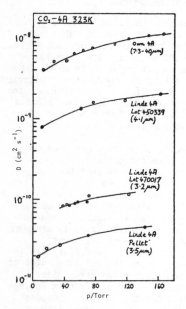

Fig. 7 Comparison of diffusivities for different 4A zeolites (CO$_2$, 50°C).

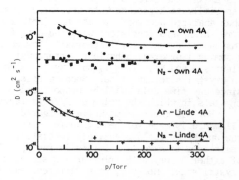

Fig. 8 Comparison of diffusivities for Ar and N$_2$ at 273K in different 4A zeolites (own crystals, 34μm, o, ●; 21.5μm ▲; 7.3μm ■ . Linde 4A pellet (3.4μm), x, +.)

Table 2: Comparison of Diffusivities and Activation Energies
for Different Zeolite Samples

Zeolite 4A	Diam. (μm)	N_2		C_2H_6		CH_4		CO_2	
		$D_0 \times 10^{10}$	E	$D_0 \times 10^{10}$	E	$D_0 \times 10^{10}$	E	$D_0 \times 10^{10}$	E
Linde Pellet	~3.5	2.6-0.7	6.0	0.008	5.6	0.19	6.4	0.09	-
Linde Crystals (Lot 470017)	3.2	-	-	0.002	6.3	-	-	0.24	5.0
Linde Crystals (Lot 450339)	4.1	6.3	5.8	0.06	7.2	-	-	3.4	-
Own Crystals	7.3-40	21	5.6	0.14	8.2	1.1	5.8	17	5.5

Zeolite 5A	Diam. (μm)	CF_4		nC_4H_{10}	
		$D_0 \times 10^{10}$	E	$D_0 \times 10^{10}$	E
Linde Crystals (Lot 550045)	3.6	.016	9.1	0.14	4.0
Own Crystals (95% Ca++)	27,55	6.5	6.6	24	4.6
Own Crystals (65% Ca++)	7.3,34	-	-	8	4.5

Activation energies (E) are in kcal/mole. D_0 is the corrected diffusivity at 323K ($cm^2 sec^{-1}$). Data for Linde 4A pellets are for different batches which may not be strictly comparable.

Comparison of Sorption and NMR Diffusivities

Theoretical considerations suggest that the self-diffusivity (D_S) derived from an NMR experiment should be comparable with the corrected diffusivity (D_0), calculated from the sorption diffusivity (D) according to Darken's equation; $D = D_0 \, \partial \ln c / \partial \ln q^*$, in which the non-linearity correction factor $\partial \ln p / \partial \ln c$ is obtained directly from the equilibrium isotherm[24,25]. When compared in this way reported values of D_0 and D_S show remarkable similarities in qualitative trends yet there is a discrepancy of several orders of magnitude in the numerical values[7,8]. To a large extent this discrepancy can be explained simply by differences in the zeolite samples. The NMR experiments were generally carried out on laboratory synthesized zeolite crystals whereas the earlier sorption rate data were obtained with commercial zeolite crystals (which are marketed in dehydrated form) and for which the diffusivities are very much smaller (<u>vide supra</u>). Because of the large differences between samples a reliable comparison can only be made if identical samples are used in both sorption and NMR measurements.

In previous comparisons[7,8] the only sorption data obtained on samples similar to those used for the NMR studies are the data of Ruthven and Doetsch[17] for hydrocarbons in 13X (17μm crystals). More detailed studies reveal that the assumption of isothermal behaviour was probably not valid under the conditions of these experiments so that the reported diffusivities are probably too low. The more recent sorption data obtained with larger crystals of A and X zeolites, synthesized by Charnell's method, give values of D_0 which are much closer to the NMR values (see Table 3). The NMR and sorption data for 2,2,4-trimethylpentane in 13X, which were obtained on the same zeolite crystals, are in order of magnitude agreement. There is also order of magnitude agreement for some of the other systems although, since the data refer to different zeolite samples, the comparisons are less valid. The most striking discrepancy is in the data for CH_4-4A. For this system the sorption data have been confirmed by experiments with different sizes of crystal and it seems unlikely that the very large discrepancy can be accounted for simply by the difference in samples even though the sample used in the NMR studies contained about 20% Ca^{++}.

Table 3:   Comparison of Sorption and NMR Diffusivities

System	T (K)	Sorption		NMR		Notes
		D_o	E	D_s	E	
CH_4-4A	140	$\sim 10^{-15}$	5.8	10^{-7}	<2	Sorption data are extrapolated from 273-323K. NMR data[26] are for sieve with 20% Ca^{++}.
C_3H_8-5A	473	$\sim 5 \times 10^{-8}$	3.5-4.0	10^{-7}	3.5-8.0	Sorption data are extrapolated from lower temperatures.
C_4H_{10}-5A	298	1.3×10^{-9}	4.6	4×10^{-8}	2.8	NMR values from correlation time with jump distance 12Å[27].
C_6H_6-13X	400	$\sim 5 \times 10^{-7}$ $(\Theta \to 1.0)$	-	10^{-6}-10^{-7}	3.5-8.0	NMR data from Lechert et al.[28,29] show strong dependence of E on Θ.
nC_7H_{16}-13X	406	$\sim 10^{-7}$ $(\Theta \to 1.0)$	-	10^{-6}-10^{-7}	3.0	NMR data show strong dependence on Θ[30].
2,2,4-trimethyl-pentane-13X	403	$\sim 3 \times 10^{-8}$ $(\Theta \to 1.0)$	-	10^{-7} $(\Theta \to 1.0)$	4.0	Same zeolite sample in both studies (48-60µ) from Shdanov. Strong dependence of D_s on Θ.

Values of D_o and D_s are in cm^2 sec^{-1} and E in kcal. Sorption data for 5A are for our fastest diffusing sample (27 and 55µm crystals, 95% Ca^{++}). NMR data for C_3H_8-5A and 2,2,4-TMP-13X are from unpublished data kindly provided by Dr. J. Karger. Sorption diffusivities for the 13X systems were obtained by non-isothermal analysis of the uptake curves (eqn. 1 and 2) and are subject to considerable uncertainty.

Accurate comparisons between sorption and NMR diffusivities are rendered difficult by the inherent limitations of the techniques. Diffusivities of less than about 10^{-7} - 10^{-8} $cm^2 sec^{-1}$ cannot be accurately measured by the NMR (self diffusion) method and this is close to the maximum that can be reliably determined from sorption rate experiments with crystals of about 50 µm, the largest available to us. Nevertheless, there is a small range over which both techniques are applicable and by a judicious choice of sorbate and temperature range a more reliable comparison should be possible in the future.

Chromatographic Studies

Chromatographic measurements provide a simple alternative to conventional sorption methods of studying both the kinetics and equilibria of sorption. The method offers advantages of speed, simplicity and versatility but suffers from the disadvantage that it is sometimes difficult to eliminate the

contributions from axial dispersion and other mass transfer resistances.
Where direct comparisons are possible, chromatographic and sorption rate
measurements show generally good agreement[31]. The large discrepancy in the
diffusivity of n-butane in 5A, reported by Hashimoto and Smith[32], has been
shown to be due merely to a difference in the definition of the micropore diffusivity[6,31].

The chromatographic method has recently been extended to the study of
diffusion in a binary adsorbed phase[33]. The chromatographic response for a
binary system (2 adsorbable components) is of the same form as for a single
component system with appropriately re-defined rate and equilibrium constants.
The mean (μ) and variance (σ^2) of the response peak are given by:

$$\mu = (L/v)\{1 + K(1-\varepsilon)/\varepsilon\} \tag{7}$$

$$\frac{\sigma^2}{2\mu^2} = \frac{D_L}{vL} + \frac{1}{k_o} \, (\frac{L}{v}) \, (\frac{1-\varepsilon}{\varepsilon}) \cdot \frac{1}{\mu^2} \tag{8}$$

where $\quad K = XK_2 + (1-X)K_1; \frac{1}{k_o} = \frac{XK_2}{k_2} + \frac{(1-X)K_1}{k_1}$ $\tag{9}$

The rate constants for each component may be expressed in terms of the contributions from film, macropore and micropore diffusional resistance:

$$\frac{1}{k_1} = \frac{r^2}{15D_1} + K_1 \left\{ \frac{R}{3k_f} + \frac{R^2}{15\varepsilon_p D_p} \right\} \tag{10}$$

with a similar expression for k_2. The contributions from film and macro
resistance may be minimized by the use of small adsorbent particles and under
these conditions the dispersion of the binary response peak is controlled by
the combined effects of axial dispersion and micropore diffusion of both
components. The individual contributions of the two components cannot be
separated but the overall rate coefficient for the binary system may be compared with the value predicted from eqns. 9 and 10 using the values of D_1 and
D_2 determined for the individual components (using a helium carrier) and duly
corrected, according to the Darken equation for differences in the isotherm.

Such studies have been carried out for a range of light gases in both 4A
and 5A zeolites. The single component diffusivities from the chromatographic
measurements agreed well with the values determined previously by gravimetric
methods. The binary rate coefficients (k_o) estimated from the single component
data by the method outlined above also showed good agreement with the experimental values. In particular, the strong variation of k_o with composition
for systems such as CH_4-N_2 in 4A sieve are correctly predicted. Representative
data are shown in figures 9 and 10. Because of the experimental scatter the
possibility of small differences between intrinsic diffusivities in the
mixture, compared with the pure components, cannot be excluded but the results
show that, at least for these systems, any such differences are relatively
minor. The diffusivities in the mixture can be approximately predicted on the
assumption that both components diffuse independently with the same intrinsic
diffusivities as for the single component systems, with due correction for differences in the driving force due to differences in the equilibrium isotherm.
These results were all obtained at relatively low sorbate concentrations and it
is possible that deviations from this simple pattern of behaviour may become
more pronounced at higher concentration.

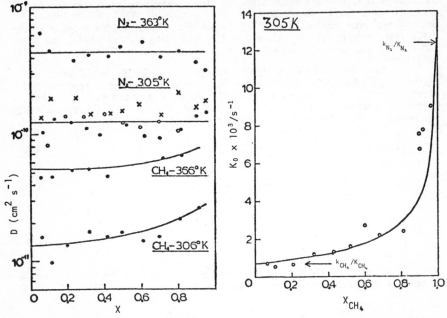

Fig. 9 Single component chromatographic diffusivities for N_2 and CH_4 in 4A (Linde pellets) (He carrier, •; H_2 carrier, x; i-butane carrier, o) X is mole fraction of sorbate in gas phase.

Fig. 10 Variation of overall rate coefficient for CH_4-N_2 in 4A (Linde pellets) at 305K. Points are experimental, curve is calculated from single component data according to eqns. 9 and 10.

Sorption Rate Measurements in Binary Systems

The behaviour of some binary systems is more complex. Taylor[34] and Loughlin have studied counter-diffusion of C_2H_6-C_2H_4 and C_2H_6-CO_2 in 4A zeolite and cyclo-C_3H_6-C_3H_8 in 5A zeolite. Under the conditions of their experiments both C_2H_4 and CO_2 diffuse rapidly compared with C_2H_6, so that the rate of the exchange process was controlled entirely by the diffusion of C_2H_6. In the 4A sieve the diffusivity of C_2H_6 decreased dramatically with increasing loading of either CO_2 or C_2H_4 but the diffusional activation energy remained constant. Representative experimental data are shown in figures 11 and 12. These results can be quantitatively explained on the assumption that CO_2 and C_2H_4 are preferentially adsorbed at the type II cations, thus effectively blocking a fraction of the windows in the 4A sieve. Unfortunately a direct verification of this hypothesis by comparative studies with the 5A sieve, which contains no type II cations, was not possible because diffusion of C_2H_6 in the unloaded 5A sieve, under experimentally convenient conditions, was too fast for accurate measurement. For other systems such as cyclo-propane-propane in 5A and C_2H_6-N_2 in 4A, no such effect was observed.

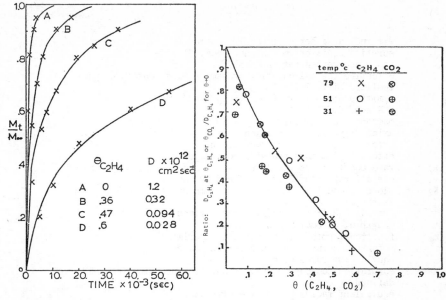

Fig. 11. Uptake curves for ethane sorption on Linde 4A pellets for varying coverage of presorbed ethylene. Solid lines indicate theoretical uptake curves as obtained from Fick's Law.

Fig 12. Variation in the reduced diffusivity of ethane with presorption coverage of ethylene and carbon dioxide.

Conclusions

Application of the non-isothermal model extends considerably the range of systems for which diffusional time constants can be derived from transient sorption rate measurements and reveals that, for some rapidly diffusing systems such as hydrocarbons in 13X and monatomic gases in 5A, the isothermal approximation may not have been valid in some of the earlier studies[17,34] leading to erroneously low apparent diffusivities. For slower diffusing systems such as hydrocarbons in Linde 5A and monatomic and diatomic gases in Linde 4A the non-isothermal analysis essentially confirms the validity of the isothermal approximation.

Studies of sorption rates in a variety of different samples of zeolites 4A and 5A reveal surprisingly large differences in diffusivity between otherwise similar zeolite crystals. In general the diffusivities for the commercial zeolites, both crystals and pellets, are smaller than the values for our own crystals although the difference is much greater for some commercial samples than for others. Despite the large differences in diffusivity the dependence on temperature and sorbate concentration was essentially the same for all samples. The key variable appears to be the initial dehydration procedure[22] which, in the case of the commercial samples, is outside the control of the investigator. Much of the apparent discrepancy between sorption and NMR diffusivities appears to arise simply from differences in the zeolite samples.

Studies of binary diffusion reveal some interesting patterns of behaviour and perturbation chromatography provides a convenient experimental method by which such behaviour can be studied.

Notation

a	external area per unit sample volume
b, b_0	sorption equilibrium constant (eqn. 5)
c	gas phase concentration of sorbate
c_s	effective thermal capacity of sample plus containing pan
D	zeolitic diffusivity
D_0	limiting diffusivity from $D = D_0(\partial \ln c / \partial \ln q^*)$
D_s	NMR self diffusivity
D_p	macropore diffusivity
D_L	axial dispersion coefficient
h	external heat transfer coefficient
k_f	external film mass transfer coefficient
k_1, k_2	rate coefficients for components 1 and 2 defined by eqn. 10
k_0	overall rate coefficient defined by eqn. 9
K_1, K_2	slope of equilibrium curves for components 1 and 2 (binary isotherm)
K	overall equilibrium coefficient defined by eqn. 9
L	length of chromatographic column
m_t/m_∞	fractional uptake in a sorption experiment
q	sorbate concentration in adsorbed phase
q^*	equilibrium value of q
q_m	saturation concentration in Langmuir expression (eqn. 5)
q_n	defined by eqn. 2
r	equivalent radius of zeolite crystal
R	radius of pellet or equivalent radius of sample of crystals on balance pan
t	time
T, T_0	temperature, steady state temperature
v	interstitial gas velocity
X	mole fraction in gas phase
ΔH	enthalpy change on sorption
Θ	sorbate concentration expressed as fraction of saturation (q/q_m)
α, β	dimensionless parameters defined in text
ρ	density of sample
τ	dimensionless time Dt/r^2
σ^2	variance of chromatographic peak
μ	mean of chromatographic response
ε	voidage of chromatographic column
ε_p	macroporosity of molecular sieve pellet

Acknowledgements. Much of the work reported here was carried out by Dr. Lap-Keung Lee, Dr. H. Yucel, Dr. R. Kumar, Mr. A. Vavlitis and Dr. R.A. Taylor and their extensive contributions as well as those of Dr. K.F. Loughlin are gratefully acknowledged. Thanks are also due to Dr. J.D. Sherman (Union Carbide) for a number of helpful discussions and for the provision of zeolite samples. Special thanks are due to Dr. J. Karger (KMU Leipzig) for providing us with the NMR data and with samples of some of the zeolites used in the NMR studies, as well as for helpful and stimulating comments on several aspects of this work.

References

1. R.A. Taylor, Ph.D. Thesis, University of New Brunswick, Fredericton, 1978 (also R.A. Taylor and K.F. Loughlin <u>A.I.Ch.E. Jl</u>. - to be published).
2. M. Kocirik, J. Karger and A. Zikanova - to be published.
3. P.E. Eberly, <u>Ind. Eng. Chem. Fund.</u>, 1969, <u>8</u>, 25.
4. A.C. Sheth, M.Sc. Thesis, Northwestern University, Evanston, Ill., 1969.
5. D.R. Garg and D.M. Ruthven, <u>Chem. Eng. Sci.</u>, 1972, <u>27</u>, 417 <u>Ibid</u>, 1974, <u>29</u>, 571.
6. D. Gelbin, R.H. Radeke, A. Roethe and K.P. Roethe, <u>Chem. Techn.</u>, 1976, <u>28</u>, (5) 281.
7. J. Karger and J. Caro, J. <u>Chem. Soc. Faraday Trans I</u>, 1977, <u>73</u>, 1363.
8. D.M. Ruthven, <u>Am. Chem. Soc Symp. Ser.</u>, 1977, <u>40</u>, 320.
9. R.M. Barrer, "Zeolites and Clay Minerals", Academic Press, London, 1978, Chapter 6.
10. J.D. Eagan, B. Kindl and R.B. Anderson, <u>Adv. Chem.</u>, 1971, <u>102</u>, 164.
11. H.J. Doelle and L. Riekert, <u>Am. Chem. Soc. Symp. Ser.</u>, 1977, <u>40</u>, 401.
12. A.A. Armstrong and V. Stannett, <u>Die Makromelekulare Chemie</u>, 1966, <u>90</u>, 145 and 1966, <u>95</u>, 78.
13. K. Chihara, M. Suzuki and K. Kawazoe, <u>Chem. Eng. Sci.</u>, 1976, <u>31</u>, 505.
14. Lap-Keung Lee and D.M. Ruthven, <u>J. Chem. Soc. Faraday Trans I</u> - in press (also Lap-Keung Lee, Ph.D. Thesis, University of New Brunswick, 1977).
15. J. Crank, "Mathematics of Diffusion", Oxford University Press, 1956, pp 86-88.
16. K.F. Loughlin, R.I. Derrah and D.M. Ruthven, <u>Can. J. Chem. Eng.</u>, 1971, <u>49</u>, 66.
17. D.M. Ruthven and I.H. Doetsch, <u>A.I.Ch.E. Jl.</u>, 1976, <u>22</u>, 882.
18. J. Karger, J. Caro and M. Bulow, <u>Z. Chemie</u>, 1976, <u>16</u>, 331.
19. M. Bulow, P. Struve, G. Finger, C. Redszus, K. Ehrhardt and J. Karger, <u>J.C.S. Faraday Trans I</u> - submitted.
20. J.F. Charnell, <u>J. Crystal Growth</u>, 1971, <u>8</u>, 291.
21. D.M. Ruthven, <u>Can. J. Chem.</u>, 1974, 52, 3523.
22. E.F. Kondis and J.S. Dranoff, <u>Ind. Eng. Chem. Proc. Design Develop.</u>, 1971, <u>10</u>, 108.
23. D.W. Breck, "Zeolite Molecular Sieves", Wiley, New York, 1974, pp 490, 615.
24. R. Ash and R.M. Barrer, <u>Surface Sci.</u>, 1967, <u>8</u>, 461.
25. J. Karger, <u>Surface Sci.</u>, 1973, <u>36</u>, 797.
26. J. Caro, J. Karger, G. Finger, H. Pfeifer and R. Schollner, <u>Z. Phys. Chem.</u> (Leipzig), 1976, <u>257</u>, 903.
27. L. Labisch, R. Schollner, D. Michel, V. Rossiger and H. Pfeifer, <u>Ibid</u>, 1974, <u>255</u>, 581.
28. H. Lechert, K.P. Wittern and W. Schweitzer, <u>Acta Physica et Chemica</u>, 1978, <u>24</u>, 201.
29. H. Lechert and K.P. Wittern - personal communication.
30. J. Karger, H. Pfeifer, M. Rauscher and A. Walter, <u>Z. Phys. Chem.</u> (Leipzig), 1978, <u>259</u>, 784.
31. D.B. Shah and D.M. Ruthven, <u>A.I.Ch.E. Jl.</u>, 1977, <u>23</u>, 804.
32. N. Hashimoto and J.M. Smith, <u>Ind. Eng. Chem. Fund.</u>, 1973, <u>12</u>, 353.
33. R. Kumar, Ph.D. Thesis, University of New Brunswick, Fredericton, 1978. (also D.M. Ruthven and R. Kumar, <u>Can. J. Chem. Eng.</u> - in press).
34. D.M. Ruthven and R.I. Derrah, <u>J.C.S. Faraday Trans I</u>, 1975, <u>71</u>, 2031.

Diffusion of C_6 Hydrocarbons in NaX and MgA Zeolites.

A. Zikánová, M. Kočiřík*, A. Bezus, A.A. Vlček

(Heyrovský Institute of Physical Chemistry and Electrochemistry of the
Czechoslovak Academy of Sciences, Prague, ČSSR)

M. Bülow, W. Schirmer

(Central Institute of Physical Chemistry of the Academy of Sciences of the
GDR, Berlin, GDR)

J. Kärger, H. Pfeifer

(NMR Laboratory of the Karl Marx University, Leipzig, GDR)

and

S.P. Zhdanov

(Institute of Silicate Chemistry of the Academy of Sciences of the USSR,
Leningrad, USSR)

Introduction.

The aim of this study is to contribute to the elucidation of discrepancies
between the effective diffusion coefficients D_{eff} and self-diffusion coefficients D of hydrocarbons in A and X type zeolites for the "intracrystalline"
region as obtained by sorption measurements and n.m.r. pulsed field gradient
technique respectively.

In order to deal with the question of these discrepancies, we must briefly
note the problem of compatibility of n.m.r. and diffusion kinetic measurements.
For intracrystalline diffusion ($\langle \Delta x^2(\delta) \rangle \ll \langle R^2 \rangle$) the compatibility of n.m.r.
and sorption data has been analysed in detail[1-3]. For intercrystalline diffusion phenomena ($\langle \Delta x^2(\delta) \rangle \gg \langle R^2 \rangle$) it is necessary to regard the zeolites
(layers, compacts and pellets) as a bidisperse medium having at least two regions
of different diffusion resistances[4]. Therefore, the question of compatibility
is much more complicated. Only in some special cases are n.m.r. and sorption
measurements governed by analogous features of molecular transport. We shall
discuss here only the case when intercrystalline mass flow occurs in the Knudsen
region. If intracrystalline diffusion does not contribute to the total transport
we may then assume the validity of Darken's equation (1) for the intercrystalline diffusion:

$$D_{eff} = D^*_{inter} \frac{d\ln c_g}{d\ln c_a}$$

$$(1)$$

where D_{eff} is the effective diffusion coefficient of the intercrystalline diffu-

sion, and D^*_{inter} is the corresponding self-diffusion coefficient of intercrystalline diffusion.

The Direct Measurement of Intracrystalline Self-diffusion.

The intracrystalline self-diffusion is most accurately measured by the
n.m.r. pulsed field gradient technique[3,5], which allows the direct observation
of the molecular mean square displacement $\langle \Delta x^2(\delta) \rangle$ during a time interval δ
of several milliseconds. The consistency of the zeolitic self-diffusion coeffi-
cients thus obtained has been proved in a number of crucial experiments[3].

A lot of information is known[3,6] on the mobility of various hydrocarbons in
zeolites of A and X types. We shall present some typical examples for the self-
diffusion behaviour of C_6 hydrocarbons in the NaX zeolite.

Figs. 1 - 3 show the concentration dependence of the coefficients of intracrystalline self-diffusion of n-hexane, cyclohexane and benzene in NaX zeolite
for different temperatures. It is evident that the difference in the micro-
dynamic behaviour of unsaturated and saturated hydrocarbons in NaX zeolites[7] is
reflected in different patterns of the concentration dependence. For the specifically sorbed benzene molecules, self-diffusion proceeds via jumps between
fixed positions so that the jump length, and therefore the self-diffusion coefficient, remains almost unaffected by increasing the concentration. Only for concentrations of more than half of the saturation capacity, can a decrease in the
self-diffusion coefficient be observed. Due to the lack of specific interaction,
however, the mutual hindrance of the molecules of the saturated hydrocarbons
leads to a continuous decrease in the jump lengths and therefore to a decrease
in the self-diffusion coefficients by more than two orders of magnitude. It is
remarkable that an analogous concentration dependence is observed in selective
self-diffusion measurements of mixtures[8] of saturated and unsaturated hydrocarbons in NaX zeolites. From Fig. 4 it can be seen that the decrease in the
mobility of the non-specifically sorbed components (n-heptane) can be affected
also by an increased amount of a second component (in this case by benzene (compare the values in circles in the vertical direction)). In contrast, up to the
medium pore filling factors the self-diffusion coefficient of the specifically
sorbed component is not influenced by the increased amount of the second component.

Uptake Experiments.

In uptake measurements the time of observation is much longer as compared
with the n.m.r. measurements. If the intracrystalline diffusion does not control
the transport exclusively, the sorbate molecules cross many times the inter-

facial area between two regions. It was necessary therefore to apply a model
of mass transport in bidisperse media and we used the bidisperse model previously developed[9]. The statistical moments treatment of this model results in the
following expression for the 1st moment E_{ω_o} of the uptake curve[10]:

$$E_{\omega_o} = \frac{L^2}{3D_{eff}} + E_{\omega_1} = K_2 Z + K_o \; ; \; Z = L^2 \qquad (2)$$

where E_{ω_1} represents the characteristic time for penetration of sorbate molecules into crystals and 2L denotes the thickness of the zeolite sample.

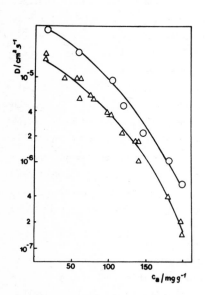

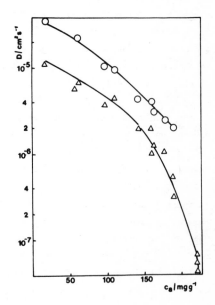

Fig. 1: Concentration dependence of self-
diffusion coefficient of n-hexane in NaX
zeolite at 293K (Δ) and 373K (0)

Fig. 2: Concentration dependence
of self-diffusion coefficient of
cyclohexane in NaX zeolite at
293K (Δ) and 393K (0).

The intracrystalline self-diffusion coefficients D_a^* estimated from E_{ω_1}
differ by 4 or 5 orders of magnitude from the n.m.r. self-diffusion coefficients

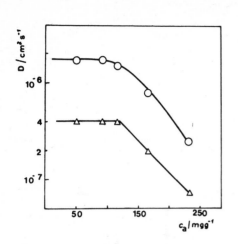

Fig. 3: Concentration dependence of self-diffusion coefficient of benzene in NaX zeolite at 360K (Δ) and 440K (0)

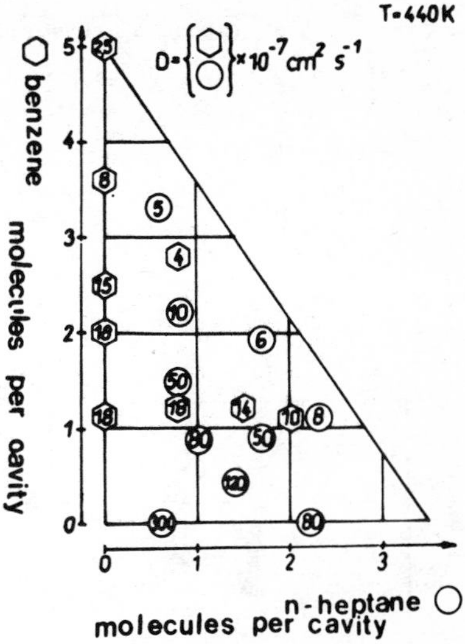

Fig. 4: Self-diffusion coefficients of n-heptane (0) and benzene (Q) in NaX zeolite at 440K

of intracrystalline diffusion. With respect to this discrepancy we formulated two hypotheses:

<u>Hypothesis A:</u> E_{ω_1} gives the true value of characteristic time of isothermal penetration of the molecules from the surroundings into the region of microporosity.This means the bidisperse model of sorption kinetics is accepted and the discrepancy should be explained by the action of the surface.

<u>Hypothesis B:</u> There is no connection between E_{ω_1} and the mass transport into the region of microporosity. Thus the bidisperse model of sorption kinetics should be rejected and E_{ω_1} can be explained by external effects such as

(i)   finite rate of sorption heat dissipation,

(ii) finite rate of sorbate supply to the sorbent.

Examination of Surface Barrier Effects.

Following the investigation of the crystal size dependence of sorption kinetics for several gaseous hydrocarbons on 5A type zeolites[3,11] we performed an analysis of this dependence for the sorption kinetics of n-hexane on Mg-A zeolite[12]. This compound was selected because the time constant of the uptake processes is high enough for the application of sorption techniques.

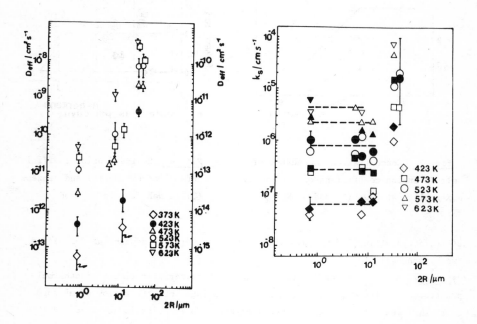

Fig. 5: Dependence of effective diffusion coefficient of n-hexane on crystal size (mean edge length) of MgA zeolites

Fig.6: Crystal size dependence of effective surface parameter k_s for n-hexane on Mg-A zeolite calculated by eqn .(2) (full symbols) and $k_s = D_{eff} \, \pi^2 / 3R(1 + \varepsilon)$ (open symbols), respectively

Fig. 5 shows the dependence of the effective diffusion coefficient for n-hexane in Mg-A zeolite on the crystal size as obtained by the constant volume - variable pressure method (excluding the intercrystalline transport and the finite rate of sorbate supply). The values of D_{eff} are calculated by means of the corresponding solution of Fick's 2nd law for a spherical particle. The form of this dependence suggests that with the exception of the largest crystals the uptake process is not limited by intracrystalline diffusion. On the contrary, the results can easily be explained by the action of some kind of crystal surface barrier. The corresponding dependence of the surface parameter k_s on the crystal size, calculated by means of the relationship

$$c_g(t) = c_{go} - \frac{\varepsilon}{1 + \varepsilon}(c_{go} - c_{gv}) \left\{ 1 - \exp\left[- \frac{3k_s}{R}(1 + \varepsilon)t \right] \right\} \qquad (3)$$

is given in Fig. 6. k_s is defined by the relation $j_s = k_s(c_a^s - Kc_g)$ in which j_s is the mass flow density through the crystal surface, c_a^s is the actual sorbate concentration just within the surface, and Kc_g is the corresponding value in equilibrium with the surrounding gaseous phase.

Examination of External Effects.

Application of the n.m.r. pulsed field gradient technique to the systems n-hexane (cyclohexane)/Na-X type zeolites has shown[4,13] that during observation times of several milliseconds the sorbate molecules can cover distances considerably larger than the crystal diameters. It must be concluded, therefore. that for these systems molecular uptake cannot be significantly influenced by barrier effects on the crystal surface. Consequently, Hypothesis A must be rejected in this case.

(i) Thermal Effects: We tested Hypothesis B on the system cyclohexane/Na-X type zeolite (Linde) compacted into thin plates. The uptake curves were measured gravimetrically under constant pressure at the surface of the sample. The thickness of the plates ranged from 0.02 cm to 0.1 cm (crystal size $\sqrt{\langle R^2 \rangle} \simeq 1.46\ \mu m$). Sets of the experiments were performed at the following temperatures: 303K, 378K, 393K, 408K and 423K. At every temperature four pressure steps were performed in the region 1.33 to 816 N m^{-2}. Fig. 7 shows a typical example of the dependence of E_ω on L^2. The experimental data were fitted by eq. (1) and the values of E_{ω_1} which ranged from 5 to 15 s were significant. If the bidisperse model is adequate, then E_{ω_1} is related to the coefficient of intracrystalline diffusion D_a and to the characteristic dimension of spherical crystals by the relation[4]:

$$E_{\omega_1} = \frac{\langle R^2 \rangle}{15 D_a} \tag{4}$$

Using the value of D_a estimated from n.m.r. self-diffusion coefficients D for the system under consideration[6] by means of Darken's formula we obtain $E_{\omega_1} \approx 10^{-4} s$. Such a short time cannot be measured by uptake experiments. One possible explanation was suggested in our previous paper[14] on the basis of the finite rate of heat dissipation. From the statistical moments solution of the nonisothermal model we obtained for E_{ω_0} the expression:

$$E_{\omega_0} = K_1 Z^{1/2} + K_2 Z \; ; \; Z \equiv L^2 \tag{5}$$

The plot of E_{ω} against Z should be therefore fitted by a parabola passing through the origin, instead of a straight line with an intercept E_{ω_1}, as given for the bidisperse model (see Fig. 8).

Due to the scattering of the experimental points it was impossible to decide between these two models. We developed therefore further our solution of the nonisothermal model[14] using a treatment similar to that given by Ruthven for a spherical particle[15]. We obtained for the fractional uptake $\gamma(\tau)$[16]:

$$\gamma(\tau) = 1 - \sum_{n=1}^{\infty} \left[\frac{2 \exp(-q_n^2 \tau) \left(\frac{tgq_n}{q_n}\right)^2}{\left(\frac{tgq_n}{q_n}\right)\left[q_n^2\left(\frac{tgq_n}{q_n}\right) + 1\right] + \frac{2}{\beta} + 1} \right] \tag{6}$$

where q_n is the nth root of the equation

$$\tag{7}$$

and

$$tgq_n = \frac{-q_n^2 + \alpha}{\beta q_n} \quad ,$$

$$\tau = \frac{D_{eff} t}{L^2} \quad , \quad \alpha = \frac{hL}{D_{eff} \rho_p \kappa_p} \quad , \quad \beta = \frac{(-\Delta H_{ad})}{\kappa} \left| \left(\frac{\delta c_a}{\delta T}\right)_p \right| \quad . \tag{8 - 10}$$

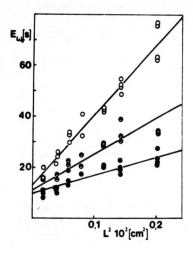

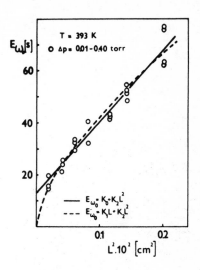

Fig. 7: Plot of E_ω vs. L^2 for the sorption of cyclohexane on Na-X type zeolite at 393K and different pressure steps Δp (Nm^{-2}): $\circ$... 1.33-53.2; ◑... 53.2-182.2; ●... 182.2-363

Fig. 8: Plot of E_ω vs L^2 for the sorption of cyclohexane on Na-X zeolite at 393K. Pressure step Δp (N m^{-2}) 1.33 - 53.2

o experimental points

———— fitting by the bidisperse model

---- fitting by the nonisotherm.

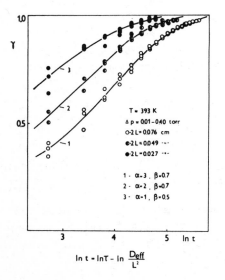

Fig. 9: Plots of γ vs $ln\ t$ for the sorption of cyclohexane on Na-X type zeolite

———— fitting by the nonisothermal model

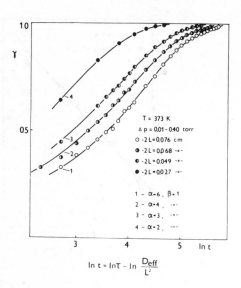

The symbols used denote:

h ... heat transfer coefficient

$(-\Delta H_{ad})$ heat of sorption

κ_p .. heat capacity of the
sorbent

ρ_p .. density of the sorbent

c_a sorbate concentration in
the crystals related to the
mass of the sorbent $(mmol\ g^{-1})$

To test the validity of the
model we had to estimate the
parameter β from independent
equilibrium measurements. Then
we estimated α by a comparison
of the experimental γ vs. lnt
plots with the corresponding
theoretical curves. Fitting
of the uptake data is illustrated in the Fig. 9 and 10.

$$\ln t = \ln \tau - \ln \frac{D_{eff}}{L^2}$$

Fig. 10: Plots of γ vs. lnt for the sorption
of cyclohexane on Na-X type zeolite
—— fitting by the nonisothermal model.

The results of the fitting are summarized in the Table 1. However, we have only
included the data for lower pressure steps; for the higher pressure steps we
found a systematic deviation from the model.

The consistency of the model has been confirmed by the following results:
1. D_{eff} obtained for plates of different thickness, though scattered, do not
exhibit any systematic trend with L.

2. Parameter α as obtained by fitting decreases with decreasing L.

3. The heat transfer coefficient h is practically constant.

4. E_{ω_0} as calculated by

$$E_{\omega_0} = \frac{L^2}{D_{eff}} \left(\frac{1}{3} + \frac{\beta}{\alpha} \right) \tag{11}$$

is in good agreement with the experimental values of E_ω. Consequently the
parabolic dependence E_ω vs L^2 reflects the true mechanism and the non-zero
intercept obtained when applying the bidisperse model probably has no physical
meaning for the system cyclohexane/Na-X.

To compare the uptake data with n.m.r. self-diffusion coefficients we cal-

Diffusion of C_6 Hydrocarbons in NaX and MgA Zeolites 67

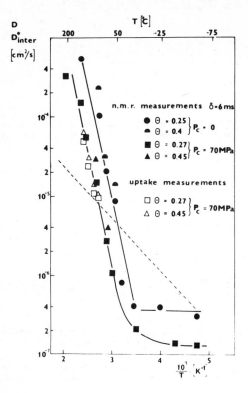

Fig. 11: Plots of D and D^*_{inter} vs
1000/T for cyclohexane/Na-X crystals
(p_c= 0) and for cyclohexane Na-X compacts
(p_c = 70MPa)
--- Arrhenius extrapolation of D from the
region of intracrystalline diffusion.

culated D^*_{inter} using eq. (1) (see Fig. 11). The results of the uptake measurements (empty symbols) are close to the Arrhenius plot of the n.m.r. self-diffusion coefficients for compacted zeolite (filled symbols). Both n.m.r. and uptake measurements were performed with the same plates, compacted under the pressure 70 MPa.

The agreement between n.m.r. and uptake data is better than we expected. On the other hand there is no reason for a large discrepancy, as both types of measurement give consistent results, and the conditions of the experiments were very close to each other. The only difference consisted in the influence of the heat effect on the uptake measurements, but this effect we corrected for using the non-isothermal model.

The most important point is the selection of the characteristic dimension of the diffusion process. In accordance with the results of the uptake measurements we have taken the half thickness of the plate and not the radius of the crystals.

(ii) Finite Rate of Sorbate Supply to the Sorbent: From our measurements under constant volume- variable pressure conditions we have found that valves and other tubing may represent significant resistances to the gas supply to the surface of the sorbent. Such an external effect acts in a similar way on the form of uptake curves as the surface barrier. On the other hand, a non-zero value of E_{ω_1} in the plot E_ω vs L^2 may result. We must analyse carefully the effect of the apparatus in sorption experiments. For our gravimetric measurements we estimated the contribution of this effect to E_{ω_o} to be less than 2 seconds.

Table 1. Results of the kinetic measurements

T /K	c̄_a /mmol g⁻¹	2L /cm	$D_{eff} \cdot 10^5$ /cm² s⁻¹	$D^* \cdot 10^5$ inter /cm² s⁻¹	α	β	$E_\omega°$ calc /s	$E_\omega°$ exp /s	$\frac{\beta/\alpha}{1/3+\beta/\alpha} \cdot 100$ %	$h \cdot 10^4$ /cal cm⁻² K⁻¹ s⁻¹	ΔT adiab /K
363	0.61	0.076	1.16	0.97	6.0	1.0	61.4	60.3	33	4.0	30.4
		0.068	1.46		4.0	1.0	46.0	49.3	43	3.7	30.5
		0.049	1.10		3.0	1.0	35.6	34.4	50	3.2	29.4
		0.027	0.95		2.0	1.0	15.8	15.7	60	4.2	26.1
	1.35	0.076	9.02	1.05	1.0	1.0	21.1	24.4	75	5.5	7.5
		0.068	7.98		1.0	1.0	19.3	22.8	75	5.4	7.5
		0.049	9.10		0.5	1.0	15.1	16.7	86	4.6	7.3
378	0.45	0.076	1.26	1.04	6.0	1.0	56.6	57.3	33	4.3	21.3
		0.049	0.83		4.0	1.0	41.3	40.0	43	3.2	20.6
	1.10	0.076	6.82	1.67	1.0	1.0	27.9	25.5	75	4.1	11.0
		0.049	6.51		0.5	1.0	21.1	—	86	3.3	10.1
393	0.26	0.076	1.62	1.97	3.0	0.7	49.8	51.1	33	2.8	15.1
		0.049	1.45		2.0	0.7	27.7	32.2	52	2.8	14.5
		0.027	1.36		1.0	0.5	11.0	16.5	60	3.0	12.4
	0.81	0.076	6.68	3.21	1.0	1.0	28.4	31.5	75	4.0	14.2
		0.068	6.22		1.0	1.0	24.7	28.9	75	4.1	14.3
		0.049	7.80		0.5	1.0	17.6	22.1	86	3.9	13.7
408	0.14	0.076	2.29	2.76	3.0	0.3	26.9	27.6	23	3.9	10.4
		0.068	2.07		3.0	0.3	24.1	25.9	23	3.9	10.5
		0.049	1.85		2.0	0.3	15.4	19.4	31	3.5	9.8
	0.55	0.076	6.29	5.35	1.0	1.0	30.2	32.6	75	3.7	13.3
		0.068	5.98		1.0	1.0	25.7	27.9	75	3.8	13.4
		0.049	6.99		0.5	1.0	19.6	23.5	86	3.4	12.9
423	0.95	0.076	18.9	6.71	0.4	1.0	21.3	21.3	88	4.6	6.5
		0.068	18.3		0.4	1.0	17.9	22.7	88	4.9	6.0
		0.049	25.1		0.2	1.0	12.5	11.9	94	5.0	6.2
	0.33	0.076	2.97	2.80	3.0	0.7	27.2	28.3	41	5.1	9.1
		0.049	2.26		2.0	0.7	17.7	17.8	51	4.3	8.8
		0.027	1.48		1.0	0.5	10.0	8.6	60	2.7	8.2
	0.67	0.076	4.76	2.81	2.0	1.0	24.8	26.1	60	5.6	7.1
		0.068	4.56		2.0	1.0	21.1	23.7	60	5.9	7.1
		0.049	5.02		1.0	1.0	15.6	—	75	4.2	6.8

Conclusions

The application of the n.m.r. pulsed field gradient technique gives a deep insight into the microscopic behaviour of sorbate-sorbent systems and enables the interpretation of intracrystalline mobility in terms of molecular properties. This is illustrated by comparative study of three different C_6 hydrocarbons and one binary mixture. On the other hand the n.m.r. technique has shown that most of the published data concerning diffusion into zeolites and obtained by the sorption technique requires a critical reconsideration. It has become evident that a number of effects which previously were considered to be unimportant in determining overall sorption kinetics require a thorough analysis.

For many cases of sorption kinetics in heterogeneous media, where a comparison of n.m.r. and uptake results is possible, we can take as a starting point the assumption that the n.m.r. technique gives true values of intracrystalline diffusion coefficients. This enables us to formulate models for sorption kinetics in bidisperse media.

In connection with the discrepancies mentioned above we formulated two hypothesis. The Hypothesis A(concerning the existence of surface barriers)was tested and proved to be important for 5A type zeolites. It may be supposed that a barrier also exists in the case of smaller molecules where a discrepancy is obtained by direct comparison of experimental n.m.r. and sorption data.

Concerning Hypothesis B we produce in summary several pieces of evidence that for the case of X type zeolites the term $E_{\omega 1}$ does not contain any information on intracrystalline kinetics (barrier and/or intracrystalline diffusion). It should be pointed out that the model of bidisperse sorption kinetics in zeolitic compacts and beds must be reconsidered.

References.

1. H. Pfeifer, *Phys. Rep. (Phys. Letters C)*, 1976, <u>26</u>, 293.

2. M. Bulow and J. Karger, "Thesis to the Promotion B", Karl-Marx-Universität Leipzig, 1978.

3. J. Karger and J. Caro, *J.C.S. Faraday I*, 1977, <u>73</u>, 1363.

4. J. Karger, M. Bulow, P. Struve, M. Kocirík and A. Zikánová, *J.C.S. Faraday I* 1978, <u>74</u>, 1210.

5. H. Pfeifer, in "N.M.R.-Basic Principles and Progress" (Springer, Berlin, Heidelberg, New York), 1972, <u>7</u>, 53.

6. J. Karger, H. Pfeifer, M. Rauscher and A. Walter, J.C.S. Faraday I, in press.

7. H. Pfeifer, Proc. "RAMIS 77", Poznan, 1977, p. 107.

8. P. Lorenz, M. Bulow, J. Karger and H. Pfeifer, Izv. Akad. Nauk SSSR, in press

9. E. Ruckenstein, A. S. Vaidyanathan and G.R. Youngquist, Chem. Eng. Sci., 1971, 26, 1305.

10. M. Kocirik and A. Zikanova, Ind. Eng. Chem. Fundam., 1974, 13, 347.

11. J. Karger, J. Caro and M. Bulow, Izv. Akad. Nauk SSSR, 1977, 2666.

12. M. Bulow, P. Struve, G. Finger, C. Redszus, K. Ehrhardt, W. Schirmer and J. Karger, J.C.S. Faraday I, in press.

13. J. Karger, M. Bulow and N. van Phat, Z. physik. Chem. (Leipzig), 1976, 257, 1217.

14. M. Kocirik, J. Karger and A. Zikanova, J. Chem. Tech. and Biotech. in press.

15. L.-K. Lee and D.M. Ruthven, Proc. "6th CHISA, 1978", Prague, 1978.

16. M. Kocirik, M. Smutek, A. Zikanova and A. Bezus, Collect. Czech. Chem. Commun ., in press.

Discussion on Session 1:   Diffusion Processes

Chairman: Professor K.S.W. Sing (Brunel University, U.K.)

(Throughout this section reference numbers correspond to those in the particular paper under discussion).

Paper 1.   (R.M. Barrer).

Prof. Riekert: Darken considered the relation between self-diffusivity of components in binary alloys on the one hand and the binary diffusivity in the same alloy on the other hand; this treatment also applies to other binary systems. In these cases the self-diffusivity and binary diffusivity can be measured independently and Darken's relation can thus be verified experimentally.

Barrer later applied Darken's results to diffusion (or rather flow) of matter in solids such as zeolites. However, here the situation is somewhat different: the problem involves defining a quantity (the intrinsic diffusivity), which cannot be measured in any other way.

The relation cannot thus in this case be falsified - it must be true. Theory does not require the so-defined "intrinsic diffusivity" to be constant as a function of composition(loading). Also there is no theoretical basis to relate self-diffusivities (as obtained from NMR) to intrinsic diffusivities as defined by the Darken-Barrer relation, as shown by Barrer and Ash. (Besides, it appears questionable whether a local gradient of chemical potential in a crystal can be derived from an equilibrium isotherm).

So after giving these problems some thought, the question remains: what kind of physical quantity is the "intrinsic diffusivity", and is it really helpful?

Prof. Barrer: Flow or diffusion through the plane at x and for simplicity in the x-direction only in a large single crystal can be expressed in two equivalent ways:

$$J \;=\; -D \, \frac{dC}{dx} \;=\; -(BC) \, \frac{d\mu}{dx}$$

where D is the differential intrinsic diffusivity, B is a corresponding mobility, C is concentration, μ is the chemical potential of diffusant and J is the flow through unit area normal to the x-direction.   The second of these relations can be re-written as

$$- \left(RTB \, \frac{d\ln p}{d\ln C} \right) \frac{dC}{dx}$$

where p is the equilibrium pressure expected when the concentration is C.   It is assumed that this pressure measures the activity, and is obtainable from the equilibrium isotherm. Accordingly the differential intrinsic diffusivity is related to RTB by the Darken-type expression

$$D = RTB \frac{d\ln p}{d\ln C}$$

There is no requirement that either D or B should be independent of concentration, though they are independent of dC/dx and $d\mu/dx$ respectively. Next, in terms of the irreversible thermodynamic formulation of isothermal diffusion it is possible to relate RTB to the self-diffusivity D*, as shown by Ash and Barrer[22] and by Karger[23], and to the straight and cross-coefficients L_{AA} and L_{AA*} of the tracer experiment. Approximately only it appears from the results of Barrer and Fender[3] that RTB $\sim$ D*. I do not regard D as undefined; it is indeed defined by $J = -DdC/dx$, and moreover it is the coefficient measured experimentally, as for example in the work of Tiselius[1] with water in single crystals of heulandite. In the practical sense it is more important than D*. Like D*, D outside the Henry's law range of sorption can, as noted above, be concentration dependent, a property which complicates the evaluation of D from sorption kinetics.

Prof. Ruthven: The questions raised by Prof. Riekert are of fundamental importance. There is convincing evidence that the true driving force for diffusive transport is the gradient of chemical potential rather than the concentration gradient. The contrary view that diffusion is a randomising process for which concentration is the true driving force (see Danckwerts, P.V., in "Diffusion processes", vol. 2, p. 545, edited by J.N. Sherwood, A.V. Chadwick, W.M. Muir and F.L. Swinton, Gordon and Breach, London, 1971), overlooks the fact that, in the presence of a concentration gradient, the *a priori* jump probability will not be the same in all directions.

Formulation of the transport equation in terms of a chemical potential driving force leads to

$$J_x \propto -\frac{d\ln a}{d\ln C} \frac{\partial C}{\partial x} \tag{1}$$

or by comparison with Fick's equation

$$D = D_o \frac{d\ln a}{d\ln C} = D_o \frac{d\ln p}{d\ln C} \tag{2}$$

where we have introduced the additional assumption of an ideal vapour phase. As Riekert points out, equation (2) is merely a definition of the coefficient D_o, (the corrected or intrinsic diffusivity). It provides no evidence concerning the concentration dependence of D_o, neither does it imply that D_o is necessarily equal to the self-diffusivity and, in that sense, it is not strictly correct to refer to equation (2) as the "Darken relation". However, equation (2) does show that the Fickian diffusivity is a complex quantity which consists of the products of a transport coefficient and the thermodynamic correction factor $d\ln a/d\ln C$. (Since chemical potential is given by $\mu = \mu^o + RT\ln a$ and in an ideal system the activity (a) is proportional to concentration (C) it is evident that the factor $d\ln a/d\ln C$ is simply the ratio $\Delta\mu$ (real) / $\Delta\mu$ (ideal)). In discussing transport behaviour it is logical to eliminate this thermodynamic factor and focus attention on D_o rather than D since, in comparing different systems it is of considerable interest to know whether the differences in diffusional properties reflect real differences in mobility or just differences in the equilibrium isotherms. These considerations are especially important for adsorbed species since for such systems the thermodynamic factor may be very large.

The behaviour of a binary system (adsorbent plus one adsorbed species) can be fully described by the Fickian diffusion equation with a concentration dependent diffusivity and for such a system the use of a corrected diffusivity does not improve our ability to model the transport properties. However, the behaviour of multicomponent systems (two or more absorbed species) cannot be properly described without recourse to the thermodynamic formulation, and the same is true of non-isothermal binary systems.

Paper 2 (R. Haul)

Prof. Sing: Professor Haul has mentioned that the porous silica contains a narrow pore size distribution. How has this been assessed? What is the effect of a change in the interparticle porosity on the magnitude of the diffusion coefficient?

Prof. Haul: An analysis of nitrogen and benzene desorption branches was made for the pore size distribution using the modelless method of Brunauer. A narrow distribution was obtained around 7.2 nm for sample 3. A V-t plot gave no indication of the presence of micropores.

The packing of the relatively large absorbent particles has practically no effect on inter-particle diffusion; however, it does affect the heat conduction properties and thus influences the sorption process. This is the reason for the difference between the diffusion coefficients obtained from NMR signal intensity and those obtained from gravimetric measurements (t → ∞), curves (d) and (e) in fig. (4), respectively.

Paper 3   (D.M. Ruthven)

Prof. Riekert: I was really impressed to hear from Professor Ruthven that experimentally obtained diffusivities can vary by orders of magnitudes for the same system, when zeolite samples from different lots of preparation are compared. Is it then worthwhile investigating the details of dependence of diffusivities on degrees of loading? Would it not be better just to try to obtain a correct order of magnitude only?

Prof. Ruthven: The large differences in diffusivity between different preparations of the same zeolite make it difficult to compare kinetic data from different laboratories, but I do not believe that this complication invalidates the detailed study of factors such as the concentration dependence of diffusivity. Despite the large differences in absolute diffusivities, the trends with concentration are essentially the same for the different samples. This is illustrated in figures 7 and 8 of the paper, although in these particular examples the concentrations are relatively low and the concentration dependence over the relevant pressure range is not especially large.

Prof. Haul: In the discussion so far, diffusivities from field gradient NMR and sorption methods have been mainly considered. It is well known that analyses of relaxation times can give more detailed information on the elementary processes of diffusion of adsorbed molecules. For instance, in our case (see paper 2, R. Haul) we can deduce from the temperature dependence of the T_R relaxation times that the cyclohexane molecule would appear to glide along the surface without isotropic rotation.

Prof. Ruthven: I agree that a great deal of additional information can be obtained from relaxation measurements. However, to derive a diffusivity from relaxation data requires an assumption concerning the mean jump distance. In order to avoid this additional uncertainty we have used, where possible, the self-diffusivities derived directly from pulsed field gradient measurements for comparison with the sorption values.

Dr. Sherry: Can you describe, or name the method employed, to make the large crystals that you used in your sorption studies?

Prof. Ruthven: The large crystals of both A and X zeolites were prepared by Charnell's method[20].

Paper 4.   (A. Zikanova, M. Kocirik, A. Bezus, J. Karger, H. Pfeifer M. Bulow, W. Schirmer, S.P. Zhdanov)

Prof. Barrer: When plugs are prepared from powders by compaction there is evidence that compaction is greater near the outer faces of the compacts. This means that properties, such as porosity, are dependent upon distance into the plug. Do you think that distance dependence of diffusion behaviour could influence any of the results you have described?

<u>Dr. Kocirik</u>: We examined the pore size distribution by mercury porosimetry, and we found a slight difference in the pore size distribution between thick and thin plates. To describe the intercrystalline diffusion we used the distributed pore model (E. Lippert, Thesis, Prague 1974). According to this approach the intercrystalline diffusion coefficient of a non-adsorbing species (based on the permeability) depends, after some simplification, on $\langle r \rangle$ and $\langle r^2 \rangle$, where

$$\langle r \rangle = \int_{r_{min}}^{r_{max}} r \, f(r) \, d\ell nr$$

and $f(r)$ is the pore size distribution ($= dV/d\ell nr$). $\langle r^2 \rangle$ can be estimated in a similar way to $\langle r \rangle$. For our thickest plate we found $\langle r \rangle$ to equal 445nm, whereas for the thinnest plate, the value of $\langle r \rangle$ was 395 nm.

In our experiments it was necessary to use conditions under which Knudsen diffusion prevailed. Therefore

$$D_g = D_K \propto \langle r \rangle$$

and

$$100 \left\{ (\overline{D}_K)_{thick} - (\overline{D}_K)_{thin} \right\} \Big/ (\overline{D}_K)_{thin} \sim 12\%$$

Barred symbols refer to values averaged over the plate thickness. The effect on the local diffusion coefficients will be even more pronounced. For the purpose of our comparative study of sorption and NMR diffusion coefficients however, we decided not to introduce this effect into the model for simultaneous heat and mass transport. For the solution of the non-stationary problem it would be an unnecessary complication.

On the other hand, both NMR and sorption measurements were carried out with the same plates and under similar conditions and we do not expect the fact that the plates are not quite uniform to be a serious drawback.

The Adsorption Separation Process
J. W. Carter
The University of Birmingham, Birmingham B15 2TT, U.K.

Developments in the applications of molecular sieve adsorbents over the last
15-20 years were reviewed at the Third[1] and Fourth[2] International Conferences
on molecular sieves. A major proportion of these applications are gas and
liquid purification processes removing water vapour from various streams
including natural gas for liquefaction, helium extraction, ethane recovery and
other purposes, from air for cryogenic distillation, hydrogen for catalyst
protection, cracked-gas for olefine production, ethylene, propylene and
acetylene after salt cavern storage and particularly natural gas with high acid
gas contents[3]. Removal of carbon dioxide and/or various sulphur compounds
is also frequently necessary. For these and other separations the adsorbents
are normally used in a granular form and on a regenerative basis undergoing
a very large number of adsorption/desorption cycles. The granules are usually contained in a cylindrical pressure vessel as a fixed bed supported on a
suitable grid. The vessel is provided with inlet and outlet fluid flow connection such that the gas or liquid being treated passes through the adsorbent bed
without by-passing or short-circuiting. To allow for continuous processing at
least two such vessels are used with suitable valving and pipework so that one
of them can be on-stream while the adsorbent in the used one is being regenerated.

Because of this unsteady state nature fixed bed adsorption was a neglected unit
operation so that theoretical analysis has been the domain of mathematicians
until recently. Useful solutions to the unsteady state equations describing the
process of adsorption from a fluid-flow passing through a fixed bed of adsorbents were first available to us through the work of Hougen and Marshall[4]
and Rosen[5]. These were limited to isothermal conditions and a linear isotherm but are of continuing use. A pseudo-steady state method due to Michael
and reported by Treybal[6] assumed a constant shape mass transfer zone (MTZ)
moving through the bed. This applied to non-linear isotherms but was still
for isothermal conditions. For non-isothermal conditions an approximate
method was proposed[7] which assumed that a heat transfer zone (HTZ) preceded the MTZ. This was essentially the situation reviewed in 1960[7]. The
availability of high-speed digital computers helped subsequently to rectify it by
enabling the simultaneous heat and mass balances and rate equations to be
solved using numerical techniques. These were used to study systems in which
the effects of the heat of adsorption were significant, in particular the adsorption of water vapour[8-11]. Then, short-cut methods of calculation were developed[12,13] to economise on computing time and cost without excessive loss of
accuracy.

This brief outline shows the continuous improvement in the use of theoretical
models for the prediction of both the isothermal and non-isothermal adsorptive
process in fixed beds. However, the authors of such publications assumed
that the adsorbent was first in a fresh, or fully adsorptive, condition throughout the bed, whereas in practice continuous fixed bed adsorption is a cyclic
process and depends upon satisfactory regeneration before a bed can be reused. Early papers on desorption or regeneration[14-17] were limited and used a
semi-empirical heat balance method of calculation. Recent papers[18-22] used
fixed bed equations to study the effect of favourable isotherms on the shape of
isothermal adsorption and desorption curves. Using adsorbate free fluid

under conditions comparable to adsorption these papers showed that desorption
times were longer than adsorption times for a bed equally saturated throughout. To ensure continuous cyclic operation it is therefore necessary to speed
up desorption in dual adsorber systems and this is normally done by using a
flow of hot regeneration fluid (thermal swing). Heating is necessary because
the heat of desorption is a large proportion of the heat requirements when high
adsorption loading occurs on the adsorbent beds. Only in pressure swing
adsorption, which is typified by low differential loading on the adsorbent, may
the desorption process be considered to approach isothermality. Also, in
dual adsorber systems, an adsorbent bed is rarely evenly saturated throughout its length because it is taken off-stream when the breakthrough of impurity
occurs. Thus, the direction of flow of the desorption fluid through the bed may
be important. Normally it would be in the reverse direction to adsorption
because this ensures that the "pure" side of an adsorbent bed is always fully
regenerated and kept free of the impurities. In many cases the process fluid
is treated at a high pressure (e.g. 2200 psi[3]) which may cause problems of
pressure blow-down and build-up with gas losses if regeneration is done at low
pressures where it is more efficient. Also, in liquid treating, liquid hold-up
on the adsorbent bed after draining has to be evaporated before proper regeneration can take place. This can impose significantly increased heating requirements and other problems.

Adsorption/Desorption Cycles

For continuous operation the adsorption and desorption parts of the cycle
cannot be considered separately because the performance during adsorption
is determined by the previous desorption (regeneration) conditions. Various
types of adsorption/desorption cycles that may be used can be shown on a
simple isosteric equilibrium diagram. This is given in figure 1 for a single
adsorbate entering the adsorber in a
process gas stream at a partial
pressure p_3 to be reduced p_2 or less.
A complete cycle of operation of a
small portion of adsorbent at, say,
the outlet end of the bed can be
followed using this diagram.

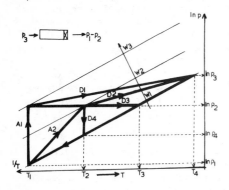

Fig.1: Adsorption/Desorption Cycles

When freshly regenerated adsorbent
with a residual loading W_1 and cooled
to a temperature T_1 is contacted by
the impure gas, adsorption may follow
an isothermal route (A_1) or a non-
isothermal route (A_2). The latter will
occur if the amount of heat evolved due
to adsorption is significant and will
raise the adsorbent temperature

resulting in a lower effective breakpoint adsorption capacity (W_2-W_1) versus
(W_3-W_1). If only untreated process gas, containing impurity at partial
pressure p_3, is available for regeneration it is necessary to heat the adsorbent
to a temperature T_4 to remove adsorbate down to the residual of W_1. However,
if some of the treated gas is available with the impurity reduced to p_2 a lower
regeneration temperature T_3 can be used. These are desorption routes D_1,
D_2 and D_3. If the desorbed impurity can be removed economically from the
regeneration flow by condensing or other methods, this flow can then be recycled to the main inlet flow to the adsorber for re-purifying to p_2 as long as
regeneration is carried out at the working pressure. Thus there need be no

loss of valuable process fluid, although if treated gas is used for regeneration this will impose an additional load on the adsorber and therefore increase its size and cost whether the gas is recycled or not. If regeneration gas at a much lower partial pressure p_4 is available, a lower temperature still can be used which can be advantageous in reducing the range of temperatures to which the adsorbent is subjected and thus help prolong its useful life. This gas may be provided by expanding a portion of the treated high pressure flow to a lower pressure but would normally involve re-compression if it is to be recovered.

Dual Adsorber Circuits

The flow circuits for continuous cyclic operation therefore depend on the method of regeneration[23] and a limited selection is given in figure 2. The

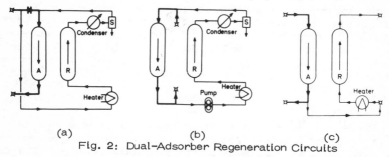

(a)   (b)   (c)
Fig. 2: Dual-Adsorber Regeneration Circuits

circuits in figures 2(a) and 2(b) regenerate at the working pressure but use respectively a proportion of the untreated or treated gas flows heated to a suitable temperature in each case. Circuit 2(b) is most useful to obtain high adsorption efficiency (i.e. high product purity) by combining the use of treated gas for regeneration with a high regeneration temperature. The circuit in 2(c) uses treated process gas flow reduced to a lower pressure so that an even lower partial pressure of impurity is obtained in the regeneration flow. Alternatively, an entirely independent source of regeneration gas flow can be used. For simplification purposes, these diagrams do not show the valves and pipework required for changing over the flows from one adsorber to the other.

Calculation of Adsorption in Fixed Beds

Early methods of design and prediction of performance were essentially empirical or semi-empirical, and dependent on experience. Such methods still exist partly because of the need to make quick comparisons using altern- ative adsorbents and designs and because of the difficulties in many cases of describing the processes occurring by mathematical models. Also many known models are expensive in computing especially if alternative designs are to be compared. The early analytical approaches[4,5] were used with some success in predicting previously published small and large-scale results[7]. For single component adsorption there are publications on isothermal[2,9,19-21, 24-26] and adiabatic[8,9,11,25,27-29] fixed beds.

Isothermal Adsorption.

Where adsorbed phase diffusion is an important rate mechanism a solid diffusion model has been widely used in the past. Recent work[26] at 1.2 atmospheres pressure compared the adsorption of carbon dioxide at partial pressures (p) up to 0.4 mm mercury in nitrogen and helium by deep beds of molecular sieves 5A and 4A (Grace-Davison). Details of the adsorbents were:

	M.S.5A	M.S.4A
Mesh size	8-12	8-12
External granule area m^2/m^3	1572	1572
Granule density kg/m^3	1120	1120
Bed density kg/m^3	729	760
Granule peak pore radius $m \times 10^6$	0.0625	0.050

and the Langmuir isotherms and rate data:

M.S.5A/CO_2/N_2

$$W^* = \frac{0.0275p}{1+0.865p} + 6\%$$

$$D/b^2(h^{-1}) : 0.402$$

M.S.4A/CO_2/He

$$W^* = \frac{0.1124p}{1+3.455p} + 6\%$$

$$0.040$$

where W^* = Weight adsorbate/unit weight of adsorbate free adsorbent at
equilibrium, D = effective adsorbate diffusivity and b = mean effective
adsorbent granule radius. The experimental and calculated breakthrough
curves for various bed depths upto 1.245m (4ft.approx.) are shown in
figures 3 and 4. The calculation procedure was to use the numerical

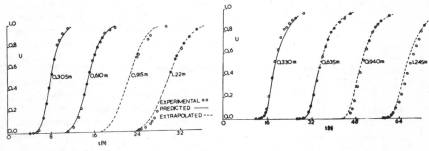

Fig.3: Breakthrough Curves:
N_2/CO_2/MS5A. 5m/min.

Fig. 4: Breakthrough Curves:
He/CO_2/MS4A. 5m/min

solution of the partial differential equations to determine the value of D/b^2
which predicted accurately the first breakthrough curve. The value then
predicted later ones very accurately. To save computing time a simple
extrapolation technique was developed for subsequent bed-lengths based on the
assumption that for steady influent conditions each particular concentration
travelled at a constant velocity through the bed. Constant velocity lines for
dimensionless gas-phase concentrations U = 0.05, 0.5 and 0.85 are given in
figures 5 and 6 for the two systems. U is the ratio of the effluent concentration to the steady inlet value. The extrapolated breakthrough curves using
this technique are also plotted in figures 3 and 4 showing close agreement
with experiment. The linear bed velocities in the N_2/CO_2/5A and He/CO_2/
4A experiments are identical (5.0m/min)but the figures clearly show the
much lower velocity of the MTZ, smaller width and rate of expansion for the
latter system. This is essentially due to the much more favourably shaped
isotherm and its higher adsorptive capacity.

Pore diffusion in granular adsorbents can be the principal rate controlling
mechanism[8,9,20]. Of particular importance is the work of Garg and Ruthven
[20] concerning macropore and micropore diffusion and their relative signifi-

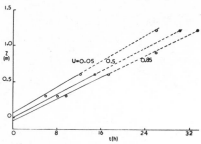

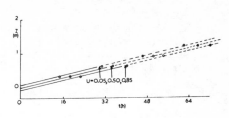

Fig.5: MTZ Development: Fig. 6: MTZ Development:
$N_2/CO_2/MS5A$. 5m/min $He/CO_2/MS4A$. 5m/min

cance in molecular sieve adsorbents. In connection with work on the coarse
purification plant for the helium coolant for the Dragon Nuclear Reactor the
author examined[30] the importance of macropore diffusion control for the
$MS4A/CO_2/He$ system in beds up to 1.855 m deep at 25°C, 1.25 atm pressure
and 0.1885 mm Hg CO_2 partial pressure. The experimental, calculated and
extrapolated results are given in figure 7. The adsorbent (Grace-Davison)

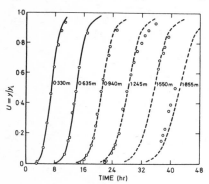

details and isotherm are as just given.
The bed velocity in this case was approximately 17m/min. The effective
experimental pore diffusion coefficient
Dpe of the carbon dioxide in the granules was 1.44×10^{-6} m^2/s with a mean
granule radius of 0.0012m. Using
Wheeler's method[31] Dpe was calculated
to be $1.55 \times 10^{-6} m^2/s$ using the peak
macropore radius determined by
mercury penetration measurements.
From experimental work on the same
system by Lange[32] for Euratom and
Brown Boveri Krupp, but at 40 atmospheres pressure and using Union
Carbide 1/16in 4A M.S., the author
determined the experimental value of

Fig. 7: Breakthrough Curves:
$He/CO_2/MS4A$. 17 m/min.

Dpe = 0.085×10^{-6} m^2/s and a calculated value of 0.11×10^{-6} cm^2/s from the
peak macropore radius of $0.225 \times 10^{-6}m$. The lower Dpe in this case is due
essentially to the higher helium pressure. The effect of higher pressure is
compared in figures 8 and 9 giving calculated breakthrough curves and MTZ
development for the same experiment as in figure 7 upto 0.635m in the bed at

1.25, 5 and 10 atmospheres respectively. The same linear bed
velocity was maintained in each
case. The figures show the earlier
breakthrough times and greater
MTZ widths caused by lower values
of Dpe as pressure increases. The
effects on initial MTZ widths and
rates of expansion are summarised
in table 1.

Table 1: Effect of Operating Pressure
on Mass Transfer Zone

Pressure (atm.abs)	Initial zone width (m)	Rate of zone expansion (m/m bed)
1.25	0.17	0.19
5	0.23	0.27
10	0.32	0.40

One of the problems in numerical computation of the solutions for a pore
diffusion model results from the non-linear concentration dependent coeffici-

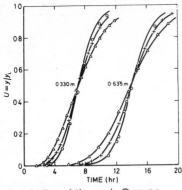

Fig.8: Breakthrough Curves:
1.25, 5 and 10 atm. 17m/min.

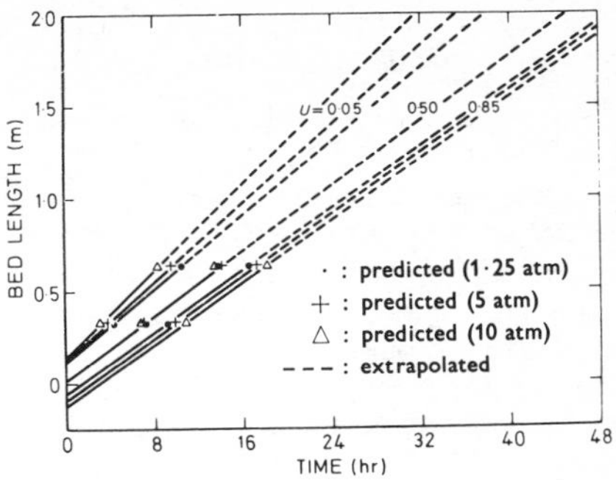

Fig.9: MTZ Development:
1.25, 5 and 10 atm. 17 m/min.

ents in the finite difference form of the equation for diffusion within the granule. This is because of the need to apply the equilibrium relationship at each radial increment and significantly lengthens the iterative procedures and therefore the time of the calculation. The calculation for the solid diffusion model is much quicker because the concentration (and temperature) dependent equilibrium applies only at the fluid/granule interface. This was therefore modified by using a quasi pore diffusion model[8] based on the approximate relationship:

$$D \simeq \frac{Dpe \, \rho_f}{K \, \rho_g}$$

between the effective solid diffusion coefficient D and Dpe. $K = W^*/y$ and is a mean Henry's Law coefficient at the gas phase composition y(weight adsorbate/unit weight adsorbate free gas). ρ_f and ρ_g are fluid and granule densities respectively. At a fixed pressure, if gas phase concentration and temperature do not vary widely then DK is substantially constant so that:

$$DK = DoKo \quad \text{and:} \quad D = DoKo/K$$

Do and Ko can be determined experimentally at known conditions, conveniently the inlet concentration and temperature to the bed. Since K at other conditions can be calculated from adsorption equilibrium data, then D at these conditions can then be calculated throughout the fixed bed as concentration (and perhaps temperature) and thus K varies with time and distance. This is supported by the excellence of isothermal and non-isothermal predictions using such data for water vapour adsorption on granular activated alumina and molecular sieves[9,11]. An appreciably reduced computing time compared with the pore diffusion model was obtained.

Non-Isothermal Adsorption . The heat of adsorption which is evolved in the adsorption process is frequently sufficient to raise significantly the adsorbent temperature and consequently reduce the maximum adsorption loading that is possible. The temperature rise will be insignificant only if the adsorbate concentration in the fluid is low or the fluid sensible heat capacity is high and is most likely to occur in treating liquids or high pressure gases. Cooling systems in an adsorbent bed and pre-chilling the gas to slightly above freezing point to remove a condensable adsorbate such as water[10] can reduce considerably the temperature rise due to the heat of adsorption and show improved performance and economies in capital and running costs.

Pan and Basmadjian[28] proposed a simple criterion which enabled two distinguishable types of non-isothermal operation to be identified in fixed beds. If the following inequality applies:

$$(W/y)_i > C_g/C_f$$

where subscript i represents conditions at the inlet end of the bed and C_g and C_f are the specific heats ($Kgcal/Kg^oC$) of the adsorbent granules and the flowing fluid respectively, a heat transfer zone (HTZ) will precede the main mass transfer zone (MTZ) so that the latter will occur on adsorbent already heated to the temperature in the HTZ. This has been called pure thermal wave formation. However if the criterion has the reverse inequality viz:

$$(W/y)_i < C_g/C_f$$

it is impossible for a pure thermal wave to pull ahead of the main adsorption zone so that a combined heat and mass transfer wave front occurs. This criterion is useful but approximate because it is based on the assumption of adsorption and heat transfer equilibria. The kinetics of these transfer processes will have local effects on the criterion and a numerical investigation for non-equilibrium conditions has been made by Cooney[33]. Other work has studied the performance of adiabatic colums under equilibrim using simple wave theory by following adsorption/desorption paths along the characteristics of the partial differential equations[27,29].

A very high inlet concentration of adsorbate, such that the temperature rise of the adsorbent due to the heat of adsorption reduces W so that $(W/y)_i < C_g/C_f$, is a prerequisite for combined wave formation. Pan demonstrated this experimentally[28] by reducing the feed concentration of CO_2-He mixtures adsorbing non-isothermally on a bed of $1/8"$ 5A M.S. pellets. It is relevant to note that earlier work by Getty[34] showed this phenomenon very clearly indeed when drying airflows of various inlet humidities in beds of activated alumina. This showed two distinct break-points for low inlet humidities as expected from the author's experience. The first occurred for a relatively short time before the pure thermal zone had penetrated the bed, raising the exit frost-point from a low initial value (about -85^oF) to slightly higher plateau values. The latter increased (-80^oF to -50^oF approximately) as the inlet humidity, and therefore the temperature of the bed behind the pure thermal zone increased. The time to the second break-point also decreased until at the highest humidity complete breakthrough occurred from the time of the first break point, that is when the thermal wave first penetrated the bed, giving a combined wave front. It would appear that such high temperature rise conditions are an inefficient use of an adsorbent bed. So called pure thermal wave behaviour in adiabatic adsorption is shown in figures 10, 11 and 12 for the adsorption of water from an airflow by a fixed bed of granular activated alumina 0.457m (1.50ft) deep[9]. The inlet air flowrate was $0.258Kg/m^2s$ at 21.1^oC and $0.005Kg/Kg$ humidity. Figure 10 shows the experimental and closely predicted concentration and temperature breakthroughs and figures 11 and 12 the predicted gas phase concentrations and temperatures throughout the bed at various times. The much faster development of the HTZ and the temperature plateau it forms is clear. The fall in temperature behind the plateau as the concentration in the MTZ rises and a consequent reduction of rate evolution of latent heat occurs is also apparent. It should be noted that the equilibrium and diffusional rate data used in these predictions were determined by separate experiments on a small but representative sample of the adsorbent, as were the data for other non-isothermal fixed bed experiments

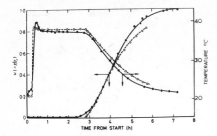

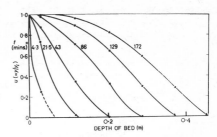

Fig. 10: Adiabatic Breakthrough
Concentration and Temperature

Fig. 11: Adiabatic Air Humidities
in Bed.

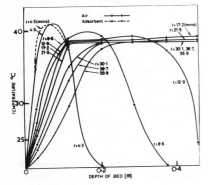

Fig. 12: Adiabatic Air and Bed
Temperatures

adsorbing water vapour on molecular sieves[11].

Multi-component Fixed Bed Adsorption

It is very rare that process adsorption involves the removal by the adsorbent of one component only, except perhaps in some instances when drying gases. Because of the different degrees of adsorbability of mixed impurities, and also perhaps of the carrier, zones of the various adsorbates form throughout the bed. The least strongly adsorbed one will precede the others and may be desorbed by them if the adsorption process proceeds for a long enough time. In many cases it may therefore be necessary to design the fixed bed operating cycle to prevent breakthrough of a lighter component. For example, using 5A molecular sieve to sweeten natural gas[35] it was necessary to design to prevent breakthrough of the hydrogen sulphide zone which preceded the water zone but succeeded the main carbon dioxide zone.

Other investigational work[36-39] has been done on two component adsorbate systems. For more than one adsorbate it is necessary to describe the equilibria by mixed adsorption equations and the research workers in these references have used the Langmuir relationship or an adaption of it. For the simultaneous adsorption at $0^\circ C$ of water vapour and carbon dioxide from a flow of helium by 4A type molecular sieves, the experimental breakthrough curves of both adsorbates were very closely predicted using rate and equilibrium data determined from single adsorbate experiments[36]. A pore diffusion model was used. The data are given in Table 2 and the experimental points and predicted outlet concentrations (U) in figure 13. The latter are the continuous lines. The discontinuous line is the calculated adsorbed phase carbon dioxide concentration (V̄) at the bed exit. The bed depth was 3.43 cm and helium flowrate $0.0774 Kg/m^2s$. Four zones formed in the bed and are shown schematically in figure 14. U and V̄ refer to the dimensionless gas and average solid phase concentrations respectively. Subscripts 1 and 2 refer to water and carbon dioxide respectively. In zones I and II both adsorbates are present but only the least strongly adsorbed, carbon dioxide, in zones III and IV. This example demonstrates the general problems in multi-component adsorbate systems of the formation of zones of different adsorbates and of

Table 2: Equilibrium and Rate Data for Single Adsorbates, Water-Carbon
Dioxide-4A Molecular Sieves 0°C

	Water	Carbon Dioxide
W* (p_1 and p_2 are mm mercury):	$\dfrac{11.1p_1}{1+46.7p_1}=2\%$	$\dfrac{0.435p_2}{1+5.84p_2}=4\%$
Dpe(cm²/sec):		
Exptl.	0.0631	0.0083
+Calc'd-Wheeler	0.0238	0.0147
Wakao	0.0115	0.0074

+The values of Dpe were calculated by the methods of Wheeler[31] and Wakao[40]
using a peak pore radius of 0.050×10^{-6}m in the granule determined previously[24]

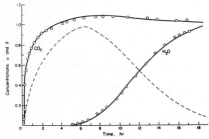

Fig. 13: Binary breakthrough curves:
He/H_2O/CO_2/MS4A.

determining their time of appearance at
the exit of the bed and is of particular
importance in preventing the escape of
impurities. Although some impurities
may have entered below the maximum
allowable concentration, perhaps from
a previous treatment process, their
outlet concentration will be higher after
desorption by a stronger adsorbate.
The problem of differential desorption
will also occur in the regeneration
process where it can create problems
of recovery or disposal associated with
the design of the regeneration equipment.

Calculation of Thermal Regeneration of Fixed Adsorber Beds
The main variables affecting the residual impurity levels are the temperature
difference of the adsorbent between adsorption and regeneration and the impurity concentration in the regeneration fluid (see figure 1). While some flexibility is possible in thermal swing design to accommodate slight differences in
impurity levels, the adsorbents do impose some limitations. For example a
3A sieve is normally regenerated at maximum temperatures of 200 to 230°C
to prevent too rapid a cyclic decay in adsorptive capacity. Also, to achieve
high purity during adsorption a low residual impurity level on the outlet section
of the bed is necessary. To do this it is usually sufficient to arrange the
regeneration heating flow counter current to the adsorbing flow and the cooling
flow co-current to prevent premature adsorption on the outlet sections of the
column. This avoids forcing high adsorbate concentrations through less
saturated portions of the bed and ensures that no traces of impurity are
adsorbed on downstream surfaces, adsorber vessel and pipework, etc., that
may desorb and spoil high purity product in the adsorption cycle.

Co-current Heating.   If very high purities are not required in the product
flow then reverse flow heating need not be used. Both regeneration heating
and cooling may then be carried out in the same direction as adsorption
without the complication and cost of any extra pipe work and valves for flow
reversals. However, in these circumstances it is important that the heating
period is completed and not shortened in any way. This is to ensure that the
desorption MTZ is pushed completely out of the adsorbent bed. If the adsorp-

tion phase begins before this, product flow which is purified in the first part of the bed will then continue the unfinished desorption lower down and probably come out less pure than at the bed inlet. It is also advantageous to have regeneration heating co-current with the adsorption flow to use the main desorbate wave to desorb impurities that can foul the adsorbent. In the drying of natural gas containing pentanes and higher, Wunder[16] expected a considerable amount of these heavy hydrocarbons to adsorb at the drier outlet. Co-current regeneration was used because if countercurrent heating were employed coking of the heavy hydrocarbons could occur, due to contact with hot reactivation gases, and cause a serious decrease in desiccant capacity. With co-current reactivation the gases were cooled to the water desorption temperature of 65–120°C before contacting the heavy hydrocarbons. In addition the water, as it moved down through the bed, stripped the hydrocarbons from the desiccant minimising the possibility of coking still further.

Use of Simultaneous Heat and Mass Transfer Equations. Although the heat balance method of calculation has been used extensively it is limited to situations for which experimental efficiency factors are available. In addition it gives little or no information about the temperatures and impurity concentrations at the bed exit and in the bed itself. The investigation and comparison of alternative regeneration conditions is therefore not possible by this simple method and has led to investigating thermal-swing regeneration by solving the partial differential equations previously used for non-isothermal adsorption.

The effectiveness of industrial adsorption depends on the high loading of an adsorbate from a dilute mixture or solution onto a solid phase of high capacity. This enables it to be desorbed to give much higher concentrations at which it can be separated by a simpler technique, such as condensation, for recovery or removal from a closed system, or it can be discarded to stack or to flare using as low a flow as possible of purge gas. During the regeneration of an adsorbent bed with such loadings significant temperature and concentration profiles are formed and some work on this has been done using heat and mass transfer equations[41–43]. Figure 15 shows the outlet air flow concentration and temperature during the adiabatic adsorption on a 0.228m (0.75ft) deep bed of granular activated alumina. These are the accurate predictions from a previous experimental and theoretical study[8,9]. The inlet air flow rate was 0.258 kg/m^{2}s at 21.1°C and a humidity of 0.0048 kg/kg.

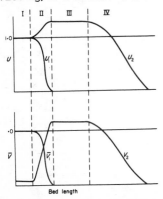

Fig. 14: Zones Generated in bed during Simultaneous H_2O and CO_2 Adsorption

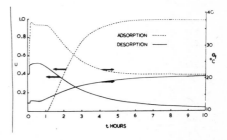

Fig. 15: Adsorption and Desorption Curves. 21.1°C Inlet.

The rate coefficient for this adsorbent was $KD/b^2 = 3.25h^{-1}$ and the heat of adsorption (or desorption) 2907 kJ/kg. The equilibrium equation was:

$$W^* = 0.129(y/y_{satn_{\Theta_g}})^{0.46}$$

where:

W^* = kg adsorbate/kg adsorbent at equilibrium

y = air humidity kg/kg

$y_{satn_{\Theta_g}}$ = air humidity saturated at adsorbent granules temperature Θ_g

Figure 15 also shows the calculated outlet concentration and temperature with time for the desorption of a fully saturated bed using the same flow rate of air at zero humidity at the same temperature, 21.1°C. Figure 16 shows the desorption curves under the identical conditions except the regeneration air enters at 93.3°C. The differences are clearly apparent in that in the second case desorption of water is much quicker, the peak outlet humidity reaching three times the inlet adsorption value whereas it reached only half the inlet value for desorption with 21.1°C air.

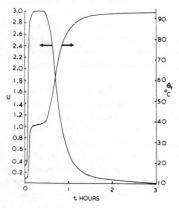

Fig. 16: Desorption Curves. 93.3°C Inlet

After adsorption to the break point in Figure 15 the calculated distribution of water on the activated alumina fell almost linearly from the equilibrium value at the inlet to zero at the outlet. Figures 17 and 18 give the desorption curves for this distribution using the same air-flow conditions as in Figure 16 but with the regeneration flow co-current and countercurrent respectively to adsorption. In Figure 17 the co-current example, the adsorbate peak at the inlet is forced through the bed and re-adsorbed in the less loaded regions. This shows in the figure as a minor peak in the outlet temperature due to the heat of re-adsorption. This is a loss of heat which has to be replaced by further regeneration air thus lengthening the process and reducing its efficiency. In figure 18, with countercurrent heating, regeneration is quicker because no

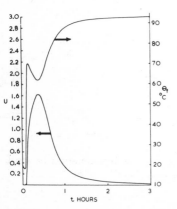

Fig. 17: Desorption Curves. Partially Saturated Bed. 93.3°C Inlet. Same Direction as Adsorption.

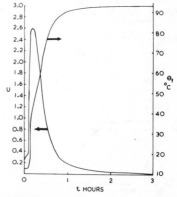

Fig. 18: Desorption Curves. Partially Saturated Bed. 93.3°C Inlet. Reverse Direction to Adsorption.

such loss of heat occurs and also the peak outlet humidity is higher. Although these curves are realistic in the temperature range investigated, for normal thermal swing regeneration temperatures of 150–250°C a more accurate equation for the concentration and temperature dependent equilibrium would be desirable.

Thus desorption is significantly non–isothermal in a heavily loaded adsorbent bed and if the regeneration gas flow is not pre–heated its desorbate capacity is reduced below that of the adsorption inlet flow. The regeneration time under these conditions will be much longer than the adsorption time, or a much higher flow will be needed to regenerate in the same time. The higher the temperature of the regeneration heating flow, the greater is its desorbate carrying capacity. Therefore there is normally no problem of regeneration being carried out within the adsorption time so that dual adsorber plants can operate continuously. The principal constraint is then the maximum temperature to which the adsorbent may be exposed.

A recently published method of predicting a limiting desorption time used Rosen's fixed adsorption bed analysis including a linear isotherm[44]. This is based on isothermality throughout the bed at a known adsorption isotherm ie. known temperature. With thermal swing regeneration isothermality may be approximated under certain conditions but the temperature may not be known. Also, the analysis is based on a bed equally loaded throughout at the start of desorption and would be less accurate for partially saturated beds.

Isenthalpic Methods. It had been noted by the author when commissioning adsorption plant that in a majority of thermal swing desorptions the outlet gas temperature rose to a pleateau, at which most of the desorption of adsorbate occurred, before rising to a higher temperature approaching the inlet value. This had also been noted by others[14]. The author has calculated these "hygroscopic wet–bulb" conditions approximately by means of an enthalpy balance using the adsorption isostere to determine the adsorbate equilibrium vapour pressure as a function of inlet temperature. The calculation also takes into account the inlet adsorbate partial pressure and was checked on a high pressure carbon dioxide adsorption drier using the regeneration circuit in Figure 2(b).

The calculation was also done for the regeneration of 4A molecular sieves containing 10% adsorbed water using carbon dioxide at 300°C and 350°C as the regenerating gas. This was for conditions that were being explored for the adsorbent beds used in the coarse purification plant for the Dragon nuclear reactor coolant gas[30]. The effects of pressures between approximately 1 and 50 atmospheres were studied. These are shown in figures 19 and 20. Figure 19 gives the calculated "plateau" water vapour pressures (p mm Hg) and gas humidities (X mols/mol). While the former increase with pressure, rising from 100 mm to 1330 mm mercury in the range from 1 to 50 atmospheres for an inlet temperature of 350°C, the gas humidities fall. Thus more gas is required to desorb a particular amount of water as pressure increases. Also, condensation problems may arise before the condenser when regenerating at high pressures (the dewpoint temperature at 50 atmospheres is 116°C. See figure 20), so that careful attention to the design of the regeneration circuit would be necessary. The higher inlet temperature (350°C) gives an appreciable increase in humidity but figure 20 shows the higher temperatures obtained in the regenerating gas as the pressure increases so that the thermal efficiency of regeneration heating decreases. Thus, when the heat of desorption is

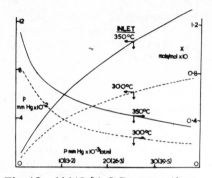

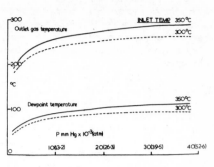

Fig. 19: 4AMS/H_2O Regeneration
by hot CO_2. Effect of Total Pressure on
Outlet Vapour Pressure and Humidity.

Fig. 20: 4AMS/H_2O Regeneration
by hot CO_2. Effect of Total Pressure on Outlet Gas and Dewpoint
Temperatures.

a significant amount of the regeneration heating load, the most efficient use of
the sensible heat of the regenerating gas occurs at the lowest pressure and
highest temperature. Under these conditions more water is desorbed per unit
mass of the gas but may have to be balanced against the loss of gas in a
pressure reduction operation. In addition to a limiting maximum temperature
to which the adsorbent may be exposed, at high pressures there may be a
hydrothermal ageing problem due to high aqueous vapour pressures.

Regeneration Heating of Liquid Adsorption Plant . Because liquid phase
treating by adsorption to remove a range of impurities including dissolved
water, sulphur compounds, some reaction products, etc. is increasing it is
important to discuss the regeneration of these adsorbers. Before regeneration
can begin the adsorber has to be drained to recover as much of the organic
liquid as possible. The remainder which is held-up between or inside porous
adsorbent granules will then have to be evaporated either before or during
regeneration heating. Then the adsorbent can be raised finally to the required
temperature for desorption of the impurities. The magnitude of the problem
of evaporation of liquid hold-up may be illustrated by the hold-up data given
in Table 3 for a granular molecular sieve.

Table 3: Liquid Hold-up on Beds of 3mm 3A Molecular Sieves

	i-Butyl alcohol	Benzene	Trichlor-ethylene
g liquid per 100g dry M.S	29.7	33.2	55.4
Liquid S.G.	0.805	0.879	1.466
ml liquid per 100g dry M.S.	36.9	37.8	37.8

The evaporation heating load can thus be highly significant, in fact frequently
greater than the desorption heating load for the adsorbed impurities. Two
types of liquid hold-up occur, namely low vapour pressure and high vapour
pressure systems. These have been investigated on a laboratory and on a full

plant scale to check out the theoretical predictions using the simultaneous heat and mass transfer equations. Figures 21 and 22 show the calculated temperatures and concentrations respectively for two such systems.

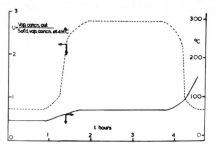

Fig. 21: 6m Bed 4AMS/C_8 Liquid Hold-up Regeneration. Outlet Gas Temperatures and Vapour Concentrations. 315°C Inlet.

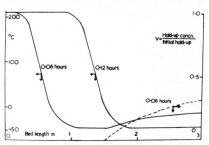

Fig. 22: 3m Bed 3AMS/C_4 Liquid Hold-up Regeneration. Gas Temperatures and C_4 Hold-up in Bed. 210°C Inlet.

Figure 21 gives the calculated outlet gas temperature and dimensionless vapour concentration (U) during the regeneration heating at one atmosphere of a 6m deep 4A molecular sieve bed with a C_8 hydrocarbon liquid hold-up. U is the ratio of the calculated outlet vapour concentration to the saturated value at 41.1°C. The regeneration circuit used re-circulated inert gas so that the hold-up liquid could be recovered by condensation. The curves show similar symptoms to a normal desorption, the outlet temperature rising to a plateau of 63°C with the outlet vapour concentration rising to approximately 4.0 times the inlet concentration (ie off the condenser). Thus appreciable recovery of the hold-up is possible. This plateau precedes the heat transfer zone (HTZ) so that after evaporation the temperature rises to approach the inlet value of 315°C. If the desorption heating load for the impurities is also significant, such as it may be for adsorbed water, then the hold-up plateau will be succeeded by the water plateau before the outlet temperature rises to the maximum value.

Figure 22 gives the temperature and dimensionless liquid hold-up (V) distributions at different times (0.06 and 0.12 hours) in a 3m deep 3A molecular sieve bed with C_4 hydrocarbon liquid hold-up. V is the ratio of the calculated liquid hold-up on the bed to the initial value after draining. The inlet regeneration gas is at 210°C at approximately 3 atmospheres. The evaporation of the volatile liquid is very rapid and complete by 0.12 hours. Because the evaporation MTZ moves through the bed much quicker than the HTZ the latter is preceded by temperatures down to -45°C, eventually forming a low temperature plateau. This results from the latent heat of evaporation being taken from the adsorbent bed and the cooled regeneration gas flow. This situation may be avoided in some instances by evaporating at higher pressures but this is not always possible. It may then be necessary to use low temperature steels for construction of the adsorber vessels and associated pipework. Internal lagging may be worth considering to prevent the vessel walls reaching such low temperatures. It would also reduce the total regeneration heating load. Systems investigated in this work included halogenated and aromatic hydrocarbons, saturated and unsaturated paraffinic hydrocarbons and other organics.

Other Problems of Thermal Regeneration

The thermal ageing of adsorbents can reduce adsorptive capacity[9] but rates of adsorption can also seriously be affected by chemical reactions such as the polymerisation of adsorbed olefins[45] and adsorbent attack by acids adsorbed as trace impurities or from the hydrolysis of halogenated solvents during regeneration. Such an attack on activated alumina granules reduced the effective adsorbed phase diffusion coefficient for water to one third whilst only reducing the adsorptive capacity to 90% of the original value. This reduced the useful breakpoint time from 8 hours to 2 hours[46]. During the high temperature regeneration ($300^{\circ}C$) of molecular sieve driers for natural gas containing upto 3500 v.p.m. oxygen, reaction with hydrocarbons occurred forming water and carbon dioxide[47]. These can reduce drier capacity and impair operating efficiency. The increase in pressure drop in a molecular sieve adsorber for high water vapour and carbon dioxide concentrations was explained by adsorbent granule breakdown caused by the formation of acid condensate during regeneration. The reduction in crushing strength values is shown in Table 4 of granules of a molecular sieve wetted with distilled water and water saturated with carbon dioxide compared with the strength of granules.

Table 4: Average Crushing Strengths of 4mm 4A Molecular Sieves

Condition	Strength (kg)
Dry	2.62
H_2O	1.52
H_2O/CO_2	1.23

Conclusions

Although considerable advances have been made in modelling and calculating the adsorption process in fixed beds they have yet to be made in calculating the desorption or regeneration process. This depends on temperature and concentration dependent equilibrium and rate data which are not readily available. In particular the regeneration of adsorbent beds containing multiple adsorbates or with liquid hold-up has scarcely been studied. Special problems are associated with thermal swing regeneration at high pressures including natural gas driers. The need for a greater flow of regenerating gas has to be balanced against the de-pressurising and re-pressurising effects. Adsorbent deterioration can occur in thermal swing regeneration for a number of different reasons. Environmental testing is desirable if not necessary to check for decay and its effects on performance.

References

[1] W.M. Meir and J.B. Uytterhoeven (Editors), "Molecular Sieves", Advances in Chemistry Series 121, American Chemical Society, Washington D.C., 1973.

[2] J.R. Katzer(Editor), "Molecular Sieves-II", ACS Symposium Series 40, American Chemical Society, Washington D.C., 1977.

[3] P.N. Kraychy and A. Masuda, Oil and Gas J., 1966 (Aug.8), 64, 66.

[4] O.A. Hougen and W.R. Marshall, Chem.Eng.Progress, 1947, 43, 197.

[5] J.B. Rosen, I.E.C., 1954, 46, 1590.

[6] R.E. Treybal, "Mass Transfer Operations", McGraw-Hill, N.Y., 1955.

[7] J.W. Carter, Brit.Chem.Eng., 1960, 5, 472, 552, 625.

[8] J.W. Carter, Trans.Instn.Chem.Engrs., 1966, 44, T253.

[9] J.W. Carter, Trans.Instn.Chem.Engrs., 1968, 46, T213, T221.

[10] J.W. Carter, "Scale-up in the design of fixed bed adsorption plant", Proc. Symposium "Dropping the Pilot", Institution of Chemical Engineers and the Journal British Chemical Engineering, 21st March 1969.

[11] J.W. Carter and D.J. Barrett, Trans.Instn.Chem.Engrs., 1973, 51, 75.

[12] J.W. Carter, H. Husain and D.J. Barrett, Dechema Monographien, 1974, 73, 17.

[13] J.W. Carter and H. Husain, Chem.Ing.Tech., 1977, 49 (Nr. 6), 517.

[14] R.G. Cappell, E.G. Hammerschidt and W.W. Dresner, I.E.C., 1944, 36, 779.

[15] D.A. Hougen and F.W. Dodge, "The Drying of Gases", J.W. Edwards, Ann Arbour, 1947.

[16] J.W.J. Wunder, "Design of a Natural Gas Drier", Presented at the Annual Meeting, Chemical Institute of Canada, Edmonton, Alberta, May 1962.

[17] J.W. Carter, Chem.Proc.Eng., 1966, 47(9), 70.

[18] I. Zwiebel, R.L. Graiepy and J.J. Schnitzer, A.I.Ch.E. J., 1972, 18, 1139.

[19] D.R. Garg and D.M. Ruthven, A.I.Ch.E. J., 1973, 19, 852.

[20] D.R. Garg and D.M. Ruthven, C.E.Sci., 1974, 29, 571.

[21] D.R. Garg and D.M. Ruthven, C.E. Sci., 1975, 30, 436.

[22] G. Bunke and D. Gelbin, C.E. Sci., 1978, 33, 101.

[23] H.L. Brooking and D.C. Walton, The Chemical Engineer, June 1972, 13.

[24] J.W. Carter and H. Husain, Trans.Instn.Chem.Engrs., 1972, 50, 69.

[25] T.Vermeulen and G. Klein, Chem.Eng.Prog. Symp. Series, 1973, 69(134), 65.

[26] J.W. Carter, H. Husain and D.J. Barrett, Rev.Port.Quim., 1974, 16, 90.

[27] N.R. Amundson, R. Aris and R. Swanson, Proc.Roy.Soc., 1965, 286A, 129.

[28] C.Y. Pan and D. Basmadjian, C.E. Sci., 1970, 25, 1653.

[29] C.Y. Pan and D. Basmadjian, C.E. Sci., 1971, 26, 45.

[30] C. Harper, The Chemical Engineer, July 1972, 271.

[31] A. Wheeler, Catalysis, 1955, 2, 105.

[32] G. Lange, Brown Boveri/Krupp, Private Communication.

[33] D.O. Cooney, I.E.C. Process Des.Dev., 1974, 13, 368.

[34] R.J. Getty and W.P. Armstrong, I.E.C. Proc. Des.Dev., 1964, 3, 60.

[35] C.W. Chi and H. Lee, Chem.Eng.Prog.Symp.Series, 1973, 69(134), 95.

[36] J.W. Carter and H. Husain, C.E.Sci., 1974, 29, 267.

[37] J. Shen and J.M. Smith, I.E.C. Fundam., 1968, 7, 100, 106.

[39] W.J. Thomas and J.L. Lombardi, Trans.Instn.Chem.Engrs., 1971, 49, 240.

[40] M. Wakao and J.M. Smith, C.E. Sci., 1962, 17, 825.

[41] J.W. Carter, A.I.Ch.E. J., 1975, 21, 380.

[42] K.E. Olsen, D. Luss and N.R. Amundson, I.E.C. Proc.Res.Dev., 1968, 7, 96.

[43] G. Bunke, K-H Radeke and D. Gelbin, Chem.Tech., 1977, 29(2), 92.

[44] D.M. Ruthven, A.I.Ch.E. J., 1978, 24, 540.

[45] A.J. McDonald, Ph.D Thesis, University of Birmingham, 1968.

[46] J.W. Carter, Ph.D. Thesis, University of Birmingham, 1966.

[47] G. Bancroft, K.R. Clark and G. Corvini, Hydrocarbon Proc., September 1975, 54, 203.

The Practical Application of Pressure Swing Adsorption to Air and Gas Separation

J W Armond

BOC TechSep, Angel Road, London N18 3BW

1. INTRODUCTION

Pressure swing adsorption (PSA) is a process for separating gas mixtures by short time cycle adsorption/desorption on fixed beds of adsorbent using gas pressure variation as the principal operating parameter. Generally no steps are taken to insulate a plant from the surroundings and the adsorbent beds operate effectively at ambient temperature.

The technique of separation by PSA has been known for at least 20 years, patents dating from 1957 (Ref. 1 and 2). Since then the interest in the process has been widespread with the number of patents published reaching into the hundreds. In spite of this wide interest the PSA plants operated commercially have principally been confined to a few specific separations as follows:

a) Air Drying

The so called "heatless drier" is now widely used for drying instrument air and probably some half of the air driers now sold employ this principle. In these the drying is carried out at an upper pressure in the region of 7 bar (g); the regeneration of the adsorbent bed is effected by countercurrent purging with a portion of the dry air product at a lower pressure close to atmospheric. Activated alumina and silica gel are commonly used adsorbents.

b) Production of Hydrogen

Compared with the other components of hydrogen-rich feedstocks, hydrogen is very weakly adsorbed at ambient temperature and lends itself particularly well to production at high purity by the PSA route. To date there are over 100 commercial plants - the largest in the region of 100 tons per day (50 000 Nm^3/hr)[*] - in operation or under construction throughout the world,

[*]Nm^3 signifies that the volume of gas is measured at 273.15 K and 760 Torr.

producing hydrogen,in some cases at extremely high purity,
from a variety of feedstocks such as refinery off-gas,
cracked ammonia or hydrogen-rich gas from steam reforming.
A product purity of 99.999% can be reliably achieved
where required. Typically purification is at 14-17 bar(g)
with bed regeneration by venting and purging with
pure hydrogen product close to atmospheric pressure.
Activated carbon and zeolite molecular sieve are commonly
employed as the adsorbents.

c) Production of Oxygen and Nitrogen from Air

This separation is not practical with the traditional
adsorbents such as activated carbon, silica gel and
alumina. However, the tailor-made materials of the
carbon molecular sieve and zeolite molecular sieve
classes have provided selectivity between oxygen and
nitrogen and permitted development of processes for
producing nitrogen or oxygen from air. The process
cycles employed have greater variety than in heatless
drying and hydrogen purification; thus in oxygen
producing processes employing zeolite the swing of
pressure may be between a few bars positive and
atmospheric pressure in one type of cycle and between
atmospheric pressure and 150 torr in another cycle. A
nitrogen producing process using carbon molecular sieve
can have an upper pressure in the range 1 to 7 bar (g)
and a lower pressure of 100 torr.

2. PROCESS DESIGN CONSIDERATIONS

Some of the major design parameters of a PSA plant are
discussed here. As far as possible these will be illustrated
in respect of a typical cycle which produces oxygen from air
using 5A molecular sieve in two adsorbent beds. A typical
operating sequence for this cycle, which is depicted in
Fig. 1, is as follows:

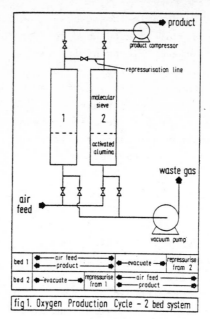

fig 1. Oxygen Production Cycle - 2 bed system

While bed 1 is on stream adsorbing nitrogen from air to give oxygen product bed 2 is regenerating by evacuation to 150 torr; this is followed by countercurrent repressurisation to 500 torr using product quality oxygen from the outlet of bed 1. A distinguishing feature of the cycle is that air feed is drawn into the plant at slightly sub-atmospheric pressure by virtue of the vacuum pump removing waste gas and by the product compressor drawing off oxygen product. The air flow rate into the plant is on a demand basis and varies several-fold during a cycle, being highest when the regenerated bed begins to repressurise and lowest immediately before repressurisation.

In order to prevent the moisture in the inlet air from reaching the molecular sieve a layer of dessicant, such as alumina or silica gel is located at the inlet end of the bed and the PSA process is selected so that the alumina or silica gel section acts as a heatless drier at the same time as the zeolite separates the nitrogen and oxygen.

The overall process solves the problem of obtaining product
at any chosen elevated pressure using only two pumps without
the wasteful operation of compressing the whole of the feed
air to the required product pressure. It also has the
advantage that by the use of a feed pressure close to
atmospheric the process operates with a better nitrogen/
oxygen separation factor than would apply at a higher feed
pressure. Defining separation factor as:

$$\frac{\text{Nitrogen adsorptive capacity}}{\text{Oxygen adsorptive capacity}} \times \frac{\text{Oxygen partial pressure}}{\text{Nitrogen partial pressure}}$$

Table 1 gives separation factors for air in the range 0 to
4 bar (g) based on the zeolite adsorption isotherms shown
in Fig. 2. The oxygen yield of this process at 15°C is 42%
for 90% oxygen product and 50% for an 80% oxygen product.

<table>
<tr><td colspan="5" align="center">TABLE 1</td></tr>
<tr><td colspan="5" align="center">Nitrogen-Oxygen Separation Factors on 5A Molecular Sieve at 25°C</td></tr>
<tr><td rowspan="2">Total Pressure
bar(g)</td><td colspan="2">Partial Pressure
torr</td><td colspan="2">Adsorptive Capacity
ml/gm</td><td rowspan="2">Separation
Factor</td></tr>
<tr><td>N_2</td><td>O_2</td><td>N_2</td><td>O_2</td></tr>
<tr><td>0</td><td>602</td><td>158</td><td>6.32</td><td>0.60</td><td>2.77</td></tr>
<tr><td>1</td><td>1204</td><td>316</td><td>11.10</td><td>1.15</td><td>2.54</td></tr>
<tr><td>2</td><td>1806</td><td>474</td><td>14.96</td><td>1.72</td><td>2.29</td></tr>
<tr><td>3</td><td>2408</td><td>632</td><td>18.03</td><td>2.24</td><td>2.12</td></tr>
<tr><td>4</td><td>3010</td><td>790</td><td>20.65</td><td>2.75</td><td>1.97</td></tr>
</table>

By using a three-bed version of the process Fig. 3, with
full part cycles allocated to feed and evacuation, the
utilisation of the vacuum pump is increased and power
consumption improved. The addition of a second cut operation,
in which the last fraction of gas from a bed taking air
feed - which has a higher oxygen content than air but is
below product quality - is used as the first feed to the
next bed, allows the former bed to be more fully saturated
with nitrogen and hence increases the oxygen yield compared
with the two bed cycle.

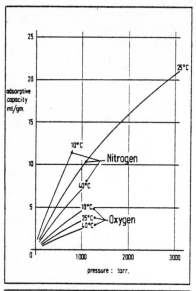

fig 2. Adsorption Isotherms: O_2-N_2 on 5A mol. sieve

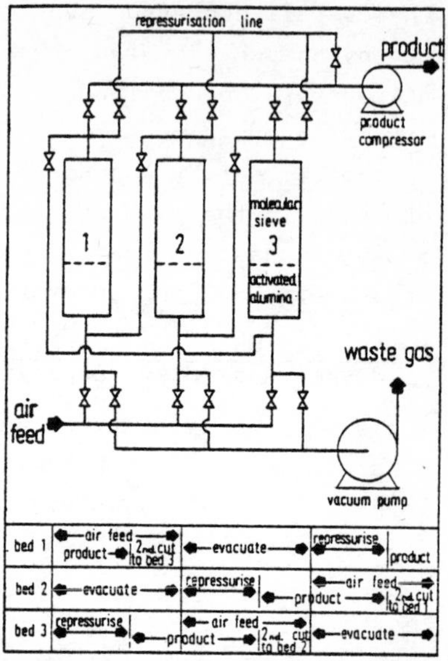

fig 3. Oxygen Production Cycle - 3 bed system

The choice between the 2 and 3 bed oxygen process depends upon the relative importance to the customer of capital cost and power consumption. The two bed process is more expensive on power than the three bed by a factor of about 1.6; however the two bed plant being simpler and more compact is cheaper to build at the smaller sizes. With scale factors of about 0.65 for plant capital cost and 1.0 for power it becomes economic to take advantage of the better power utilisation of the 3 bed plants at larger sizes. On these considerations the maximum size for a two bed plant is in the region of 50 Nm^3/hr oxygen product.

2.1 Cycle Time

For a given plant output, reducing the cycle time reduces the bed size and thus the cost and size of the plant. The minimum practical cycle time is determined by some of the parameters discussed in this section. Typically for the oxygen plant described, operating from 0 bar(g) to 150 torr, the part cycle time is one minute. For a

PSA plant operating from 15 to 0 bar (g) such as a
hydrogen plant the part cycle time is typically 6-7 mins.

2.2 Adsorbent Choice

There are several factors determining the choice of
adsorbent for a process:

a) Selectivity. The adsorbent must have the highest
   possible selectivity between the gases to be separated.
   Whilst selectivity is in many cases on the basis of
   differences in equilibrium adsorptive capacities,
   which includes the case of true molecular sieving
   action, there is also the alternative of selectivity
   on the basis of the rate of adsorption. This is very
   well illustrated by the carbon molecular sieves, the
   pore sizes of which can be adjusted in the manufacturing process to exhibit large differences in the rate
   of adsorption between different species of molecule.
   Fig. 4 illustrates the basis of the selectivity
   between oxygen and nitrogen using zeolite and carbon
   molecular sieves, the zeolite separating by the
   difference in the equilibrium adsorptive capacity
   whilst the carbon sieve separates by differences in
   the rates of adsorption.

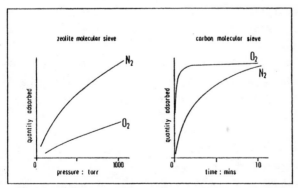

fig 4. Carbon & Zeolite Molecular Sieve Selectivity for Oxygen-Nitrogen

The PSA process may be applied to the recovery of
either the most strongly, or most rapidly, adsorbed
component, or the least strongly, or least rapidly,
adsorbed component of a gas mixture. For recovery
of the most strongly adsorbed component it is usual,
after the adsorption step, to purge away unwanted
components before reducing the pressure to draw off
the product. In such a process it is difficult to
purge away the last traces of impurity so that very
high purity product is not usually obtainable.
Furthermore, relatively small leakages through valves
during the product withdrawal stage can seriously
reduce product purity. In contrast to this, the
process for the recovery of the least strongly
adsorbed component is capable of yielding product of
very high purity from the outlet of an adsorbent bed
at the same time as feed gas is passed in at the
opposite end; furthermore leakage past valves is
never in a direction to contaminate pure product.
This latter mode of operation is favoured where high
to very high purity product is required and in the
case of the separation of air the opposite selectivities exhibited by the zeolite and carbon sieves
towards oxygen and nitrogen enable this cycle
feature to be used for plants producing oxygen using
zeolite molecular sieve and nitrogen using carbon
molecular sieve.

b) Adsorptive Capacity. Unlike temperature swing
   adsorption the highest possible adsorptive capacities
   are not essential for PSA separations and are less
   important than good selectivity. For example, a
   molecular sieve 5A adsorbent for separating oxygen
   and nitrogen has adsorptive capacities at $25^{\circ}C$ and
   0 bar(g) of only 2.65 and 7.65 ml/gm for oxygen and
   nitrogen respectively (0.36 and 0.90% by weight) and
   yet forms the basis of an economically viable process.

c) Physical Properties. Adsorbents must be available
   as particles of the correct size range in order to
   control pressure drop. They must also have good
   resistance to attrition to minimise dust formation.
   Any adsorbent we use is subjected to attrition

testing on a comparative basis. Where necessary,
pressure drop data is measured.

2.3 Power Consumption

A vital aspect of the competitiveness of a PSA plant
is its power consumption. Once the process is chosen
the power needed to compress the feed or product can be
minimised by the best choice of pumps and by ensuring
that pipework pressure drops are low. However, in a
process using a vacuum for the lower pressure, apart
from the correct choice of pumps, considerable power
saving can be achieved by the correct choice of adsorbent
particle size and bed dimensions.

For example, in the oxygen process described we have to
evacuate an adsorbent bed from 0 bar (g) to 150 torr
and the bed pressure drop is found to have a significant
effect. For a given bed volume and a given pump, the
effect of bed dimensions and particle size on pumping time
and power consumption is shown in Table 2. These results
have been calculated using a computer programme which
solves a set of non-linear partial differential equations
describing the dynamic behaviour in terms of mass balance,
component balance and pressure drops (Ref. 3). The
method was checked carefully against measured evacuation
performance before being adopted for plant design.

TABLE 2

Evacuation of a Bed of Zeolite Molecular Sieve with a Fixed Pump

Starting Pressure: 0 bar(g)
Final Pressure: 150 torr

Particle Size							
1-2 mm Balls	Bed Length	3	4.5	6	7.5	9	ft
	Bed Diameter	4.243	3.464	3.000	2.683	-	ft
	Time	57.6	59.7	62.9	67.4	-	secs
	Power	0.452	0.468	0.494	0.530	-	kWh
	Effect on Plant Output	1.02	0.988	0.938	0.875	-	-
3-5 mm Balls	Bed Length	3	4.5	6	7.5	9	ft
	Bed Diameter	4.243	3.464	3.000	2.683	2.449	ft
	Time	56.7	57.6	59.0	61.0	63.6	secs
	Power	0.444	0.452	0.463	0.479	0.500	kWh
	Effect on Plant Output	1.04	1.025	1.000	0.967	0.929	-

As would be expected, increasing the particle size and
decreasing the bed aspect ratio improves power consumption. Increasing the particle size further would give
lower power consumption; however we have found that
using adsorbent in a ball form the separating efficiency
of the process deteriorates at sizes above 3-5 mm
probably due to mass transfer within the particles
becoming a significant factor. In practice, for the
oxygen process described, we have generally standardised
on the use of 3-5 mm balls or equivalent particles of
other shapes together with molecular sieve bed lengths
typically not below 4 feet. This combination provides
an acceptable flow distribution without an excessive
power penalty and keeps the overall height of the plant
within reasonable bounds. The height is important where
a plant has to be transported as completed skid mounted
unit from the factory to the operating site.

2.4 Effect of Ambient Temperature

The adsorbent isotherms of nitrogen and oxygen on 5A
molecular sieve alter with temperature, Fig. 2, in such
a way that both adosrptive capacity and selectivity
between oxygen and nitrogen decrease with increasing
temperature, Table 3. As PSA plants usually operate in
the open air and are subject to changes of ambient
temperature, the extent of the temperature effect needs
to be known for design purposes.

TABLE 3

Effect of Temperature on Nitrogen-Oxygen Separation Factor

| Temperature °C | Adsorptive Capacities ml/gm | | Separation Factor |
	Nitrogen at 602 torr	Oxygen at 158 torr	
10	9.62	0.81	3.13
25	6.32	0.60	2.77
40	4.50	0.48	2.50

In the case of the 2-vessel oxygen process exampled
here, the measured change in output over the range
15-40°C for constant cycle time and 150 torr vacuum is
shown in Fig. 5.

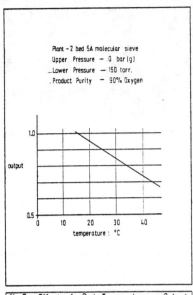

Plant – 2 bed 5A molecular sieve
Upper Pressure — 0 bar (g)
Lower Pressure — 150 torr.
Product Purity — 90% Oxygen

fig 5. Effect of Bed Temperature on Output

3. COMMERCIAL PLANTS

PSA plants are of simple construction using carbon steel
vessels and pipework, standard valves, pumps and other
components. Provided they are not too large plants leave
the factory complete in the form of one or more transportable
skid mounted units as shown in Figs. 6 and 7. In the case
of the oxygen and nitrogen plants they are, whenever
possible, precommissioned before delivery to site.
Precommissioning is not normally possible in the cases of
feed gases other than air.

In addition to nitrogen, oxygen and hydrogen plants and
heatless driers, PSA units have been manufactured for helium
recovery from diving gas, nitrous oxide recovery from

admixture with nitrogen and carbon monoxide recovery from
steam reformer gas.

Fig. 6. 350 Cubic Meter
         per hour Oxygen
         Plant.

Fig. 7. 300 Cubic Meter
         per hour Nitrogen
         Plant.

REFERENCES

1. Skarstrom, C.W. - B.P. 850 443

2. Guerin de Montgareuil, P and Domine, D - French
    Patent 1 223 261

3. Sebastian, D J G - A Mathematical Model for Pressure
    Swing Adsorption. Proceedings of the 7th IFIP Conference
    on Optimisation and Modelling at Nice, September 1975.

The author wishes to thank BOC Limited for permission to
present this paper.

Molecular Sieves for Industrial Separation and

Adsorption Applications

C.W. Roberts

Laporte Industries Limited, Widnes, Cheshire, U.K.

The choice of subject matter for this paper presented several problems. Firstly, the industrial applications of molecular sieves have been reviewed many times, eg [1], and yet another review did not seem appropriate. Secondly, for a manufacturer, there are constraints on the topics which can be discussed. For example, many manufacturing details cannot be divulged because these must remain proprietary information. Also, secrecy agreements between manufacturers and users are common, and many process details cannot be discussed. Thirdly, only a short account was possible; thus it was necessary to be very selective in choice of material. However, perusal of the recent academic, review and patent literature led to the conclusion that the practical problems associated with molecular sieve design for general and particular applications had not been properly summarised. It is the purpose of this paper therefore to attempt to fill that gap.

The topics selected are listed in Table 1.

Table 1

Selected Topics

1. Some aspects of manufacture
2. General product optimisation
3. Experimental techniques

Firstly, some aspects of manufacture :- A survey of commercially available molecular sieves for the major separation and adsorption applications [2] reveals that most compositions still consist of zeolite A or X crystals bound with naturally occurring clays. A simplified manufacturing route to such products is shown in figure 1.

Figure 1

Manufacture of Clay Bound Molecular Sieves

Simplified Flow Diagram

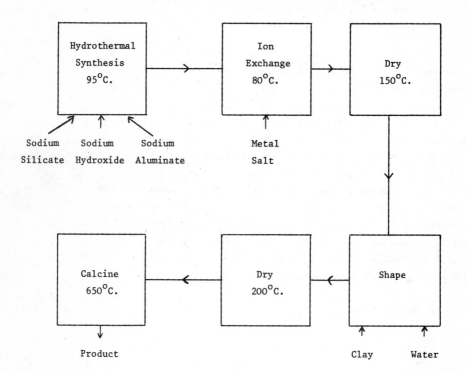

For the sake of simplicity, the necessary filtration and washing
stages after hydrothermal synthesis and ion exchange have been omitted.
Sieving operations are also not shown. Clearly if the sodium zeolite is
required, no ion exchange is necessary. Also in some circumstances, the
ion exchange step is performed on the calcined shaped pieces rather than
on the crystalline product from the hydrothermal synthesis. This is
usually because the thermal stability of the desired product is insufficient
to withstand the final high temperature calcination, the purpose of which is

to dehydroxylate the clay binder irreversibly. Also, in certain circumstances, a fully dehydrated product is not called for.

The chemical engineering involved is fairly conventional and the temperatures shown are typical. Careful control of all stages is required to ensure consistent product quality. However, the two stages which prove particularly troublesome in practice are the hydrothermal synthesis and shaping processes. These will be dealt with in more detail. These special problems are listed in Table 2, together with some of the more common clay binders.

Table 2

Special Manufacturing Problems

1. Hydrothermal synthesis
2. Shaping, especially choice of binder
    eg   Sepiolite
          Attapulgite
          Kaolinite
          Montmorillonite

Firstly, hydrothermal synthesis :-   Production of crystalline zeolite A is relatively simple, both on the laboratory and industrial scale, and a pure crystalline product can be formed for example in four hours at 95°C in a conventional agitated reactor. The original route to zeolite X, however, called for a lengthy ageing period under quiescent conditions and presented considerable problems on the large scale. Without these special precautions the usual product is zeolite P, which has a small pore size and is of no value as an industrial adsorbent.

In 1964, the concept of certain sodium metasilicates in promoting selective formation of zeolite X in agitated reactions was published[3]. At first, this so-called activity was observed in certain silicate manu- facturers' products but not in others. The source of this activity was unknown, but it was always associated with a hydrated rather than an anhydrous metasilicate. Subsequently, special methods of hydration of the anhydrous material to produce active products were patented, but the source of the activity remained a mystery. More recently, parallel industrial[4] and academic investigations[5,6] have shown that this activity to promote formation of zeolite X, rather than zeolite P, is associated with the adventitious presence of aluminium in commercial hydrated metasilicates, and there is good evidence that the source of activity is an alumino-silicate seed material. At the present time, the precise nature of this seed is unknown, but it is significant that the same seed is also of value for synthesis of zeolite A. A common factor between A and X, of course, is the sodalite cage structure.

Other methods of seed manufacture have been patented, eg [7], and it seems likely that most molecular sieve manufacturers use seeding processes. In addition to allowing easy synthesis of high purity A and X type zeolites, adjustment of the nature and amount of seed permits control of crystal size and state of aggregation. Naturally, in common with other crystallisation processes, these factors are controlled by many other variables, such as agitation, temperature, and overall reactant composition. It is found that synthesis control by seeding is the most valuable technique. The scanning electron micrographs shown in Plate 1 illustrate the variation in zeolite A crystal size which may be obtained by addition of seed. In this case, seed addition has decreased the average crystal size by about a factor of three.

Secondly, the shaping process :-   At least three shaping processes are employed. These are extrusion, extrusion followed by rolling to form spheres, and simple granulation to form spheres. All three methods result in zeolite crystals cemented together by the clay binder. The binders commonly used are sepiolite, attapulgite, kaolinite and mont-morillonite. Often mixtures of two or more clays together perhaps with other substances such as alumina and silica are found. Organic additives, such as lignosulphonates[8], may be used to modify the rheology of the mix. Important considerations in the choice of clay binder are, of course, the physical properties of the final product. In the early days of molecular sieve manufacture, these were probably the prime criteria.

Plate 1

Effect of Seed Addition on Crystal Size for Zeolite A

Seeded with Commercial Sodium Metasilicate Pentahydrate

No Seed

For the purpose of achieving adequate physical properties, binder levels in the range of 10 to 25% in the finished product are commonly used. However, manufacturers have now learned that choice of binder type for each application is very important. This is a consequence of the nature of naturally occurring clays which themselves are high surface area materials. Each type has unique sorption and catalytic properties. In addition there is little doubt that, during the thermal treatments following the shaping process, chemical reaction occurs between the zeolite and binder surfaces leading to a thin layer of a new phase. As well as these complications there are many binder type variants, for example, regarding particle size and purity. The scanning electron micrographs shown in Plate 2 are fractured molecular sieve type A pieces bound with sepiolite and kaolinite. These photographs show the intimate contact between binder and zeolite crystals. The characteristic fibrous structure of sepiolite and the platy structure of kaolinite are apparent.

The situation is therefore very complex, and the effect of binder type on product performance can be dramatic. The disadvantages of some binders are shown in Table 3.

Table 3

Disadvantages of Some Binders

Cause	Effect
Blinding of crystal surfaces	Poor kinetic properties
High surface area	Poor selectivity
Catalytic activity	'Coking' in some applications

These effects are perhaps not surprising, and are to an extent predictable. For example, calcined kaolinite and montmorillonite are active hydrocarbon conversion catalysts. Sorption on attapulgite and sepiolite, which are high surface area clays, can impair selectivity in some separation processes.

The foregoing discussion might seem to indicate that the use of clay binders is always disadvantageous. However, whilst several so-called binderless products are commercially available, the market is still

Plate 2
Zeolite A Containing Two Common Binders

Kaolinite

Sepiolite

dominated by clay-bound materials. Many factors, including cost,
contribute to this situation, but the main reason is that by proper
choice of binder for each application these difficulties can usually
be overcome. In addition, there are instances in which the use of
certain clays confer advantages compared with binderless products.
However, details in this area must remain proprietary information.

Some product optimisation problems will now be discussed. It is
difficult to deal with these in a general way because some applications
call for extraordinary solutions. However, there are some general
principles which apply to virtually all situations.

Manufacturers and users of sieves are concerned about two factors –
the initial performance, and the deterioration of this performance
during service. For all separation and adsorption applications, the
prime consideration is the dynamic capacity. Optimisation of this allows
use of the smallest (and least expensive) user process installation. For
a given application, effective sieve capacity may be broken down into
equilibrium capacity and kinetic behaviour. Initial performance and rate
of deterioration depend critically upon the nature of the sieve manufacturing process. The efficiencies of most adsorption and separation
processes depend upon rates of diffusion in both the zeolite micropores
and extracrystalline macropores. The effect of additional diffusional
resistance at the zeolite binder interface has been referred to already,
and this effect does not seem to be generally recognised. The principal
factors affecting initial performance are listed in Table 4.

Table 4
Principal Factors Affecting Initial Performance

Equilibrium Capacity	Kinetic Behaviour
Zeolite purity	Crystal size
Cationic composition	Cationic composition
Binder level	Crystal aggregation
Thermal history during	Crystal packing
manufacture	Binder type
	Piece size and shape
	Thermal history during
	manufacture

This list applies, of course, to a given zeolite framework composition which, in practice, is substantially fixed in the case of zeolite A types, but is variable for zeolite X.

Framework composition, zeolite purity, crystal size and state of aggregation can all be optimised by control of the hydrothermal synthesis. Additional important factors affecting zeolite purity are the washing processes which follow the hydrothermal synthesis and ion exchange stages. For simplification reasons, these are not shown in figure 1. For example, incomplete removal of the highly alkaline mother liquor from the hydrothermal synthesis stage can cause serious degradation during subsequent processing, particularly during drying and calcination. Cationic composition is easily controlled and the general effect of this on the properties of zeolites A and X is well known. However, for certain specialised applications, optimisation of cationic composition continues. Binder level and types have been mentioned already. The effect of piece size is well recognised and its choice depends upon the application. Generally, the largest possible piece size is chosen to minimise pressure drop problems, but requirements vary widely. In practice, for spherical products, mean piece diameters range from about 3.5 m.m. for gas phase applications to 0.5 m.m. for some liquid phase processes. Crystal packing depends not only upon choice of shaping process, but also upon binder type and quantity, water level, and any additive which may be used to affect the rheology of the mix. Extremes of packing density are shown in Plate 3. These variations occurred within the same piece, and probably arose from poor blending of the zeolite and binder powders. The greater concentration of binder in the more tightly packed structure is apparent.

The effect of thermal treatment during processing deserves special attention. Drying of the zeolite powder, shaped pieces, and final calcination subject the product to steaming conditions. This situation can lead to serious degradation of equilibrium and kinetic properties. Zeolite X is particularly susceptible, and requires very careful thermal processing to minimise loss of crystallinity. In particular, thick beds must be avoided, and good purge conditions maintained. A high silica sodium X is much easier to process than a low silica product.

Plate 3

Variation in Crystal Packing for Zeolite X

Tight

Loose

Whilst steamy processing conditions tend to destroy the crystalline structure of zeolite X, the effect on A types is much more subtle. Sodium A in particular undergoes a pore closure effect, but the crystalline structure remains intact[9]. An example of the practical outcome of this phenomenon is in the use of sodium A for removal of carbon dioxide from gas streams. A product which has undergone excessive pore closure will behave badly for this duty due to slow rate of carbon dioxide sorption.

Loss of initial performance during use poses even greater problems for the sieve manufacturer. Again it is difficult to generalise, but there are two very common problems which can apply to both A and X types in many applications. Typically in a fixed bed adsorption process for example, lengthening of the mass transfer zone occurs resulting in loss of effective capacity. This may be due to deterioration of sieve kinetic behaviour or loss of equilibrium capacity or both. The two common causes of performance loss are listed in figure 2.

Figure 2
Two Common Causes of Loss of Sieve Performance

1. Steam damage   (Loss of crystallinity
                               (
                               (Pore closure

2. 'Coking'   (Intracrystalline
                               (
                               (Extracrystalline

Firstly, steam damage :- Because so many molecular sieve applications involve a drying duty followed by thermal regeneration, this is an extremely common problem. Thus, there is a direct correlation between the problems of damage by this mechanism during manufacture and during use. In fact, damage during use can be regarded as a continuation of the processes occurring in thermal treatment during manufacture. The comments made in that context also apply to deterioration during use. An example is the removal of water and carbon dioxide from natural gas by sodium A, prior to liquefaction. In this case,

steam damage can occur at the inlet end of the bed which performs
the water sorption duty. With continued use, therefore, progressive
pore closure takes place along the bed and mass transfer zone lengths
for both water and carbon dioxide increase.

Secondly, 'coking' :-   The nature of this phenomenon depends
very much upon the application. The term is normally applied to deposition of carbonaceous material which occurs during some hydrocarbon
separation processes. It is clearly not a problem with many other
applications - air purification prior to liquefaction for instance.
This deposition of carbonaceous matter may be intracrystalline or extracrystalline, or both, depending again upon the duty. 3A sieve for
drying cracked gas for example has a finite useful life. In addition to
deterioration by steam damage, coking can occur. Product discharged
after some years service contains carbonaceous matter and it can be
demonstrated that by careful burn-off the rate of water uptake is at
least partly restored. In practice, in this case, regenerative burn-off
is not incorporated in the process. For this application of course, 3A
sieve is chosen to exclude the unsaturated components in the stream.
However, the degree of exclusion depends critically upon sieve design.
If insufficient pore closure to exclude readily polymerisable unsaturates
has not been achieved during manufacture, then intracrystalline coking
will occur. Extracrystalline coking depends to a large extent upon
choice of binder. All the binders listed previously have been used and
montmorillonite, in particular, is not favoured for this application[10].

Some of the more common experimental techniques used in product
optimisation will now be discussed. Many more than those listed have
been used, but there are certain key methods without which little
progress can be made. Investigation can be divided into three problem
areas. These are :-

> Zeolite crystal characterisation
> Fresh product characterisation
> Product degradation investigation

Firstly, zeolite crystal characterisation :- The more important
techniques are listed in Table 5.

Table 5

Zeolite Crystal Characterisation

Usual Techniques

Problem	Technique
Purity	Powder X-Ray Diffraction
	Equilibrium Sorption (Vacuum Microbalance)
Chemical Composition	X-Ray Fluorescence
Size	Scanning Electron Microscopy (S.E.M.)
Aggregation	S.E.M. and Sedimentometer

These techniques are of course fairly commonplace, except perhaps that the excellence of X-ray fluorescence analysis for aluminosilicates is not generally recognised. Typically a set of complete chemical analyses for eight elements can be obtained within a working day. The add-up is usually within the range 99 - 101%.

Secondly, fresh product characterisation :- The usual techniques are listed in Table 6.

Table 6
Fresh Product Characterisation
Usual Techniques

Problem	Technique
Single Component Equilibrium Sorption	Vacuum Microbalance
Single Component Sorption Rate	Flow Microbalance
Multicomponent Equilibrium Sorption	Microflow Technique
Overall Multicomponent Behaviour eg. Exhaustion Capacity, Mass Transfer Zone Length	Process Simulation in Pilot Plant

Only a few comments are necessary. Electric microbalances are preferred to spring microbalances or volumetric techniques. The reason is that rates of sorption, as well as equilibrium values, are obtained directly and simply on a pen recorder. Multicomponent studies usually involve complex mixtures and elevated pressures and may be in the liquid phase. Thus a microflow technique is used as far as possible, often merely to minimise feed fluid requirements. Conventional methods of macropore characterisation such as mercury porosimetry are purposely excluded. This is because, unlike other adsorbents such as alumina, the macroporosity is fixed by deliberate choice of crystal size and binder type. Thus, for a given piece density, the outcome is largely predictable.

However, a useful tool not listed here is gas chromatography. This is particularly valuable for selectivity optimisation. An example is the use of 5A for normal paraffin separation processes. In this case, the presence of small traces of a large pore zeolite such as X can spoil product purity. Very small traces are difficult to detect by X-ray diffraction and gravimetric sorption methods. One solution then is to pack a conventional gas chromatograph column with the 5A adsorbent and observe the way in which pulses of appropriate probe hydrocarbons are eluted.

The real proof of the performance of a product is pilot plant simulation of the user's process. Often this can only be done by the user himself for process secrecy reasons or feedstock availability. However, for the well established applications it must be done by the sieve manufacturer.

For fixed bed gas and liquid purification processes, the general technique is to scale down in diameter as far as is possible to avoid undue wall effects, but to maintain sufficient length to contain the anticipated mass transfer zones. This type of experimentation is both expensive and time consuming. However, in many cases, it is essential. The reason is that feeds are often so complex, no reliable prediction of the multicomponent equilibrium or kinetic situations can be made otherwise. Results of this type of experimentation, of course, appear as breakthrough curves. From these, the appropriate exhaustion capacities and mass transfer zone lengths can be calculated by the wellknown Treybal method[11].

Finally, the most intractable problem of all will be discussed briefly. This is prediction of useful service life. The major problems are re-listed in Table 7, together with some of the possible solution techniques.

Table 7
Product Degradation in Service
Usual Techniques

Problem	Technique
Steam Damage	Accelerated static steaming followed by flow microbalance measurements
	Cycling in pilot plant
'Coking'	Static exposure to real or simulated fluid composition, followed by flow microbalance rate measurements
	Cycling in pilot plant

In this area, manufacturers must rely upon feedback from users, but in designing new products, this approach cannot be risked. There is also a severe time limitation; thus small scale accelerated tests must be employed.

There are many dangers, of course, with accelerated tests. For example, in connection with the steaming problem, unrealistic temperatures and steam vapour pressures are often used. This can lead to a change in degradation mechanism giving misleading results. However, in general, it seems that a product which behaves well in an accelerated test also does so under more realistic conditions[12]. On the other hand, a product which is perfectly adequate for normal service might behave badly in an accelerated test and be rejected.

Clearly, the nature of the sorbate used in the flow microbalance rate measurements depends upon the sieve type. For 3A, the most useful probe molecule is water. For 4A, in the context of pore closure by steam damage, both ethane and carbon dioxide are useful. 5A presents a special problem because rates of uptake of normal hydrocarbons by this technique tend to be too rapid to be measured properly. In this particular case, Freon 13 ($CClF_3$) diffuses much more slowly into 5A and has proved to be more useful. Again however, the real solution is expensive and time consuming pilot plant work.

Thus, molecular sieve manufacturers are faced with considerable problems. For all but the simplest applications, there are very difficult chemical and engineering problems to be overcome. The applications area is particularly troublesome and there is no real substitute for the user process simulation approach.

References

1. R.A. Anderson, "Molecular Sieves II", (Editor J.R. Katzer), A.C.S. Symposium Series 40, 1977, p. 637.

2. Laporte Industries Ltd., Unpublished Work.

3. British Patents 1,082,131; 1,145,995; 1,171,462; 1,171,463.

4. Laporte Industries Ltd., Unpublished Work.

5. E.F. Freund, J. Crystal Growth, 1976, 34, 11.

6. N.A. McGilp, Ph.D. Thesis, University of Edinburgh, 1976.

7. U.S. Patent 3,808,326.

8. British Patent 994,908.

9. D.W. Breck, "Zeolite Molecular Sieves", Wiley-Interscience, 1973, p.490.

10. British Patent 1,347,517.

11. R.E. Treybal, "Mass Transfer Operations", (Second Edition), McGraw-Hill, 1968, p.550.

12. Laporte Industries Ltd., Unpublished Work.

Oxygen Enrichment of Air on Alkaline Forms of Clinoptilolite.

I.M. Galabova*   G.A. Haralampiev

Department of Physical Chemistry,
Higher Chemical Technological Institute,
Sofia, Bulgaria.

Introduction.

Sieving properties and high selectivity have made zeolites some of the most advantageous sorbents in the last decades but there are also separation processes on zeolites requiring no sieving effect and only moderate selectivity. They have attracted growing interest in recent years. Separation of air components is an example of such a process. It produces oxygen enriched air which can be used in the chemical industry, non-ferrous metallurgy and in solving various problems of water and air pollution[1].

Synthetic zeolites are convenient sorbents for air enrichment[2,3] but natural ones are potentially promising as well[4,5]. Natural clinoptilolite shows fairly good selectivity for nitrogen in nitrogen-oxygen mixtures, the selectivity being enhanced by ion exchange to the potassium form[6].

The aim of the present paper is to prepare alkaline forms of clinoptilolite by means of ion exchange and to characterise the samples obtained in physico-chemical terms as sorbents for air components in particular. Alkaline cations were chosen since potassium is one of them and they have simple electronic configurations.

Ion exchange to a given cationic form brings specific changes in the stability, adsorption behaviour, selectivity and other important physical properties. Changes in the field strength in zeolite pores caused by ion exchange results in changes in the interaction of the field with the nitrogen quadropole moment, *i.e.* in a new selectivity for nitrogen. In view of this, each alkaline form of clinoptilolite should exhibit a specific selectivity for air components, this selectivity being a function of the properties of the ingoing cation.

Experimental and Results.

The original material used for ion exchange was clinoptilolite from Oligocene 1st horizon in the Northeastern Rhodopes, Bulgaria.[7] Composition of the unit cell, determined by chemical analyses, was $Ca_{1.88}Na_{0.9}K_{1.8}Mg_{0.13}Al_{6.8}$

$Si_{29.3}O_{72} \cdot 15.5H_2O$. A purity of between 90 and 95% was estimated[8] for this original sample.

Ion exchange was carried out at $25^{\circ}C$[8]. The average particle size of the clinoptilolite was >0.1 mm, and the solutions were the chloride salts of Li, Na, K, Rb and Cs of strength 0.5 g equiv dm^{-3}. The solution to solid ratio was 8 cm^3 g^{-1}. After ion exchange, samples were washed with distilled water and air-dried at room temperature. All sorbents were equilibrated at $25^{\circ}C$ and under a partial pressure of water of 14.1 mm Hg before being subjected to chemical and X-ray analyses.

<u>Ion exchange</u>: Results from the chamical analyses are given in Figure 1, with the cationic composition of the sorbents presented in a Gibbs triangle. This triangle is used in such a way[8] that a convenient reading of the degree of ion exchange is made possible. It also demonstrates clearly the negligible change in potassium content during ion exchange.

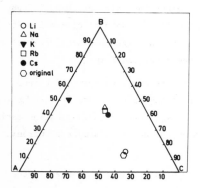

Ion exchanged form	Apex of triangle		
	A	B	C
Li form	K	Li	Cs+Na
Na form	K	Na	Ca
K form	Na	K	Ca
Rb form	K	Rb	Ca+Na
Cs form	K	Cs	Ca+Na
Original	K	Na	Ca

Figure 1. Cation composition of original and ion exchanged samples.

Legend for the choice of apex in Gibbs triangle.

For this purpose, the ingoing cation is always presented on a chosen apex of the Gibbs triangle, say apex B[8], whilst potassium is shown on apex A. The attached legend helps the readings for each sorbent and overcomes the complications with four-component systems in case of Li, Rb and Cs forms.

Full degree of ion exchange is directly read off on edge BC, and the vertical position of each sorbent gives a clear comparison of the exchangeability of each of the ingoing cations. The actual (corrected) degree of ion exchange for each ingoing cation is determined by taking into account the content, if any, of this cation in the original sample. The corrected degrees of ion exchange thus obtained for Na and K are 23% and 31% respectively.

The above data confirms the limited degree of ion exchange up to 40-46% found[8] for clinoptilolite when using ion exchange solutions of much higher concentration than those employed in the present work.

Distribution of all sorbents, including the original one, in a narrow band of almost constant content of potassium, proves the poor exchangeability of the potassium cation[8,9]. This is due to the higher coordination of K in the structure as compared to the other cations[10].

The lowest degree of ion exchange is found with the Li-form, and is probably connected with the high hydration of the lithium cation.

X-ray Analyses and Thermal Stability Checks: X-ray analyses were carried out on a TURM 61 diffractometer with CoK_α radiation. A horizontal goniometer HZG-3 was used (speed 0.5° min^{-1}) with a Greiger-Müller counter (maximum count rate 60×10^3 min^{-1}). Figure 2 presents the x-ray patterns of the sorbents.

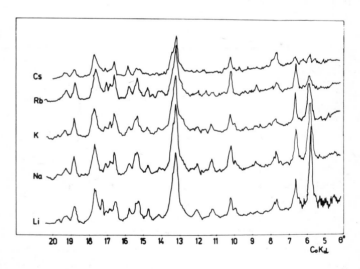

Figure 2

Ion exchange affects only (020) planes[11]. The intensities of diffraction peaks corresponding to (020) planes show an obvious decrease from the Li to Rb samples, and become negligible for the Cs sample. This behaviour of (020) planes is entirely different from that of the (200) planes, which remain unchanged for all ion exchanged forms save Cs where a total intensity decrease took place.

Analogous effects, observed with other cations[8], account for a different
population of these two crystallographic sets of planes. This is in a good
agreement with the new structural data for clinoptilolite: planes (020) contain
exchangeable cations whilst planes (200) do not. Intensity measurements were
taken in the internal 5 to 7.5°θ at a lower speed ($\frac{1}{4}$°/min) using original
clinoptilolite as the external standard and "sweeping" repeatedly through
θ to minimise the error.

Sorbents were activated before adsorption by heating to 400°C. At this
temperature their thermal stability was checked for different durations of heating (4, 9 and 24 hours). The X-ray technique and isobaric adsorption of water[12]
were both used to confirm the preservation of the structure during heating.
Isobaric adsorption was carried out under the conditions used for equilibration
of the samples. Table 1 provides data on the isobaric adsorption. The
duration of heating did not significantly affect the sorbents, except the one
enriched in Li for which a decrease in the adsorption capacity of about 25%,
and an alteration of the lattice parameter $\underline{b}$ were found (b_{4hrs} = 18.02 Å:
b_{24hrs} = 17.18 Å). A considerable decrease in the intensity of the X-ray

Table 1

		Li	Na	K	Rb	Cs
	Ingoing cation	Li	Na	K	Rb	Cs
X/mmol g^{-1}	400°C 4 hrs	6.9	7.0	6.7	5.8	5.3
	400°C 24 hrs	5.1	6.8	6.5	6.0	5.0

pattern of the Li-sorbent was observed after heating for 24 hours and a loss
of crystallinity of about 30% was established.

The check of the thermal stability of the sorbents was a necessary preliminary characterisation so as to make their adsorption comparison reliable.
The above data allow any differences found in the adsorption behaviour of the
samples (except for that of the lithium) to be ascribed to the nature of the ingoing cation and not to irreversible effects of the heating.

Adsorption of the Air Components: Adsorption of the air components was carried
out in an apparatus[6] containing an adsorption column of 12 mm diameter and length
400 mm. The system allowed for pressures up to 6 kg cm^{-2} and vacua enabling
regeneration up to 98%. The effect of contact time of adsorption, τ, has been
investigated before[6,8]. Both the pressure of adsorption and the desorption

time τ_d, were varied in the present investigation. Mean oxygen concentration, C_{O_2} final pressure P_f and corrected oxygen production β[8], were chosen[6] as parameters of the oxygen enrichment process and they were used for characterisation and comparison of the sorbents. Table 2 presents data on the sorbents examined under standard conditions[6,8] *i.e.* initial pressure 4 kg cm^{-2}, contact time 60 s and time of desorption 30 s.

Table	2		
Ingoing cation	C_{O_2} %	β/cm^{3}g^{-1}	P_f/kg cm^{-2}
Li	25.0	0.21	2.20
Na	28.8	0.50	1.80
K	36.0	0.85	1.00
Rb	29.6	0.68	1.65
Cs	27.1	0.37	1.75

Discussion

Comparison of the mean oxygen concentration and oxygen productions of the sorbents gives the following sequence:

$$K > Rb > Na > Cs > Li \qquad \ldots\ldots (1)$$

A direct relationship between the above data and the radii of the ingoing cations is not apparent.

The low enrichment ability of the Li-sorbent should be partially attributed to its unsatisfactory thermal stability. However, bearing in mind that the decrease in the thermal stability of the Li-sorbent is only about 25 - 30%, allowance for this does not change its position in the sequence (1).

Parameters of the oxygen enrichment process are affected differently on increasing the pressure, the differences being specific for the different sorbents. Mean oxygen concentrations remain unchanged for the Li-, Na- and Cs-sorbents, whilst marked changes are observed for the K- and Rb- sorbents (Fig.3). The slopes of the linear plots of concentration against pressure for the K- and Rb- sorbents are: dc/dp = -4.25 kg^{-1}cm^2 for the K- sorbent, and dc/dp = -2.57kg^{-1} cm^2 for the Rb- sorbent.

The difference in the effect of pressure on oxygen concentration for the different sorbents is of practical importance. According to the data obtained,

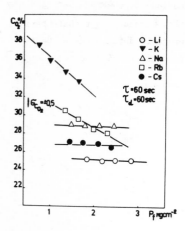

Figure 3. Mean oxygen concentration, C_{O_2}, as a function
of pressure, P_f.

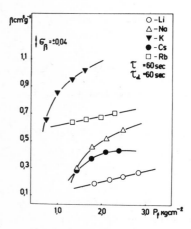

Figure 4. Oxygen production
β, as a function of pressure,
P_f.

the oxygen concentration is only a function
of pressure for the K- and Rb sorbents,
and low pressures are preferred with these
two sorbents.

Oxygen production, β, has always
been found to rise on increasing the
pressure of adsorption   (Fig. 4). The
analytical function which describes the
relationship between pressure for the K-
sorbent is

$$1/\beta = 1.48 + 0.084/P_f$$

Differences in oxygen production and mean
oxygen concentration found for the different alkaline forms of clinoptilolite show
differences in their selectivity. Physico-
chemical characteristics of each cation,
such as polarisation and magnetic perm-
eability, could play a significant role[13]
in the interaction between the nitrogen
quadrupole moment and zeolitic strength
field. These characteristics could be[14]
a convenient basis for understanding the
differences in the adsorption behaviour
of different clinoptilolite cationic forms
towards air components. In the case of
alkaline forms of clinoptilolite, the
picture is rather complicated. The dependence of oxygen production and mean oxygen concentration on polarisation sum[14]
reveals a sharp maximum in the case of the
K- sorbent (fig. 5a).

The lower value of β for the Rb- and
Cs- sorbents is most probably due to a
partial blocking of the 10-membered rings
in the channels[11]. Location of rubidium
or caesium cations at the positions of the displaced sodiums and calciums[10]
would leave rather small apertures with dimensions close to the molecular diameters of nitrogen and oxygen. Furthermore, it is not likely that cations as
large as Rb^+ and Cs^+ would site exactly in the positions of the displaced and

small Na^+ and Ca^{2+} cations. Even very small displacements of Rb^+ and Cs^+ sitings towards the centre of the channel would result in a partial blocking off of nitrogen and oxygen molecules.

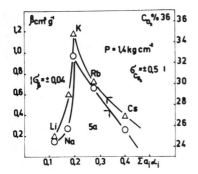 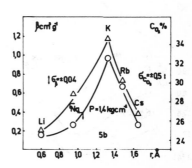

Figure 5. Mean oxygen concentration and oxygen production as functions of polarisation sum $\Sigma a_i \alpha_i$ and cation radii.

The dependence of oxygen production and mean oxygen concentration on the radius of the ingoing cation has a similar character (Fig. 5b). The similarity in the dependences in Fig. 5a and Fig. 5b is a result of the monotonous change in the polarisability with radii of the alkaline cations.

Separation coefficients are determined for $P_f = 1$ kg cm^{-2} and used as an additional and commonly employed characteristic of the selectivity. The coefficients depend on the radii of the ingoing cation. The curves go through a maximum, which also corresponds to K- sorbent (Fig. 6a).

In a search for additional explanations for the maxima observed, the computed unit cell parameters were examined. The same value for parameter $\underline{c}$ was found for all the sorbents, but the $\underline{a}$ and $\underline{b}$ parameters changed with the radius of the ingoing cation (Fig. 6b). The dependence exhibited a minimum which corresponds to the maxima in Figures 5a, 5b and 6a. This implies that the complex phenomenon of selective adsorption of nitrogen on clinoptilolite arises at least partially through a deformation of the unit cell.

The present investigation characterises the alkaline forms of clinoptilolite as promising sorbents for oxygen enrichment of air, the potassium sorbent

being the best. Using this sorbent, a pilot plant for oxygen
enrichment of air has been put into operation with a capacity of 30
m^3 (STP) of pure oxygen per hour.

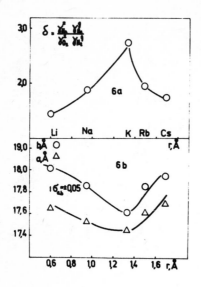

Figure 6. Dependence of separation coefficient, δ , and unit cell
parameters $\underline{a}$ and $\underline{b}$ on cation radii.

References.

1. D.W. Breck, "Zeolite Molecular Sieves",   Wiley - Interscience
   publication.   John Wiley & Sons, New York, 1974, 689.

2. D.W. McKee, US Pat. 3.140.932, 1964.

3. L.B. Batts, US Pat. 3.564.816. 1968.

4. H. Minato, T.Tamura, "Natural Zeolites", Pergamon Press, Oxford, 1978
   509.

5. I.M. Galabova, G.A. Haralampiev et al; Bulgarian National Bureau for
   Inventions No.18223, 1973.

6. I.M. Galabova, G.A. Haralampiev, B. Alexiev, "Natural Zeolites",
   Pergamon Press, Oxford, 1978, 431.

7. B. Alexiev, Compt. Rend. Acad. Bulg., 1968, 21, 1293.

8. I.M. Galabova, Geochem, Miner. Petrol., 1979 (in press).

9. I.M. Galabova, Z. Naturforsh., 1979, 34a (in press).

10. K. Koyama, Y. Takeuchi, Z. Crystall., 1977, 145, 216.

11. A.B. Merkle, M. Slaughter, Amer. Miner., 1968, 53, 1120.

12. E.E. Senderov, N.I. Khitarov, 'Tseol. ikh sinteza i usloviya obr. v prirode',
    Nauka, Moscow.

13. A. Kirkwood, A. Muller, Proc. Roy. Soc. A, 1936, 154, 634.

14. I.M. Galabova, G.A. Haralampiev, Proc. Sci. Sess. Kinetics and Mech.,
    Higher Chem. Technolog. Inst., Sofia, 1975, 287.

Discussion on Session 2:   Separation Processes

Chairman: Dr. A.L. Smith (Unilever Research,U.K.)

(Throughout this section, reference numbers correspond to those in the particular paper under discussion).

Paper 5. (J.W. Carter)

<u>Mr. Roberts</u>: Do you ascribe the carbon dioxide mass transfer zone differences you have observed between 4A and 5A products entirely to the isotherm situations for the two products?

<u>Dr. Carter</u>: No. The work of 4A was done with carbon dioxide in a flow of helium gas, whereas the 5A work used nitrogen as the carrier gas so that the smaller effective macropore diffusivity of CO_2 in the latter would also contribute to zone lengthening.

<u>Mr. Roberts</u>: We have found precisely the reverse situation at similar carbon dioxide partial pressures.

<u>Dr.Carter</u>: Assuming the comparison is to be made at the gas velocities in the bed, or more precisely at the same rate of feed of the adsorbate to the mass transfer zones, then their behaviour will depend upon isotherm capacities, shapes and also effective diffusivities. Therefore, the reverse behaviour with other source adsorbents is quite possible. Also, in our case we believe the nitrogen was co-adsorbed on the 5A, making the isotherm shape less favourable, so that this may be another reason why your experience differs from ours.

Paper 6   (J.W. Armond)

<u>Mr. Roberts</u>: You have mentioned 5A and carbon molecular sieves for this application. Which adsorbent type is now dominant in practice?

<u>Mr. Armond</u>: For each gas mixture to be separated by pressure swing adsorption (PSA) the adsorbent type has to be selected for the best efficiency. Thus PSA hydrogen plants principally employ activated carbon and zeolite 5A, heatless driers use activated alumina and silica gel, whilst oxygen producing plants use zeolite 5A with, in some cases, an alumina pre-drying stage, and nitrogen producing plants employ molecular sieve carbon.

In terms of adsorbent tonnage, I think the most commonly employed is activated carbon, followed in descending order of magnitude by zeolite 5A, activated alumina and finally carbon molecular sieve.

<u>Dr. Vaughan</u>: How do the economics of oxygen and nitrogen generation using PSA

compare with cryogenic separation? At what tonnage are cryogenic systems more
economic?

Mr. Armond: On the large scale, oxygen and nitrogen are produced more economically by the cryogenic route. Therefore, PSA cannot compete where oxygen and
nitrogen are used at pipeline distance from a large-scale producing site. In
addition nitrogen and oxygen produced by PSA are less pure than cryogenic gas
and will not be suitable for some applications. However, where oxygen or
nitrogen has to be transported in liquid form from the production site to the
point of usage the costs of liquefaction and transportation change the situation
and, provided that the purities provided by PSA are acceptable to the customer,
we estimate that PSA plants are more economical than delivered liquid oxygen or
nitrogen up to about 30 tonnes/day oxygen or 100 tonnes/day nitrogen, at which
point cryogenic plants should be installed on site.

Prof. Barrer: High area carbons are sometimes known to chemisorb oxygen. This
oxygen can only be released as oxides of carbon and so could slowly "eat away"
the carbon material. Is there then any evidence that the relative rates of
sorption of oxygen and nitrogen change following numerous sorption-desorption
cycles?

Mr. Armond: We understand that chemisorption of oxygen does occur with carbon
molecular sieves. However, for the purposes of practical plants producing
nitrogen from an air feed the effect is so small as to be insignificant in a
period of several years for a plant cycling every two minutes.

Mr. Bligh: In the PSA process what is the recovery of oxygen contained in the
product as a percentage of oxygen in the feed air?

Mr. Armond: If the two bed PSA processes described are operated at a vacuum of
150 torr and 15°C, oxygen yields of 42% at 90% oxygen purity and of 50% at 80%
purity are obtained. The three bed version of the process operating under the
same conditions shows oxygen yields of 45% at 90% purity and 52% at 80% purity.

Prof. Sing: Interesting differences are reported between the adsorptive properties of zeolites and carbon molecular sieves. Are the rate curves shown in
your paper for oxygen and nitrogen typical of results obtained with these two
sorbent systems? Is there any appreciable difference in the sorption capacities?
How do you explain the results obtained?

Mr. Armond: The curves shown in Figure 4 for rates of adsorption of oxygen
and nitrogen on carbon molecular sieve are typical for a carbon sieve tailored
for the separation of oxygen-nitrogen mixtures. At equilibrium the oxygen and

nitrogen sorptive capacities are about equal.

In simple terms, the gas molecules gain access to the main surface area of the carbon through "windows", and the art of manufacture of the carbon lies in reducing the size of these windows to the point at which the ratio of the rates of diffusion of the two or more species of molecule through these windows is as large as possible whilst at the same time maintaining absolute rates fast enough for the operation of a short cycle time PSA process.

Paper 7 (C.W. Roberts)

Prof. Ruthven: Presumably the pore closure effects to which you referred are related to a re-distribution of the cations. However, it is hard to imagine exactly what this re-distribution might be, particularly in the 4A sieve. Do you know exactly what is happening?

Mr. Roberts: In fact, we do think that the pore closure is due to a cation re-distribution, but we do not know exactly what is happening. The effect can be seen (by sorption measurements using appropriate probe molecules) with both the 3A and 4A sieves. We are investigating the matter further at present.

PENTASIL FAMILY OF HIGH SILICA CRYSTALLINE MATERIALS

G.T. Kokotailo, Mobil Research and Development Corporation
Paulsboro, New Jersey, USA

W.M. Meier, Institute of Crystallography
ETH, Zurich, Switzerland

The framework structures of ZSM-5 and ZSM-11
are members of the same family, regardless
of the chemical composition of the frameworks,
and these structures can be described in
terms of layer sequences. The usefulness of
the σ-transformation extended by the concept
of the i-transformation in describing the
layer sequences of this family is demonstrated. This family of structures is so closely
related that a single generic name, Pentasil,
is proposed to encompass all family members.

INTRODUCTION

The name porotektosilicate has recently been proposed[1] to encompass that class of siliceous crystalline materials which includes
both zeolites and materials which resemble zeolites in crystal
structure but may or may not have ion exchange capability. ZSM-5
and ZSM-11 are such materials, and in addition they are members of
a distinct family of such porotektosilicates. Examples of this
family which are not pure silica have zeolitic properties such as
ion exchange. Furthermore, they are members of a new class of
shape selective catalysts[2,3]. They have intersecting ten-membered ring channel systems. The pore size of the channels is intermediate between the eight-membered ring zeolites, such as type
A zeolite and erionite, and the twelve-membered ring zeolites of
the faujasite type (X,Y). This new family of materials also possesses unusual catalytic properties and has a high thermal stability which must at least in part be due to the high Si/Al ratio.

Methanol can be converted to hydrocarbons and water using this new
family of catalysts[2,3]. The hydrocarbons produced in this process are predominantly in the gasoline boiling range, C_4 to C_{10},
and are of better quality than the typical products obtained by
the conventional Fischer-Tropsch process. The intermediate pore
size of this family of materials leads to diffusion constraints
which limit the size of molecules entering the channels. The
mechanism of methanol to hydrocarbon conversion over the ZSM-5

class of catalysts has been described[4,5]. Other applications of these materials in catalysis have also been described[3,6].

Chen and Garwood[7] have shown that the diffusion of molecules through ZSM-5 is probably controlled by the matching of the size and shape of the molecules with that of the channel and there appears to be no "cage effect". Unique selectivity dependent on molecular chain length was noted. The size of the channels affects catalyst deactivation by inhibiting the formation of large condensed polyalkyl aromatics inside the channel system.

In view of the common structural feature of the materials as described in this paper, a generic name - PENTASIL - is proposed to describe and include all structures within this family regardless of the nature of the specific, individual atoms making up the structure.

THE PENTASIL FAMILY OF STRUCTURES

The related framework structures of ZSM-5 and ZSM-11 have been solved by Kokotailo et al[8a,b]. The crystal data are summarized in Table 1. Flanigen et al[9] have independently derived the structure of a silica molecular sieve having the ZSM-5 topology and hence it is appropriate to include this silica molecular sieve as a member of the pentasil family. The investigation of these new structures had for the most part to rely on X-ray diffraction powder data. However, the topology of the frameworks

Table 1   Summary of Crystal Data

	ZSM-5	ZSM-11
Number of T-atoms (Si + Al) per unit cell	96	96
Unit Cell	orthorhombic	tetragonal
	a = 20.1 A b = 19.9 c = 13.4	a = 20.1 c = 13.4
Maximum possible space group symmetry	Pnma	I$\bar{4}$m2
Framework density (number of T-atoms per 1000 A^3)	17.9	17.6

could thereby be firmly established. Further extensive refinements to determine structural details such as the location of non-framework atoms and precise interatomic distances are currently in progress.

The framework structures of members of the pentasil family can be described in terms of a common characteristic layer shown in Figure 1. The heavy dark outline in Figure 1 indicates the secondary building units (SBU) from which the characteristic layers of the pentasil family can be built. In the ZSM-11 structure these layers are joined such that neighboring layers are related by a reflection, σ, and therefore must be enantiomeric.

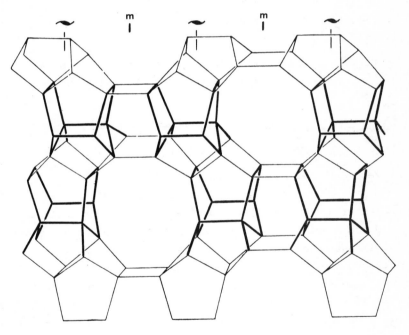

Fig.1   Characteristic Pentasil Layer

The sequence of layers in ZSM-11 can thus be expressed by σσ or AA'AA' and is shown in Figure 2a. The layers can also be linked such that neighboring pairs are related by an inversion, i, and are thus also enantiomeric. If all pairs of neighboring layers are related by inversions the framework is that of ZSM-5 and the layer sequence can be expressed by ii or AB'AB. This layer sequence is shown in Figure 2c. The layer sequences in the pentasil

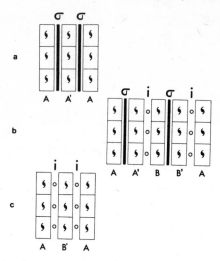

Fig.2   a   Stacking Sequence of Layers in ZSM-11
         b   Pentasil Structure with σiσi Sequence
         c   ZSM-5

structures are based on two operations:

$$\sigma : A \rightarrow A' \qquad\qquad i : A \rightarrow B'$$

Structures with other possible combinations can be readily derived.
A simple combination is σiσi (or AA'BB') in which σ and i alternate as illustrated in Figure 2b. The sequence σσii or AA'AB' represents yet another simple example. Evidently, the number of
possible combinations (σ,i) determining the layer sequence is
effectively infinite unless the repeat distance is limited by some
arbitrary value. In all these structures the layers are parallel
to (100).

As exemplified by the various pentasil structures described above,
the σ-transformation described by Shoemaker, Robson and Broussard
[10] for interrelating zeolite structure types can be extended by
introducing an analogous i-transformation. In these terms the
framework of ZSM-11 and any intermediate structures can be obtained by applying σ-transformations to ZSM-5. Conversely,

ZSM-5 and all intermediate structures are derivable from ZSM-11 by applying i-transformations. The ZSM-5 and ZSM-11 structures are the end members embracing a substantially infinite series of intermediate structures.

X-ray powder patterns, extensively used for identifications, are very similar but highly complex in case of ZSM-5 and ZSM-11[11,12]. Identification of other members of the pentasil family can be expected to pose formidable problems in view of likely complications due to possible deviations from idealized symmetry, stacking faults, and compositional variations. ZSM-11 has the highest symmetry of all members of the pentasil family. Consequently, its powder pattern has the least number of lines and these lines are present in the patterns of all the other members. Lowering of the symmetry generally results in an increase of the number of

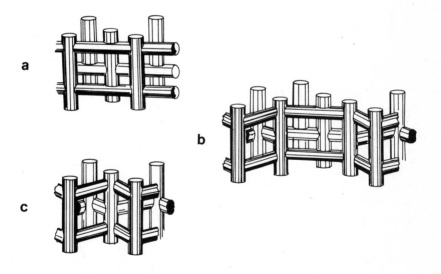

Fig.3   Channel Systems in Pentasil Structures with σσ (a), σσii (b) and ii (c) Sequences of Layers

weaker lines as well as the splitting and/or broadening of lines.

CHANNEL SYSTEMS

The pentasil family of zeolites have intersecting channel systems
with ten-membered ring openings. As shown schematically in
Figure 3, both sets of channels in ZSM-11 are straight whereas in
ZSM-5 one set is straight and the other one is sinusoidal[8a,b].
The channels parallel to the layers are invariably straight. As
can be readily established, the channels perpendicular to the
mirror planes are straight and those perpendicular to the layers
related by a center of symmetry are sinusoidal. For example, in
the structure with a σσii stacking sequence (Figure 3b) the channels perpendicular to the layers are straight in portions of the
structure where operations σ occur and bent whenever the operation i applies. On this basis, the nature of the channels in all
members of the pentasil family of structures can be easily discussed.

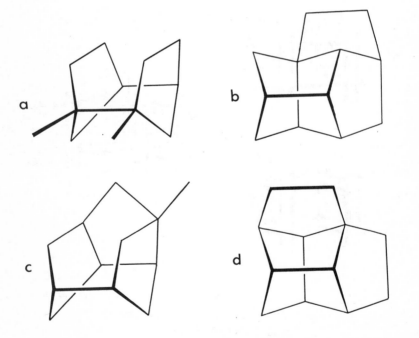

Fig.4   Non-chiral and Chiral Building
Units Comprising 12 T-atoms

SECONDARY BUILDING UNIT

As indicated above, that portion of Figure 1 which is outlined in heavy tracing constitutes the proper secondary building unit (SBU) of the pentasil family. Apart from this non-chiral secondary building unit, there are some other units comprising the same number of 12 T-atoms, and these are shown in Figure 4 together with the one adopted in this paper (a). The configuration (b) was considered to represent a secondary building unit by Flanigen et al[9] but this unit as well as (c) and (d) is chiral. This means that enantiomeric pairs of these units rather than units of one kind only are needed to build the structure. The unit shown in Figure 4a is the only non-chiral unit with 12 T-atoms (having point symmetry C_s) and should therefore be chosen.

REFERENCES

1. R.M. Barrer, private communication.

2. S.L. Meisel, J.P. McCullough, C.H. Lechthaler and P.B. Weisz, Chemtech, 1976, 6, 86.

3. S.L. Meisel, J.P. McCullough, C.H. Lechthaler and P.B. Weisz, ACS Meeting, Chicago, Ill., August 1977.

4. C.D. Chang and A.J. Silvestri, J.Catalysis, 1977, 47, 249.

5. E.G. Derouane, P. Dejaifve, J.B. Nagy, J.H.C. van Hooff, P.B. Spekman, C. Naccache and J.C. Vedrine, Comp.Rend., 1977, 284, 945.

6. N.Y. Chen, R.L. Gorring, H.R. Ireland and T.R. Stein, Oil and Gas J., 1977, 75, 165.

7. N.Y. Chen and W.E. Garwood, J.Catalysis, 1978, 52, 453.

8. a) G.T. Kokotailo, S.L. Lawton, D.H. Olson and W.M. Meier, Nature, 1978, 272, 437.

   b) G.T. Kokotailo, P. Chu, S.L. Lawton and W.M. Meier, Nature, 1978, 275, 119.

9. E.M. Flanigen, J.M. Bennett, R.W. Grose, J.P. Cohen, R.L. Patton, R.M. Kirchner and J.V. Smith, Nature, 1978, 271, 512.

10. D.P. Shoemaker, H.E. Robson and L. Broussard, Proc. of 3rd Internat. Conference on Molecular Sieves, Zurich, Switzerland, (1973), p. 138.

11. U.S. Patent 3,702,886.

12. U.S. Patent 3,709,979.

Evaluation of a New Zeolitic Catalyst
for Selective Catalytic Reduction
of NO$_x$ from Stationary Sources

Joseph R. Kiovsky, Pramod B. Koradia and Charles T. Lim

Norton Company
Chemical Process Products Division
P.O. Box 350
Akron, Ohio   44309

Introduction

The efficient reduction in the emission of nitrogen oxides from both stationary and mobile exhaust streams remains a paramount environmental problem. The emissions from stationary sources account for 55% of the man-made emissions of NO$_x$. It has been suggested that even with the application of the best available technology emissions of NO$_x$ from stationary sources, unlike SO$_2$, hydrocarbon and CO emissions, will continue to increase. By 1990 they will reach a level one and one half times that in 1975 (1).

More than 93% of the stationary source NO$_x$ emissions are contributed by combustion of fossil fuel and 5% by industrial processes (non-combustion). Although these emissions from industrial processes account for a relatively small fraction of the total emissions from stationary sources, their control has received attention due to generally higher concentration and, in many instances, locations near population centers.

Over the past decade many different techniques have been developed for controlling NO$_x$ emissions. These can be classified as either those consisting of combustion modifications to reduce the formation of NO$_x$ or those consisting of effluent treatment to remove NO$_x$.

Combustion modification techniques include flue gas recirculation, reduced air preheating, staged combustion, low combustion air and reduced heat release rate (2). The reductions achievable by these techniques are limited to less than 50% and may cause unstable flame conditions or lower thermal efficiencies. Since greater than 50% reduction is necessary to achieve the desired ambient air quality standards, flue gas treatment to remove NO$_x$ is required.

Flue gas treatment methods can be classified as dry methods or wet methods. With few exceptions dry processes only remove NO_x, while in general the wet processes are simultaneous NO_x and SO_2 removal systems. In addition, wet processes are relatively insensitive to particulates. The two primary disadvantages of the wet systems are high capital and operating costs and the formation of waste water containing nitrate anion which requires secondary treatment.

Dry processes are cheaper due to their simplicity and generally do not produce any waste stream. The primary disadvantage is their high sensitivity to inlet particulate matter. Dry methods consist of recovery of NO_x by adsorption on activated carbon or molecular sieves, radiation of the flue gas and reduction of NO_x by various reducing agents. While the adsorption processes (3, 4) can achieve high removal rates, large quantities of adsorbents and high energies of regeneration required make them economically less favorable for the utility industry. A brief discussion of non-adsorptive dry methods is given below.

In the radiation process, the electron beam converts particulates, SO_2 and NO_x into a complex powder mixture which is removed in an electrostatic precipitator. The major disadvantages are high capital and operating costs and secondary waste disposal problems.

The reduction methods are dry processes in which a reducing agent is added to the flue gas. There are three basic types of reduction processes; selective noncatalytic reduction (SNR), nonselective catalytic reduction (NCR), and selective catalytic reduction (SCR).

In the noncatalytic reduction processes NH_3 or other reducing agents are injected directly into the upper portion of the boiler to reduce the NO_x to molecular N_2. This process is the simplest of all processes as it eliminates the need for any supplemental equipment. It, however, requires a NH_3 to NO_x ratio in the range of 3.0 to 4.0 and a narrow operating temperature range. The removal level of about 70% for these processes is low compared to catalytic processes.

Catalytic reduction processes can be either selective or nonselective. Selectivity, in this respect, refers to the reaction of the reducing agent preferentially with NO_x rather than with oxygen or other oxidizers present in the gas stream.

In NCR processes, the reducing agents, such as hydrogen, carbon monoxide or hydrocarbons (CH_4) react with O_2 and SO_x in addition to NO_x and therefore a large amount of reductant, sufficient for the reduction of all of the above species must be added. In addition to the high reducing gas requirement, a number of NCR catalysts promote formation of H_2S and COS which are toxic. If CO is used as the reductant, toxic metal carbonyls may be formed. The effects of these two disadvantages can be minimized by combustion modifications to reduce excess air. However, such modifications generally lead to lower thermal efficiencies.

For selective catalytic reduction (SCR), NH_3 is used as a reducing agent which selectively reacts with NO_x to form N_2 and H_2O. Thus, the principle advantage is low reducing gas requirements. One disadvantage is the possibility of formation of ammonium bisulfate when excess NH_3 leaks through the reactor. Both of the catalytic processes (NCR and SCR) may also suffer from the decrease in catalytic activity and increase in pressure drop caused by clogging of the catalyst bed by particulates.

A 1977 survey by the United States EPA showed that there are 48 different processes for NO_x reduction in various stages of development, 26 of which are dry processes. An overwhelming majority, namely, 21, are based on SCR and use ammonia as the reductant. Process conditions and economics of the SCR processes are given in the EPA report (6).

These processes are in various stages of development from conceptual to commercial scale. The operating conditions, i.e. reaction temperature, space velocity, NH_3/NO_x and the efficiencies vary widely from one process to another. The spectrum of catalysts used includes noble metal catalysts, base metal catalysts with or without supports and activated carbon. Five of these SCR processes are designed for simultaneous removal of SO_x. Required capital investment ranges from $10/kw to $141/kw and a similar wide range is reported for operating costs.

The ideal catalyst should be highly active over a wide temperature range, not be easily poisoned by SO_2, have a long live, require a minimum amount of reductant and be relatively inexpensive. The search for such a catalyst has continued. Literally thousands of compositions have been tested and a large number of these have been patented, especially in the United States and Japan.

Some of the large differences are due to the differences in the basis of the cost estimates in terms of plant capacity. Another significant difference is the technique used to solve the problem caused by particulates. This problem is primarily controlled by the use of a hot electrostatic precipitator. In addition, most of the processes use catalyst and reactors designed to minimize the effect of dust. The catalysts used in fixed bed reactors are honeycomb, ring or pipe shaped. Those used in the moving bed reactors are in pellet form.

Synthetic and naturally occurring zeolites have been shown to possess activity for SCR. In these studies the hydrogen form of synthetic mordenite (7-9) has been shown to possess excellent activity for SCR of NO_x present in high concentration such as in the tail gas from nitric acid plants and other chemical processes. In fact, several installations of the process are presently in commercial scale use in the USA and Canada. In such cases the tail gas is normally free of SO_2 and particulates. A number of patents have been issued in Japan on the use of natural zeolites for the reduction of NO_x with or without the presence of SO_2 in the flue gas.

This paper reports the results on the activity of a proprietary zeolitic catalyst designated NC-300 developed by Norton Company for the SCR of NO_x.

Experimental
The flow diagram of the apparatus used in the evaluation of the catalyst is shown in Figure 1. It consists of an air supply line, inlets for NO, NH_3, O_2 and SO_2, arrangement for water injection, a tube furnace gas preheater and tubular reactor with gas sampling

lines and thermocouples to measure the bed temperature.

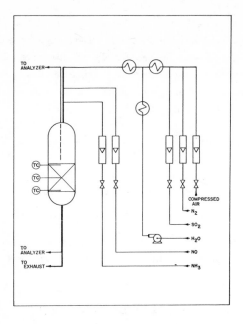

Figure 1.   The Flow Diagram of the Experimental Apparatus

The tubular packed bed reactor was made from a 316 stainless steel
tube 4 cm in diameter and 45 cm long and was equipped with two
sample ports for withdrawal of gas samples at the top and bottom
of the catalyst bed.   Three thermocouples, 10 cm apart, were placed
inside the reactor.   Pelletized catalyst of 1.5 mm diameter and 6 mm
average length was packed to a height of 20 cm supported on a stainless
steel screen.

At the start of the test, plant air is introduced into the gas preheater at the desired flow rate.   The desired O_2 concentration was
achieved by the addition of N_2 to the air stream.   In the runs
carried out to simulate power plant conditions SO_2 was added to the
air-N_2 stream.   Water was injected with a piston pump through a
heated line into the air preheater.   NO from the supply tank is
metered into the hot air before the gas stream reaches the catalyst
bed. NH_3 is introduced after steady state conditions of flow rate,

temperature and inlet concentrations of O_2, SO_2, water and NO are achieved.

The analysis of NO_x in the inlet and outlet gas stream was carried out using a chemiluminiscent analyzer made by Thermo Electron Corp. It analyzes for NO in the NO mode, then in the NO_x mode, the gas is passed through a converter to convert NO_2 to NO before being admitted to the chemiluminiscent cell. SO_2 is analyzed using a UV analyzer (DuPont Model 410).

The range of operation conditions under which the activities of the catalyst were determined are given below:

Temperature (Center of bed), °C	250 to 450
Space Velocity, hr-1	3000 to 15000
NH_3/NO_x Molar Ratio	0.6 to 1.8
Concentration of NO_x, ppm, V/V	500 or 2500
Water Concentration, Vol. %	0 to 10
Concentration of O_2, Vol. %	0 to 21
Concentration of SO_2, ppm, V/V	0 or 1000

Results and Discussion

In general, it can be stated that there were a number of combinations of temperature, space velocity and NH_3/NO_x which gave very high reduction efficiencies. The results are plotted in Figures 2 through 11 to illustrate the effects of different variables.

The effect of temperature for various NH_3/NO_x ratios is shown in Figure 2. For NH_3/NO_x ratios of 0.76 and 1.00 the reduction efficiencies increase with increasing temperatures. However, for the ratio of 0.62 there is a small decline in the reduction efficiency at increasing temperature. It should be pointed out that, based on samples taken at the top of the catalyst bed, 90% of the inlet NO_x was NO and the remaining 10% was NO_2. Thus, the stoichiometric amount of NH_3 required results in an NH_3/NO_x ratio of 0.73 (0.9 x 0.67 + 0.1 x 1.3).

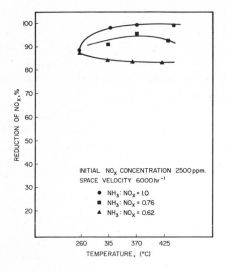 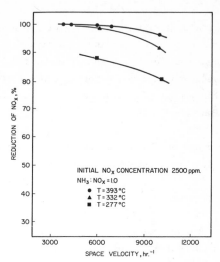

Figure 2. The Effect of
Temperature on NO_x Reduction
for Different $NH_3:NO_x$

Figure 3. The Effect of Space
Velocity on NO_x Reduction at
Various Reactor Temperatures

Effects of space velocity at various temperatures are shown in
Figure 3. In general as the space velocity is increased, the reduction efficiency decreases. This decrease of the reduction
efficiency is greater at low temperature. However, it should be
pointed out that greater than 90% reduction efficiencies are obtained at temperatures around 315°C and space velocities as high as
10000 hr^{-1}.

The effect of NH_3/NO_x ratio for various temperatures is shown in
Figure 4. It can be seen that for all temperatures reduction
efficiency increases as the NH_3/NO_x increases from 0.62 to 0.76.
For temperatures of 332°C and 393°C efficiency increases further
as the NH_3/NO_x increases to about 1.0 and then declines slightly
with further increase in the NH_3/NO_x ratio. On the other hand, at
lower temperature (277°C) efficiency decreases sharply as the
NH_3/NO_x is increased from 1.0 to 1.8.

This decrease in the NO_x reduction efficiency could be caused either by the increased chemisorption of NH_3 on the Bronsted acid sites of zeolite decreasing the sites available for NO_2 chemisorption or by the oxidation of the excess NH_3 (chemisorbed) to NO_x. Such strong chemisorption of NH_3 on H-mordenite, compared to amorphous supports has been demonstrated (10). At higher temperatures the excess NH_3 may have been oxidized to N_2 or it could have leaked through the reactor. Unfortunately the outlet gas was not analyzed for NH_3 or other nitrogen oxides.

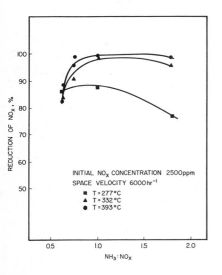

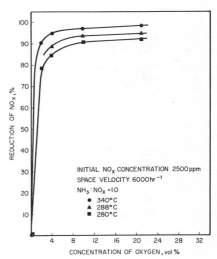

Figure 4. The Effect of NH_3:NO_x on NO_x Reduction for Different Temperatures

Figure 5. The Effect of Oxygen Concentration on NO_x Reduction for Different Temperatures

The data presented above were obtained using air as the carrier gas and therefore the O_2 concentration was about 21%. The presence of O_2 is known to enhance the reduction of NO_x (61, 62). It was observed by Bruggeman (63) that no reduction of NO occurs on H-mordenite in the absence of oxygen. A series of experiments were carried out to determine the effect of oxygen and the results are shown in Figure 5. It can be seen that no reduction occurs in the absence of oxygen on

the new zeolitic catalyst. As the concentration of O_2 is increased
from 0 to 4% the conversion increases rapidly and finally levels off
at around 10% O_2. Thus, oxygen is essential in the SCR of NO_x with
zeolitic catalysts.

This study was not aimed at determining the mechanism of the reaction
on zeolite type catalysts. However, it appears that the mechanism
proposed by Takagi et al (11), according to which the reactive species
are absorbed NH_4^+ and NO_2, would be applicable to zeolitic catalyst.
Formation of NH_4^+ ions by the chemisorption of NH_3 on zeolites is very
likely. Furthermore, the requirement of O_2 for the reduction of NO_x,
discussed above, would indicate that NO_2 is the reactive species.

Figure 6 shows the effect of concentration of water on the efficiency
of NO_x reduction from a gas stream containing 4% oxygen. It can be
seen that as the concentration of water increases the reduction at any
temperature decreases. However, at 0.76% H_2O (nitric acid plant
conditions) greater than 90% reduction is obtained at temperatures
slightly higher than 300°C (300-310°C). As the temperature increases
the effect of water becomes less significant.

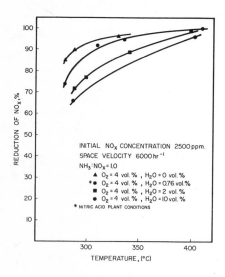

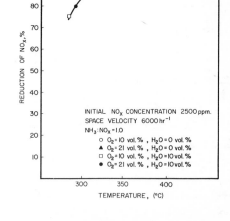

Figure 6. The Effect of Water
Concentration on NO_x Reduction
for Treatment of Gases Containing 4 Vol % Oxygen.

Figure 7. The Effect of Water
Concentration on NO_x Reduction
Under Power Plant Tail Gas Con-
ditions

Results of the runs carried out under power plant tail gas conditions, i.e. with 10% O₂ and 10% water, are shown in Figure 7. Results obtained are similar to those obtained under nitric acid plant conditions, i.e. at any given temperature the reduction obtained is lower in the presence of water. However, this difference decreases as the temperature is increased. At 340°C greater than 90% reduction is obtained.

Comparisons of the performances of Zeolon 200H (synthetic mordenite) and NC-300 catalysts under various operating conditions are shown in Figure 8 (21% O₂, 0.5% water), Figure 9 (Nitric acid plant conditions, i.e. 4% O₂, 0.76% water) and Figure 10 (power plant conditions, i.e. 10% O₂ and 10% water). It can be seen that the activity of NC-300 is higher than that of Zeolon 200H over the entire temperature range studied.

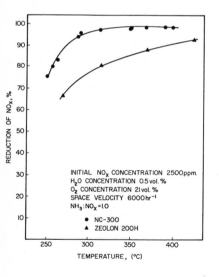

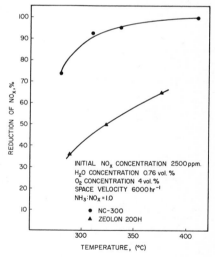

Figure 8. Comparison of NC-300 Catalyst with Zeolon 200H Catalyst

Figure 9. Comparison of NC-300 Catalyst with Zeolon 200H Under Nitric Acid Plant Tail Gas Conditions

Since power plant tail gases contain significant amounts of SO₂ (∼1000 ppm V/V), a number of runs were carried out with

Zeolon 200H and NC-300 and the results of these runs are shown in
Figure 11. In general, higher temperatures and NH_3/NO_x are required
for 90% reduction. Simultaneous removal of SO_2 in the range of 30 to
45% was observed. Outlet gases were not analyzed for the presence
of $(NH_4)_2SO_4$ or NH_4HSO_4 and therefore the mechanism of SO_2 removal
is not known. Pence and Thomas (7), in their studies with mordenite,
have detected the presence of both species in the effluent. Over the
entire temperature range the activity of NC-300 is higher than that
of Zeolon 200H.

The activity of NC-300 at 300°C is reduced from over 90% to around
50% upon exposure to tail gases containing SO_2. These runs at
300°C, after exposure to SO_2, were not carried out for a long
period to determine if this SO_2 poisoning effect is irreversible.
Activity at 370°C remains unchanged in the presence of SO_2.

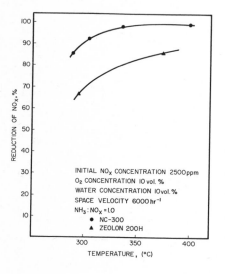

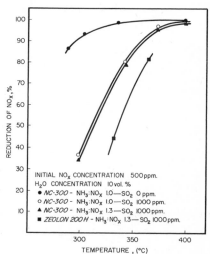

Figure 10. Comparison of NC-300
Catalyst with Zeolon 200H Under
Power Plant Tail Gas Conditions

Figure 11. The Effect of SO_2 on
NO_x Reduction Over NC-300 and
Zeolon 200H for Power Plant
Conditions

Activity of Zeolon 200H obtained in this study is lower than that
obtained by Pence and Thomas (7) with Zeolon 900H even though both

are synthetic large port mordenites.

However, comparison of their data with the data obtained here is fraught with difficulties due to differences such as their use of a condenser in the sample train. Since this reaction is highly exothermic, there is a temperature gradient along the entire length of the bed. The location of the measurement of reaction temperatures reported by them is not known. The reaction temperatures reported in the data presented here are the temperatures at the middle of the bed. Lastly the ratio of NO to NO_2 may not be the same in the two studies.

Extended life studies have not been conducted. However, in the course of the above study a sample of catalyst was in service for over 400 hours without any decline in its performance. In the process of obtaining data to determine the effect of SO_2, experiments at 373°C were repeated four times over a period of three weeks and the reduction obtained varied between 96% and 98.2% demonstrating reproducibility of the experiments.

Critical Evaluation of SCR Catalysts

As mentioned earlier literally thousands of compositions have been tested for the SCR of NO_x. The EPA study (6) reviewed the processes being developed and the information was gathered in the first half of 1977. Disclosures of new catalyst compositions have continued. Most of them are tabulated in Tables 1 through 4. Most of the information is contained in the patent literature and therefore a true comparison of various catalysts is difficult. However, it is possible to classify most of the catalysts being used in the developed processes, or likely to be used, into one of the following types:

(1)   Noble metal such as Pt, Pt-Rh, Pt-Pd on Al_2O_3 or $SiO_2-Al_2O_3$

(2)   CuO and other Cu salts on Al_2O_3 (TABLE 1)

(3) Fe_2O_3 and mixtures of Fe_2O_3 with other oxide unsupported or supported on Al_2O_3 or $SiO_2-Al_2O_3$ (TABLE 2)

(4) V_2O_5 or TiO_2 or CeO_2 or mixtures unsupported or supported

on Al_2O_3 or $SiO_2-Al_2O_3$ (TABLES 3 and 4)

(5) Base metal on activated carbon

(6) Zeolites (TABLE 5)

Advantages and disadvantages of these different types of catalysts are discussed below.

(1) Pt on Al_2O_3 catalysts have been used for decoloration, i.e. reduction of NO_2 to NO (13). The Bneron process is also based on a noble metal on honeycomb catalyst (6). In general these catalysts operate at lower efficiencies. They are prone to poisoning by SO_2 at low operating temperatures and if the operating temperatures are raised to solve the poisoning problem loss of activity occurs due to increased chemisorption of oxygen. Also they tend to promote formation of N_2O and require high NH_3/NO_x ratio (in the range of 1.6 to 2.5).

(2) CuO supported on Al_2O_3 has been used in the SCR processes developed by UOP (6). UOP has developed two processes, one for NO_x reduction only and the other for both NO_x and SO_x. When this catalyst is used for the removal of NO_x only, CuO is first converted to $CuSO_4$ by the reaction with SO_2 in the flue gas. $CuSO_4$ in turn is a selective catalyst for the reduction of NO_x. These catalysts, in contrast to Pt/Al_2O_3 catalysts have high efficiency, are not poisoned by SO_2 and operate at low NH_3/NO_x. However, the operating temperatures are on the high side (400°C) and the recommended space velocities are lower than those of some of the other processes. Also, stability of $CuSO_4$ on Al_2O_3 is doubtful. Deactivation of the catalyst by formation of Cu-aluminate may occur. Interestingly, Exxon has also developed a process based on CuO/Al_2O_3 Catalyst.

High activity has been shown for $CuCl_2$ catalyst (Table 1) at temperatures as low as 120°C. However, operation at such a low temperature is a questionable practice due to the danger of NH_4NO_3 formation.

The other low temperature catalyst consisting of mixtures of
Cu, Ni and Mn oxides supported on Al_2O_3 requires a very high
NH_3/NO_x ratio of 4.7. Use of such a high excess of NH_3 can
lead to the leakage of NH_3 or NH_4HSO_4.

(3) Iron oxide catalysts are made from a number of different raw
materials that include iron sand, spent pickling liquor, dust
from mineral sintering and blast furnace and converter slag
(Table 2). In general, Fe_2O_3 catalysts requires high operating temperatures and operate at relatively low space velocities.

(4) Table 3 shows catalysts consisting of V_2O_5, unsupported or
supported on Al_2O_3 as SiO_2-Al_2O_3.

It can be seen that vanadium catalyst in combination with other
oxides posses high activity at relatively low temperatures. The
V_2O_5-WO_3 catalyst patented by Hitachi Shipbuilding is of particular
interest (6). Composition of the catalyst is not disclosed.
The reaction temperature is in the range of 350-400°C and its
reduction efficiency drops sharply to 60-70% at 300°C. Also
the space velocity is somewhat low at around 4000 to 6000 hr^{-1}.
From the published data catalyst cost is estimated to be equal
to $10-14 per kg making it a relatively expensive catalyst.
(Basis $1 equals 300 yen). V_2O_5 imparts stability against
sulfur poisoning; however, it may also catalyze the oxidation
of SO_2 to SO_3.

Catalyst compositions based on TiO_2 and CeO_2 are given in
Table 4 . In general, these catalysts allow high reduction
efficiencies but require high operating temperatures. No
cost data are available. However, these catalysts are likely
to be expensive.

(5) Activated carbon processes are designed for the removal of both
SO_x and NO_x and they operate at relatively low temperatures of
200-250°C. However, the NO_x reduction efficiencies and space
velocities are low and requirements of NH_3 are high. Catalyst
losses by oxidation and attrition are high and therefore replacement rates as high as 50% per year are reported.

(6) Natural zeolites promoted by CuO, V_2O_5 or Fe_2O_5 possess high
    reduction efficiencies at temperatures in the range of 200-
    300°C (Table 5). Only two of the catalysts have been tested
    in the presence of SO_2. However, they require operating
    temperatures of 350-400°C. Furthermore, the stability of
    Faujasite type zeolite in the acidic environment is suspect.

Summary

A Gulf process has been in use for the reduction of NO_x from nitric
acid plant tail gas and it uses Zeolon 200H. Their tests were carried
out with tail gas containing 3-4% oxygen which is typical of the
nitric acid plant. They reported a total loss of efficiency when
the oxygen concentration is 10% or greater which is in contradiction
to the data presented above.

The reduction efficiency of Zeolon 200H for nitric acid plant tail
gas (4% O_2) is a low 35% at 300°C whereas for power plant tail gas with
10% O_2 the reduction efficiency is much higher (65%) at the same
temperature. The only difference in Gulf's operating conditions and
those of the present study is the proportion of NO_x that is present
as NO_2. In the present study, NO_2 comprises only 10% of the total
NO_x whereas in the Gulf process studies NO_2 amounts to about 50% of
the total NO_x.

In any event, NC-300 catalyst has much higher activity than Zeolon
200H under all of the operating conditions studied.

Existing catalysts suffer from various disadvantages in terms of
high cost, low life or requirement of severe operating conditions.
Thus, it is necessary to continue the search for a better catalyst.
The new catalysts developed by Norton Company show promise. It
would be a low cost catalyst and in the initial testing has shown
very high activity at moderate operating conditions. It has been
shown that it can operate in the presence of SO_2 and the high
amount of water present in the tail gases of power plants.

TABLE 1 – Catalysts Containing Copper Oxides

Composition	Temp °C	Space Velocity hr^{-1}	Copper Oxides Concentration ppm V/V NO$_x$	SO$_2$	NH$_3$/NO$_x$	Reduction %	Ref
CuO/Al$_2$O$_3$	250-400	-	-	-	-	-	14
Cu-Pyrophosphate/Al$_2$O$_3$	400	14500	126	520	1.1	91.4	15
CuCl$_2$ - V$_2$O$_5$ - Al$_2$O$_3$	≤120	-	300	200	-	90.0	60
Cu, Ni, Mn Oxides/Al$_2$O$_3$	180-250	12000	1000	-	4.7	90.0	16
CuO - TiO$_2$ - V$_2$O$_5$	325	10000	200	800	1.0	99.0	17
Cu, Mn, Fe Oxides/Al$_2$O$_3$	350	5000	-	-	-	80-90	18
Cu, Cr, Co, Ni, Zn, Mn/ SiN$_4$, FeSO$_4$	250-400	5000	300	350	1.0	90-95	19
Cu, Cr, Co, No, Zn, Mn/ Spinel Type Ferrite	400	5000	300	-	0.8	97.0	20

TABLE 2 – Catalysts Containing Iron Compounds

Composition	Temp °C	Space Velocity hr⁻¹	Copper Oxides Concentration ppm V/V		NH₃/NOₓ	Reduction %	Ref
			NOₓ	SO₂			
Fe₂(SO₄)₃	270-400	-	400	300	1.2	56-99	21
Fe₂O₃-Cr₂O₃-Al₂(SO₄)₃	200	-	1000	-	1.1	90.5	22
Fe₂O₃ from Spent Pickling Liquor	340	4000	200	50	1.3	90.0	23
Fe Sphere/Fluidized Bed	-	-	200-300	300-2000	1-1.5	75-78	24
Iron Sand	400	5000	500	2000	1.5	50.0	25
Fe₂O₃ Coated with Al₂O₃	400	5000	400	600	1.25	94.0	26
Dust from mineral sintering blast furnace (Fe & FeO)	300-350	2000	250	-	2.0	80.0	27
Converter Slag Containing Fe & Mn Oxides	300-350	1000	250	-	2.0	90.0	28
Fe₂O₃/SiO₂ - Al₂O₃	-	-	200	1000	1.1	90.0	29
Ore Slag	310	500	200	500	2.5	90.0	30
Fe₂O₃-Cr₂O₃/SiO₂-Al₂O₃	250-350	-	200	1000	1.35	90-97.5	31
Fe + W Oxides/SiO₂-Al₂O₃	350-450	-	-	-	-	90-95	32
Fe₂O₃/SiO₂-Al₂O₃	380	-	-	-	-	90.0	33
Fe₂O₃-Vanadium Sulfate/SiO₂-Al₂O₃	330	3000	760	1050	1.3	88.0	34

TABLE 3 – Catalyst Containing Vanadium Oxides

Composition	Temp °C	Space Velocity hr^{-1}	Copper Oxides Concentration ppm V/V		NH$_3$/NO$_x$	Reduction %	Ref
			NO$_x$	SO$_2$			
V$_2$O$_5$, Fe$_2$O$_5$/Al$_2$O$_3$	350	5000	150	150	1.0	90.0	35
V$_2$O$_5$/Al$_2$O$_3$	300	10000	120	–	1.0	80.0	36
V$_2$O$_5$ or base metal/ H$_2$O$_3$	330–370	–	115	–	–	99.8	37
V$_2$O$_5$ WO$_3$/Al$_2$O$_3$	180–300	10000	500	250	1.0	60–100	38
VOSO$_4$, SnCl$_4$/SiO$_2$	350	4000	750	850	1.3	91.0	39
V$_2$O$_5$, SnO$_2$/SiO$_2$	300–450	10000	750	800	1.3	79–89	40
V, Nb, Ti & Cu Oxides/ Al$_2$O$_3$	300	–	–	–	–	97.2	41

TABLE 4 - Catalyst Containing Titanium and Cesium

Composition	Temp °C	Space Velocity hr-1	Concentration ppm V/V NOx	Concentration ppm V/V SO2	NH3/NOx	Reduction %	Ref
TiO2-CeO2 (10-20 Mesh)	400	50000	300	-	1.1	90.0	42
TiO2-CeO2 - MoO3	200-400	50000	300	-	1.2	98.0	43
TiO2-V2O5/Inert Support	300	10000	200	-	1.35	99.5	44
TiO2-MnO2	450	5000	300	800	1.33	86-90	45
TiO2-WO3-V2O5-Bi2O3	250	10000	200-230	3000	1.0-1.1	82.0	46
TiO2-CuO	330	10000	300	800	1.0	90.0	47
TiO2-V2O5	323	10000	200	800	1.0	99.0	48
TiO2-V2O5	380	-	-	-	-	89.0	49
TiO2/TiO2 Ceramics	300	10000	-	-	-	99.0	50
TiO2-SiO2-ZrO2	300-400	20000	200	800	1.0	93-96	51
CeO2/Al2O3	480-500	40000	2000	-	1.0	94.0	52
CeO2, V2O5/Al2O3	450	-	3000	-	-	-	53

TABLE 5 - Zeolite Based Catalysts

Composition	Temp °C	Space Velocity hr-1	Concentration ppm V/V		NH₃/NOₓ	Reduction %	Ref
			NOₓ	SO₂			
Natural Zeolite + Fe Ore	250	5000	190	-	1.0	70.0	54
Natural Zeolites + Cu	200-300	-	190	-	1.05	95.0	55
Zeolite + Fe₂O₃	300	5000	190	-	1.1	98.0	56
Zeolite + Cu	200	5000	250	-	1.0	-	57
Natural Zeolite + V₂O₅ + Fe₂O₅	360-390	3600	110-150	4000	1.5-2.0	-	58
H-Type Zeolite + CuO	350	20000	250	3000	1.0	98.0	59
Synthetic Mordenite	200-300	15000	6520	-	0.8	-	9

References

1. G.D. McCutchen, Chem. Eng. Progress, 1977, **8**, 58.

2. W. Kenneth and C.F. Warner, "Air Pollution: Its Origin and Control," Harper & Row Publishers, Inc. 1976.

3. J.R. Kiovsky, P.B. Koradia and D.S. Hook, "Zeolon^R - SO_2 Removal Process," The Second International Conference on the Control of Gaseous Sulphur and Nitrogen Compound Emission, 1976, Salford, England.

4. J.L. Friedrich and R.M. Pai, Proceedings of NO_x Control Technology Seminar, 1976, EPRI SR-39, San Francisco, California.

5. R.K. Lyon and J.P. Longwell, ibid., 1-20.

6. H.L. Faucett, J.D. Maxwell and T.A. Burnette, "Technical Assessment of NO_x Removal Process for Utility Applications," 1977, EPRI AF-568 157.

7. D.T. Pence and T.R. Thomas, Proceedings of the Section A&E Environmental Protection Conference, Albuquerque, New Mexico, 1974, UASEC Report WASH-1332, Vol. 1, p 427-445.

8. N.R. Leist, J.J. Bockhold and D.E. Keck, "The Design and Operation of a Pilot Unit for NO_x Destruction with Ammonia and a Synthetic Mordenite Catalyst," 1976, prepared for U.S. ERDA under contract EY-76-C-05-1156.

9. J.L. Carter, M.T. Chapman and B.G. Yaokan, U.S. Patent 3,895,094, 1975, to Gulf Oil Corporation.

10. J.R. Kiovsky, W.J. Goyette and T.M. Notermann, J. Catalysis, 1978, **52** 25-31.

11. M. Takagi et al., J. Catalysis, 1977, **50**, 441-446.

12. K. Otto, et al., The Journal of Physical Chemistry, 1970, **74**, No. 13, 2690.

13. H.C. Anderson, et al., I&EC, 1961, **53**, 199.

14. H. Ito, et al., Japan Kokai 76,149,871, 1976, to Kawasaki Heavy Industries Ltd.

15. K. Nagai, et al., Japan Kokai 77,120,965, 1977, to Toho Royon Co., Ltd.

16. W. Herzog and J. Stenzil, German Offen. 2,620,378, Nov. 17, 1977, to Hoechest A.G.

17. A. Inoue, et al., Japan Kokai 77,106,389, Sept. 6, 1977, to Nippon Shokubi Kagaku Kogyo Co., Ltd.

18. S. Adachi, et al., Japan Kokai 77,02,888, Jan. 10, 1977 to TDK Electronics Co., Ltd.

19. Y. Mitsuo, et al., Japan Kokai 77,122,291, Oct. 14, 1977, to Denki Kagaku Kogyo K.K.

20. I. Sugano, et al., Japan Kokai 76, 140,873, Dec. 4, 1976, to Nippon Electric Co., Ltd.

21. S. Ohbayashi, et al., Japan Kokai 77 60,285, May 18, 1977 to Seitetsu Kagaku Co., Ltd.

22. J. Takeuchi, et al., Brit. Patent 1,447,089, June 22, 1977 to Ube Industries, Ltd.

23. M. Yoshida, et al., Japan Kokai 77 116,777, Sept. 30, 1977 to Nippon Steel Corp.

24. T. Saida, et al., Japan Kokai 77 89,560, Jan. 23, 1976 to Mitsui Toatsu Chemical, Inc.

25. S. Iguchi, et al., Japan Kokai 77 93,688, Aug. 6, 1977, to Mitsui Toatsu Chemical, Inc.

26. Y. Nishikawa, et al., Japan Kokai 77 96,992, Aug. 15, 1977, to Kyushu Refractories Co., Ltd.

27. H. Gondo, et al., Japan Kokai 76 141,770, Dec. 6, 1976 to Nippon Steel Corp.

28. H. Gondo et al., Japan Kokai 76 141,769, Dec. 6, 1976 to Nippon Steel Corp.

29. F. Sishida, et al., Japan Kokai 77 80,271, July 5, 1977, to Nikki K.K.

30. N. Yokoyama et al., Japan Kokai 77 66,891, June 2, 1977, to Mitsubishi Heavy Industries, Ltd.

31. K. Abe, et al., Japan Kokai 77 138,082, Nov. 17, 1977, to Sakai Chemical Industry Co., Ltd.

32. H. Matsuoka, et al., Japan Kokai 77 29,281, March 5, 1977, to Japan Gasoline Co., Ltd.

33. H. Matsuoka, et al., Japan Kokai 77 29,481, March 5, 1977, to Japan Gasoline Co., Ltd.

34. K. Sato, et al., Japan Kokai 77 78,751, July 2, 1977, to Asahi Glass Co., Ltd.

35. S. Adachi, et al., Japan Kokai 76 141,789, Dec. 6, 1976, to Tokyo Electric Chemical Industry Co., Ltd.

36. Japan Kokai 76 111,467, Oct. 1, 1976, to Exxon Research & Engineering Co.

37. T. Tamura, et al., Ger. Offen. 2,709,064, Sept. 15, 1977, to Research Foundation for the Development of Industries.

38. M. Inaba, et al., Japan Kokai 77 103,373, Aug. 30, 1977, to Hitachi Shipbuilding and Engineering Co., Ltd.

39. S. Sato, et al., Japan Kokai 77 75,656, June 24, 1977, to Asahi Glass Co., Ltd.

40. T. Uchino, et al., Japan Kokai 77 70,978, June 13, 1977 to Asahi Glass Co., Ltd.

41. K. Usunomiya, et al., Ger. Offen. 2,634,279, March 10, 1977, to Kobe Steel Ltd.

42. S. Uno, et al., Japan Kokai 77 42,463, April 2, 1977, to Hitachi Ltd.; Mitsubishi Petrochemical Co., Ltd.

43. S. Matsuda, et al., Japan Kokai 77 42,464, April 2, 1977, to Hitachi Ltd.; Mitsubishi Petrochemical Co., Ltd; Babcock-Hitachi K.K.

44. K. Abe, et al., Japan Kokai 77 39,589, March 26, 1977, to Sakai Chemical Industry Co., Ltd.

45. K. Saito, et al., Japan Kokai 76 142,490, Dec. 8, 1976, to Nippon Shokubai Kagaku Kogyo Co., Ltd.

46. Y. Watanabe, et al., Japan Kokai 77 76,289, June 27, 1977, to Mitsui Petrochemical Industries, Ltd.; Hitachi, Ltd.; Babcock-Hitachi, K.K.

47. A. Inoue, et al., Japan Kokai 76 144,396, Dec. 11, 1976, to Nippon Shokubai Kagaku Kogyo Co., Ltd.

48. A. Inoue, et al., Japan Kokai 77 106,389, Sept. 6, 1977, to Nippon Shokubai Kagaku Kogyo Co., Ltd.

49. T. Okano, et al., Japan Kokai 77 63,891, May 26, 1977, to Mitsubishi Chemical Industries Co., Ltd.

50. K. Abe, et al., Japan Kokai 77 48,582, April 18, 1977, to Sakai Chemical Industry Co., Ltd.

51. A. Inoue, et al., Japan Kokai 77 122,293, Oct. 14, 1977, to Nippon Shokabai Kagaku Kogyo Co., Ltd.

52. T. Okano, et al., Japan Kokai 77 123,371, Oct. 17, 1977, to Mitsubishi Chemical Industries Co., Ltd.

53. K. Takani et al., Ger. Offen. 2,727,649, Dec. 29, 1977, to Mitsubishi Chemical Industries Co., Ltd.

54. H. Abe, et al., Japan Kokai 76 144,371, Dec. 11, 1976, to Nippon Kokan K.K.

55. H. Abe, et al., Japan Kokai 77 46,385, April 13, 1977, to Nippon Kokan K.K.

56. H. Abe, et al., Japan Kokai 77 47,567, April 15, 1977, to Nippon Kokan K.K.

57. T. Maeshima, et al., Japan Kokai 76 147,470, Dec. 17, 1976, to Tao Nenryo Kogyo K.K.

58. N. Fujita, et al., *Nippon Kagoku Kaishi*, 1977, 5, 722.

59. N. Tagaya, et al., Japan Kokai 77 39,593, March 26, 1977, to Toa Nenry Kogyo K.K.

60. H. Akimoto, et al., Japan Kokai 77 78,754, 1977, to Hitachi, Ltd.; Babcock-Hitachi K.K.

61. M. Markvard and V. Pour, *J. Catalysis*, 1967, 7, 279.

62. J.R. Katzer, "The Catalytic Chemistry of Nitrogen Oxides," 1975, Plenum Press, New York.

63. A. Bruggeman, et al., 15th DOE Nuclear Air Cleaning Conference, Aug. 1978, Boston, Mass.

NMR Investigations of Mobility Mechanisms of Benzene Molecules in the Cavities of Faujasite Type Zeolites and Connections with Catalytical and Sorption Properties of these Substances

H.Kacirek, H.Lechert, W.Schweitzer and K.-P.Wittern

Institute of Physical Chemistry of the University of Hamburg
Laufgraben 24, 2000 Hamburg 13 , West-Germany

Introduction

From the effects determining the catalytic activity of porous catalysts the molecular mobility mechanism inside the pores is one which can be obtained only with some difficulty by macroscopic experiments. The knowledge of these mechanisms is, however, most interesting for the explanation of sorption processes and the catalytical properties.

One of the most powerful methods for the investigation of molecular mobility is the pulsed NMR spectroscopy. By this method the nuclear relaxation times T_1 and T_2 can be obtained which are directly related to the mobility of the molecules. The mobility is usually described in terms of the so-called correlation time τ_c, which can be regarded as the inverse of the jump frequency. The temperature function of T_1 shows a minimum for each kind of motion. T_2 usually decreases with decreasing temperature and becomes constant for molecules in fixed positions and a definite state of reorientation. At the minima of T_1 holds $\tau_c \omega_L \approx 0.6$, from which τ_c can be obtained. ω_L is the frequency of the NMR-experiment and is called Larmor frequency. Furthermore, at the minimum $T_1/T_2 = 1.6$ if the motion is isotropic and has only a single τ_c[1,2]. Any deviation from this situation leads to an increase of this ratio. Usually τ_c obeys an Arrhenius law with good accuracy. The respective activation energy can be obtained by a measurement of the T_1 minimum for different ω_L , even for cases where distributions of correlation times are present, as is usually the case in sorption systems [3-6].

The relation of the longitudinal relaxation time T_1 , which is predominately discussed in this paper, and τ_c is given by

$$\frac{1}{T_1} = \frac{3}{2} \gamma^4 \hbar^2 I(I+1) \sum_k J_k^{(1)}(\omega_I) + J_k^{(2)}(2\omega_I) \qquad (1)$$

for like nuclei and by

$$\frac{1}{T_1} = \gamma_I^2 \gamma_S^2 \hbar^2 S(S+1) \sum_k \frac{1}{12} J_k^{(0)}(\omega_I - \omega_S) + \frac{3}{2} J_k^{(1)}(\omega_I) + \frac{3}{4} J_k^{(2)}(\omega_I - \omega_S) \qquad (2)$$

for unlike nuclei. γ_I, γ_S and γ are the gyromagnetic ratios of the nuclei and I and S their spins. $\hbar$ is Planck's constant. The summation is carried out over a sufficient number of nuclei in the neighbourhood of the nucleus in question. ω, ω_I and ω_S are the respective Larmor frequencies. The $J^{(n)}$ are the socalled spectral density functions which depend on the mechanism of motion of the molecules to which the nuclei belong. For random rotational reorientations of spin pairs of a distance r_k holds [1,2]

$$J_k^{(0)} = \frac{4}{5r_k^6} \frac{2\tau_c}{(1+\omega^2\tau_c^2)} \quad , \quad J_k^{(1)} = \frac{2}{15r_k^6} \frac{2\tau_c}{(1+\omega^2\tau_c^2)} \quad , \quad J_K^{(2)} = \frac{8}{15r_k^6} \frac{2\tau_c}{(1+\omega^2\tau_c^2)} \quad (3)$$

Results and Discussion of the NMR Measurements

In the figures 1 to 3 the temperature behaviour of T_1 and T_2 in three different faujasite samples is compared.

These samples have been elected from the following points of view. The sample of Fig.1 was a NaX-faujasite with Si/Al = 1.32. In the cavities of this sample four sodium ions in the S_2-sites and about three to four ions in the nonlocalized S_3-sites can be expected to be available for an interaction with the sorbed benzene molecules. In contrast to this the sample of Fig. 2 has Si/Al =2.7 where already one free S_2-position should be present in every second large cavity. Of this sample an H-type with about 70% exchange has been prepared which was used for the experiments demonstrated in the Fig. 3.

At a first inspection can be seen that in Fig. 1 for all coverages a T_1 minimum can be observed which is shifted to higher temperatures with increasing coverage, indicating a restricted mobility with increasing number of molecules in the large cavity. To get more insight into the mechanism of motion responsible for the minima, T_1 measurements with 1,3,5-trideuterobenzene and benzene/hexadeuterobenzene mixtures have been carried out. As can be seen by a comparison of the equations 1 and 2 using the γ and I of the proton and the deuteron, the relaxation rate $1/T_1$ of the magnetic dipole-dipole interaction should be decreased to 1/13 for the intramolecular contribution and to about 1/2 for the intermolecular part in the experiments with the 1,3,5-trideuterobenzene. For the mixtures with the fully deuterated species only the intermolecular interaction is influenced. Actually, one observes, however, that neither for the experiments with the 1,3,5-trideuterobenzene nor for the mixtures is the T_1 at the minimum changed. The only possible explanation for this fact is that the minimum is caused by an

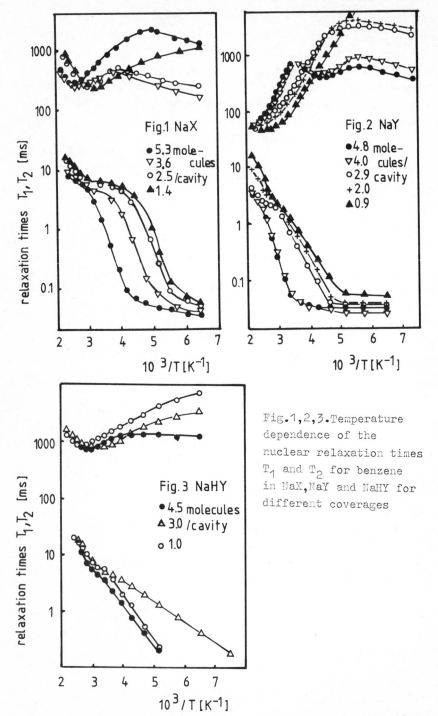

Fig.1 NaX

- ● 5.3 mole-
- ▽ 3.6 cules
- ○ 2.5 /cavity
- ▲ 1.4

Fig.2 NaY

- ● 4.8 mole-
- ▽ 4.0 cules/
- ○ 2.9 cavity
- + 2.0
- ▲ 0.9

Fig.3 NaHY

- ● 4.5 molecules
- △ 3.0 /cavity
- ○ 1.0

Fig.1,2,3. Temperature dependence of the nuclear relaxation times T_1 and T_2 for benzene in NaX, NaY and NaHY for different coverages interaction of the benzene protons with the residual paramagnetic centers. For a paramagnetic impurity of 7 ppm Fe^{3+} being present in the sample, in every twothousandth cavity an Fe^{3+}-ion can be expected. Because a benzene molecule must meet such an ion in a time comparable with T_1 at room temperature a self diffusion coefficient D of about 10^{-8} to 10^{-7} cm^2/sec can be estimated. A direct measurement of D on large crystals gave values near 10^{-7} cm^2/sec at 375 K. The activation energy has been determined to be 34 kJ/mole for 5.3 molecules in a large cavity and 14 kJ/mole for 3.6 molecules in a cavity from self-diffusion measurements as well as from the frequency dependence of the T_1 minimum mentioned above.

For the sample with the higher Si/Al = 2.7 (Fig.2) the T_1 minima are shifted to higher temperatures which are out of the range which can be measured without distilling benzene from the sample in an appreciable amount. The paramagnetic impurity of the sample in Fig. 2 is 17 ppm Fe^{3+}. Experiments with deuterated molecules show above about 300 K no difference between the T_1 of C_6H_6 and $1,3,5-C_6H_3D_3$. From this result can be concluded that the minima at the high temperatures are given again by a translation which carries the molecules from one paramagnetic center to another in a time shorter than T_1. The higher temperatures of the T_1 minima indicate, therefore, a restriction of the correlation time of the translational motion compared with the NaX. These results are in accordance with earlier results [7] of line shape investigations, showing a steady broadening of the lines at full coverage with increasing Si/Al-ratio. Above Si/Al= 2.5, where free S_2-sites can be expected, a narrow line is observed, indicating a small amount of benzene molecules with a strongly increased mobility, increasing in intensity with increasing Si/Al which corresponds to a decreasing number of S_2-sites inside the large cavities.

Going to low temperatures, for the NaX from a study of the second moments of the wide-line spectra [8] can be seen that up to four molecules be arranged in sites about 0.32 nm above the S_2-ions. The fifth and possibly a sixth molecule is most probably arranged in the window between the two cavities parallel to the plane of the window. At temperatures below 150 K only a rotation around the hexagonal axis takes place for four and more molecules in the large cavities. Below this coverage the moments indicate additional freedoms of motions.

From the relaxation behaviour of the benzene compared with the deuterated samples the inter-, the intra- and the contribution of the paramagnetic ions to the relaxation can be obtained. For 3.6

molecules in a large cavity a minimum in the intermolecular contribution can be observed at about 190 K.

The respective motions must be regarded as jumps between the possible sites being connected with tipping motions around the twofold axes and simulating a quasiisotropic motion of the reorientation. Comparing the intramolecular relaxation rate with theoretical
values obtained by Woessner [9] from model calculations a good agreement with the rate of 5.5 sec^{-1} for the isotropic reorientation
can be observed. The mentioned minimum of the intermolecular relaxation can be explained by the fact that because of the limited
space in the cavities the molecules come rather close together in
the course of the tipping motions around their twofold axes, causing an increased dipole-dipole interaction of their protons.

For high coverages the reorientation around the twofold axes cannot be separately detected. Here the high temperature decay corresponding to the rotation of the molecules in fixed position is
directly linked to the low temperature branch of the minimum of
the translational relaxation.

The values of the T_1 at low temperatures are, therefore, rather
long in spite of the increased intermolecular interaction which
must be expected by the larger number of molecules inside the cavities.

For the NaY sample a completely different behaviour can be observed. For the lowest coverage near one molecule in the cavity
the behaviour of T_1 in both samples is similar. For the coverages
up to about three molecules in the large cavity, the behaviour of
the T_1 is similar to that observed for the highest coverage in the
NaX zeolite. According to the explanation for the X-type sample,
these effects can be interpreted by benzene molecules in fixed
positions rotating around their hexagonal axis, as can be seen
by the value of T_2 50 μs which is near the value of 40 μs calculated by Woessner [9] for the intramolecular contribution of this
kind of motion. This behaviour is due to a quite strong interaction
between the benzene molecules and the ions in the S_2-positions.
This is explained by rather high electric fields which have been
calculated by Dempsey [10] above the S_2-ions for zeolites with high
Si/Al-ratios, compared with the respective fields in X-type zeolites. Between three and four molecules in a large cavity these
sites above the S_2-ions are occupied and the next molecules entering the large cavities are bound loosely. This brings the molecules closer together and increases the intermolecular interaction

which leads to a decrease of the T_1. The minima at about 200 K for
the samples with 4.0 and 4.8 molecules in a large cavity indicate
that the relaxation caused by the loosely bound molecules may be
given by a reorientational motion, similar to the tipping motions
discussed for the NaX at intermediate coverage. This is supported
by the values of T_2 which show that the molecules have to be considered to be in fixed positions below ablut 300 K. A completely
different relaxation behaviour is observed for the $Na_{30}H_{70}Y$ sample,
demonstrated in the Fig. 3.

At first the decreased temperatures of the T_1 minima for all coverages indicate a strongly increased mobility of the benzene molecules in the cavity system of the zeolite. The increased value of
T_1/T_2 at the minimum indicates an extremely broad distribution of
correlation times. For all coverages a two component behaviour of
T_1 as well as for T_2 can be observed. The T_2 of the second component lies over the whole temperature range at about 40 μs which
is characteristic for tightly bound molecules. The respective T_1
is about 10 seconds. The absolute amount of these molecules is
nearly independent on coverage and temperature showing that these
molecules are adsorbed on some kind of lattice defects, where is
not clear from these investigation whether these sorption centers
are possibly sites created by partial dehydroxylation which are
present in H-type faujasites according to the investigations of
Bolton et al. [11] The intensity of the signal in the dehydrated
sample before the sorption of the benzene suggests a degree of dehydroxylation of about 30%. The fact that the T_1 in the order of
magnitude of 10 sec can be detected separately shows that the
molecules in the bound situation do not take part in the mobility
mechanism demonstrated in the Fig.3 and have life times in their
state which are at least 10 seconds on an average. The T_1 and T_2
of these molecules is not separately shown in the Fig.3.

The molecules belonging to the relaxation times demonstrated in
the Fig.3 show in contrast to the behaviour of the NaX and the NaY
in the given temperature range no indication of a solid state behaviour. The rather long values of T_1 show in its temperature
function a slight asymmetry for the two lower coverages, indicating that at lower temperatures at least two mechanisms of mobility are responsible for the observed relaxation rates. At the
highest coverage the lower T_1 is due to an increased intermolecular interaction coming into play by the reduced mobility. The T_2
show values which are about an order of magnitude above the values

characteristic for molecules in fixed positions.
Summarizing the NMR experiments it can be stated that in the samples with cations in the large cavities the relaxation behaviour is
determined by the interaction of the benzene molecules with these
cations. Removing these cations, which can be done by exchanging
it against rare earth ions, as has been shown in an earlier investigation ^{12}, or by replacing it by protons which are situated
in positions where completely different mechanisms of interaction
with the benzene molecules must be assumed, the mobility of the
benzene molecules is increased and the distribution of correlation times, which is rather narrow in the case of a cation containing cavity, broadened.In the case of the $Na_{30}H_{70}$Y-faujasite a
small amount of rather tightly bound molecules can be observed
which belong to sites which are created by the activation mechanism of this sample.

Sorption Experiments

To look at the connections between the mobility phenomena and sorption and catalytical properties, the sorption isotherms of benzene on a large number of different zeolites and the cumene cracking
reaction have been studied. For these studies gas-chromatographic
techniques have been used. The sorption isotherms have been obtained from the elution curves of sorbate pulses of different height.
Experiments with different flow rates showed that a diffusion influence is given only at rather high flows and temperatures which
qualitatively confirms the observation of a rather good mobility
of the benzene molecules in the zeolite pores. The sorption isotherms of the three samples for a large range of equilibrium pressures are demonstrated in the Figures 4,5 and 6.
By a comparison of the Figures 4 to 6 can be seen that for a given
coverage and a given temperature, the equilibrium pressure above
the sample increases distinctly from NaX over NaY to NaHY.
The shape of all isotherms is described well by the Freundlich
type at coverages below one benzene molecule in a large cavity.
Above this coverage the description by the Langmuir isotherm is
the best. The saturation value of this isotherm is given for all
samples by about four molecules in a large cavity.
The heats of sorption are 72 kJ/mole for NaX
                          64 kJ/mole for NaY
                      and 58 kJ/mole for NaHY for coverages below
one molecule in a cavity. From one to four molecules in a cavity
these values decrease for

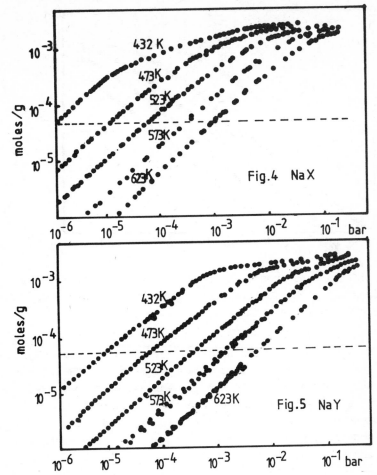

Fig.4 Sorption isotherms for benzene on NaX (Si/Al = 1.32) at different temperatures. The dotted line characterizes the coverage of one molecule per large cavity.

Fig.5 Sorption isotherms for benzene on NaY for different temperatures. The dotted line characterizes the coverage of one molecule per large cavity.

NaX from 72 to 56 kJ/mole, for NaY from 64 to 56 kJ/mole and for NaHY from 58 to 48 kJ/mole. These values show that there is a distinct correlation between the heat of sorption at low coverages and the number of cations available at the walls of the cavities as sorption centers, whereas the fieldstrength above the S_2-ions, being of some influence for the mobility inside the cavities obviously does not express itself in the sorption properties. At first glance it may seem surprising that even at

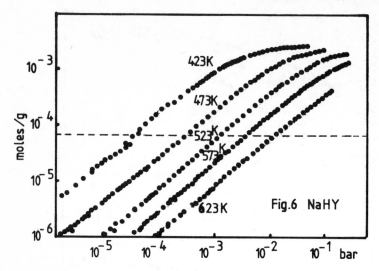

Fig.6 Sorption isotherms of benzene on NaHY at different tem-
       peratures. The dotted line characterizes the coverage of
       one molecule in a large cavity.

very low pressures the occurrence of the very tightly bound molecules in the NaHY zeolite does not show any influence. The explanation of this may be that in the sample used for the NMR experiments the dehydroxylation could not be avoided, whereas the sample
in the sorption experiment has been handled quite carefully.
From the slope of the curves in the Figures 4,5 and 6 can be seen
that the exponent of the Freundlich equation is nearly equal in
all isotherms and has values of about 0.8 .

Catalytic Experiments

To investigate whether there are connections between the mobility
behaviour, the sorption properties and the catalytic activity,
the well known cumene cracking reaction has been studied in a
fixed bed reactor.
As is well known, the NaX does not show any activity at reasonable
temperatures. For the NaY in the steady state the direct dealkylation takes place only with yields of about 0.2 %. The main product
observed in the steam is the isopropyltoluene with about 7% accompanied by a small amount of n-propylbenzene. The temperature of
the experiment was 673 K.
Going to the NaHY and to a temperature of about 473 K, the dealkylation is strongly favoured. As by-products toluene and ethylbenzene are observed to some percent.

One of the main reasons of the observed behaviour seems to lie in the fact that an appreciable part of the propene does not leave the cavity system, but undergoes consecutive reactions, supplying the intermediates necessary for the formation of the by-products. Thus, the spectrum of products leaving the zeolite is strongly dependent on the residence time of the molecules in the cavity system. If the mobility of the molecules is restricted the residence time and the chance of a reaction with the above mentioned intermediates increases. For long residence times and low concentrations of acid centers, as present in the NaY, the most of the cumene molecules will not react. The benzene resulting from the small portion of reacting molecules stays long enough in the cavity system to be involved in consecutive reactions.

One of the most important reactions of the propene and its reaction products is the poisoning of the catalyst.

In the NaHY the number of acid centers is much larger so that more benzene molecules are formed, which have a greater chance to leave the zeolite because of its increased mobility and the increased value of the equilibrium pressure. In this way the larger number of active centers, the mobility and the equilibrium pressure lead to the observed preference of the dealkylation reaction.

The type of side reactions also changes because of the higher concentration of the acid centers. The products observed with the NaY show, however, that the cations are not without any influence on the catalytic behaviour, though the overall process inside the zeolite cavities must be subject to further investigations.

References

1. N.Bloembergen, E.M.Purcell and R.V.Pound, Phys.Rev.,1954,73,679.
2. A.Abragam,"The Principles of Nuclear Magnetism" Oxford
               University Press 1961.
3. H.A.Resing, J.Chem.Phys.,1965,43,669.
4. H.A.Resing, Adv.Mol.Relaxation Processes,1972, 1, 121.
5. H.A.Resing, Adv.Mol.Relaxation Processes,1972, 1, 282.
6. H.Pfeifer, NMR-Basic Principles and Progress, Vol.7 p.53,
               Springer Berlin,Heidelberg,New York 1972.
7. H.Lechert, W.Haupt and H.Kacirek, Z.Naturforsch., 1975, 30a,1207.
8. H.Lechert and K.P.Wittern, Ber.Bunsenges.physik.Chemie,
                                          1978, 82, 748.
9. D.E.Woessner, J.Phys.Chem., 1966, 70 , 1217.
10. E.Dempsey, SCI Monograph "Molecular Sieves" London, 1968, 293.
11. A.P.Bolton and M.A.Lanewala, J.Catal.,1970, 18, 154.
12. H.Lechert, W.Haupt and K.P.Wittern, J.Catal., 1975, 43, 356.

NMR Study of the Equilibrium Exchange Rate of Water between and of the Intermolecular Proton Exchanges in the Large and Small Pores of Type Y Zeolites

W. D. Basler

Institute of Physical Chemistry, University of Hamburg,
Laufgraben 24, 2000 Hamburg 13, W.-Germany

Summary

Using H/D isotopes and pulsed H NMR, the rate by which water
molecules are exchanging between the large and small pores of
zeolite NaY has been measured and the intermolecular proton
exchange in the large and the small pores has been estimated.
An exchange rate of one molecule in 4 days has been found at
295K. As the rate is independent of water activity, it is concluded that the step of leaving the small pores determines the
rate. In the small pores, no intermolecular proton exchange has
been observed within several weeks, in the large pores it takes
place in less than milliseconds.

In zeolite CaY exchange equilibrium has been reached within less
than 15 minutes. It is concluded that removing of the cations
in the six-membered rings is the rate determining step of the
exchange. The NMR results of Y zeolites, where Na had been partly exchanged against Ca, were in agreement with conclusions of
ion exchange studies.

Introduction

Zeolites of faujasite type belong to a class of zeolites which
contain two different kinds of intracrystalline voids. In faujasites, there are large pores interconnected in three dimensions
and isolated small pores in the interior of the sodalite building units. The system of small pores or ß-cages is accessible
from the large pores or supercages through six-membered rings
of 0.25nm free diameter, but which are often occupied by cations.

From the point of view as a molecular sieve, these zeolites may
act as a double sieve, which had been observed with respect to
cations early by Barrer and Meier [1]. The same double sieve effect
may be observed using small molecules like water which can penetrate into the sodalite units. This second sieve effect may be
investigated by NMR, by which molecules in the channel system of
the supercages can be observed separately from those in the isolated small ß-cages: By the relatively high mobility of the mole-

cules in the channel system, the NMR cw-spectrum is considerably motionally narrowed resp. the free induction of pulsed NMR is decaying slowly. On the other hand, the molecules in the ß-cages are more restricted in motion and show a solid-like broad cw-spectrum resp. a fast decaying free induction. From experimental reasons, simultaneously observing and accurately separating the two kinds of molecules is only satisfactorily possible by pulse techniques.

By this method, water molecules in the ß-cages of Na-faujasites had been detected and quantitatively measured [2-4]. Further, the rate by which water molecules are penetrating into the ß-cages had been determined as a function of temperature T and activity a of sorbed water [5,6]. The time t_f to fill the ß-cages had been found to be (R = gas constant)

$$t_f \approx 10\text{min } a^{-1}\exp(+90\text{kJmol}^{-1}R^{-1}(1/T - 1/295\text{K})) \qquad (1)$$

Extrapolating t_f by eq.(1) to 475K gives $t_f \approx 10^{-3}$s for a=1. Assuming that the equilibrium exchange rate between the two pore systems is of the same order of magnitude as the rate of penetrating, it follows from general NMR theory [7] that such a rapid exchange should destroy the two-phase NMR signal and result in a uniform averaged cw-spectrum resp. free induction decay.

As this had not been observed, but the NMR signal still showed two-phase-behaviour at 475K and full water coverage, it had been concluded that the backward rate, i.e. water molecules leaving the ß-cages, determines the equilibrium exchange rate [5].

The equilibrium exchange rate can be observed by isotope labelling. In this way, Frohnecke and Fischbach have measured the exchange kinetics of Sr- and Ba-ions in Na-faujasite Linde 13X by radio-isotopes [8]. We have used H_2O and D_2O and observed the H-NMR-free induction decay. As for small loading the water is predominantly sorbed in the ß-cages [2,4], non-equilibrium distributions of water protons have been created by successive adding of H_2O and D_2O. Thereafter, the approach to equilibrium has been followed by NMR.

As a by-product, upper and resp. lower limits of the rates of intermolecular proton exchange in the large and in the small pores have been obtained.

As one of the steps of exchange is the passage through the six-membered rings, which can be modified by cation exchange, the study of Ca-faujasites should decide whether this step is con-

trolling the rate. The location of Ca-ions can be controlled by
the conditions of ion-exchange [9] and by this a most direct proof
is given that the fast relaxing fraction of the protons actually
belongs to water in the ß-cages.

Experimental

The Na-faujasites used were NaY, self-made, Si/Al=2.36 and Linde
13X and 13Y. CaY($>$98% exchange) was prepared from NaY using a
column with 0.1 n $CaCl_2$-solution for 8 days at 350K. Partly exchanged Na,Ca-faujasites were obtained by stirring in solutions
of tenfold excess of $CaCl_2$ and various concentrations over night
at ambient temperature.

The zeolites were degassed at 675K, until the pressure was below
1mPa for 12 hours. The loss of weight was $355^{\pm}5$mg/g degassed NaY
after storing at 80% rel. humidity. The zeolites were loaded
with vapor of carefully degassed distilled H_2O and D_2O; the
amount was determined by weight. The first step of sorption was
followed by storing at 370K for 24 hours to get the equilibrium
distribution for that partial loading. The second step was done
slowly in 2 hours first at 273K, then at ambient temperature.

Proton-NMR relaxation was measured at ambient temperature with
a Bruker B-KR 322 spectrometer at 60MHz as free induction decay
(FID) and by Carr-Purcell-Gill-Meiboom technique (CPGM). Signal-
to-noise-ratio was improved by accumulation (Bruker Transi-Store
with Fabri-Tek 1074); the self-made probe-head showed a residual
signal of 2 μg H corresponding 0.2 mg H_2O/g zeolite, which was
subtracted if necessary.

The degassed zeolites showed no proton signal at a detection
limit of 0.05 mg H_2O/g. After loading with water, the fraction
P_A (P_B) of protons in the large (small) pores was obtained by
decomposing the relaxation I(t) according to

$$I(t)=I_A+I_B=P_A exp(-t/T_{2A})+P_B f(t) \quad \text{with } I(0)=f(0)=1 \qquad (2)$$

The transverse relaxation time T_{2A} was in the range of ms and
given by relaxation via the paramagnetic Fe^{3+}-ions. After H_2O-
loading the fast relaxation function f(t) was found to be well
approximated by a Gauß-function

$$f(t)=exp(-t/T_{2B})^2 \quad \text{with } T_{2B}=23^{\pm}2 \ \mu s \qquad (3)$$

This corresponds to a cw Gauß-line of $19^{\pm}2$ kHz FWHH. After additional D_2O-loading the fast relaxation I_B could be separated
into two parts, corresponding to H_2O and HOD in the ß-cages:

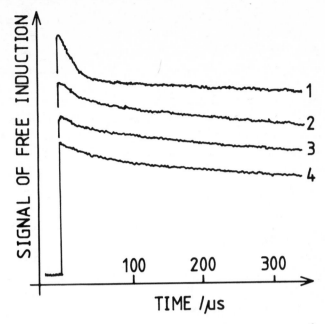

Figure 1: Beginning of proton free induction decay of 63 mg H_2O + 309 mg D_2O/g NaY. 1:1h, 2:12d, 3:33d, 4:11 months after loading.

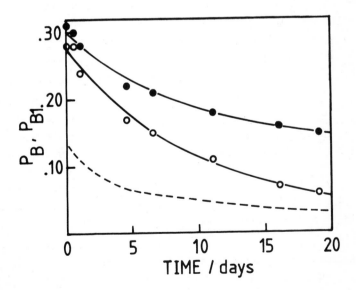

Figure 2: Fractions P_B (●) of all protons and P_{B1} of H_2O-protons (○) in the ß-cages of NaY + 63 mg H_2O + 309 mg D_2O/g as a function of time. Water activity a=0.18. ——— after eq.(10)(11), - - - - after eq.(7).

$$I_B = P_{B1}f_1(t) + P_{B2}f_2(t) \tag{4}$$

with $f_1 = f$ and $f_2 \approx \exp(-t/115 \pm 10 \ \mu s)^2$.

P_B was obtained by exponentially extrapolating I_A to t=0 in the FID resp. CPGM-technique. Decomposition of I_B was done graphically using $f = f_1$. The errors are estimated to 1-2% of total signal. The following abbreviations will be used:

N = number of H_2O-molecules per 1 large + 1 small pore loaded

A (B) = number of water molecules per 1 large (small) pore

$x_h(x_H)$= mole fraction of all (H_2O and HOD) protons in the
        small (large) pores

$x_{hh}(x_{HH})$= mole fraction of H_2O protons in the small (large) pores

N=1 corresponds to 11.25 mg H_2O/g degassed NaY.

Results and Discussions

For rotating H_2O molecules $T_2 = 22,4 \ \mu s$ is calculated for intramolecular H-H interaction, for HOD $T_2 = 135 \ \mu s$ for H-D interaction. As the experimental relaxation times closely agree (the value for HOD is only slightly greater and can be expained by minor contributions of Na^{23} and Al^{27} nuclei) and apparently do not change with varying P_{B1}/P_{B2} ratio, a possible intermolecular proton exchange in the ß-cages must be slow compared with the fast relaxation I_B and the fractions P_{B1} and P_{B2} are the true mole fractions of protons in H_2O resp. HOD in the ß-cages [7].

The non-equilibrium distribution of protons can be created by first H_2O- and then D_2O-loading or vice versa. We have studied both possibilities.

Exchange in first H_2O and then D_2O loaded NaY. After loading with 63 ± 5 mg H_2O/g (18% of full coverage) $P_B = 29 \pm 2\%$ was found, i.e. 1.6 H_2O molecules in 1 ß-cage. Then the zeolite was exposed to D_2O vapor of activity 0.18 for 2 hours; the uptake was 309 ± 10 mg D_2O/g. In these 2 hours the equilibrium of sorption between the large and small pores is attained [6] and constant activity a=0.18 is assured thereafter.

The following variation of the FID with time is shown in Figure 1 and of the fractions P_B and P_{B1} in Figure 2. From N=5.6, A=26 and B=4 one has

$$P_B = x_h B/N = 0.71 \ x_h \tag{5}$$

$$P_{B1} = x_{hh}B/N = 0.71 \ x_{hh} \tag{6}$$

If there were intermolecular proton exchange in the ß-cages,

$P_{B1}/P_B = x_h$ is expected. With eq.(5) one gets

$$P_{B1} = x_h P_B = 1.4 \ P_B^2 \tag{7}$$

The values of P_{B1} do not obey this equation (broken line in Figure 2) and it is concluded that there is no intermolecular proton exchange in the ß-cages within several weeks.

After adding D_2O, $P_{B2} \approx 3\%$ has been observed, but no change of P_{B1}. As the ß-cages are filled up ($\Delta B = 2.4$) in several minutes, in this time HOD must have been formed in the large pores. Assuming exchange equilibrium one gets $P_{B2} = \Delta B x_H/N = 0.06$ and $\Delta P_{B1} = \Delta B x_{HH}/N = 0.008$ as $x_{HH} = x_H^2$.

As the water activity is constant, the exchange rate r between large and small pores is constant, too, and it is for H_2O

$$d(Bx_{hh}) = (-rx_{hh} + rx_{HH}) \ dt \tag{8}$$

The change of x_{HH} is small compared with the change of x_{hh}. With x_{HH}=constant, the integration of eq.(8) gives

$$x_{hh}(t) - x_{hh}(\infty) = (x_{hh}(0) - x_{hh}(\infty)) \exp(-tr/B) \tag{9}$$

The same holds for HOD molecules. Therefore, with eq.(5)(6), the fractions P_{B1}, P_{B2} and $P_B = P_{B1} + P_{B2}$ all should approach equilibrium with the same time constant B/r.

Using the calculated final values $x_h(\infty) = x_H(\infty) = N/(A+B) = 0.19$ and hence $P_B(\infty) = 0.13$, $P_{B1}(\infty) = 0.025$, which are really found experimentally after some months, too, fitting the experimental points in Figure 2 gives (lines in Figure 2)

$$P_B(t) - 0.13 = (0.17 \pm 0.01) \exp(-t/18 \pm 2 \text{ days}) \tag{10}$$
$$P_{B1}(t) - 0.025 = (0.25 \pm 0.01) \exp(-t/19 \pm 2 \text{ days}) \tag{11}$$

This gives $r = 4/18.5$ days=1 molecule/4.6 ± 0.5 days, i.e. every 4.6 days on an average one water molecule is exchanged between one ß-cage and its four surrounding large pores.

Exchange in first D_2O and then H_2O loaded NaY. In a second run first 188 ± 10 mg D_2O/g NaY and then H_2O with a=0.80 were sorbed to find whether the exchange rate was dependent on the water activity. In contrast to the first run, an increase of the fast component is expected and observed (Figure 3). Using A=28, B=4, N=17 from sorption data, hence $P_B(\infty) = 0.125$ and $P_{B1}(\infty) = 0.07$, fitting the experimental values gives (lines in Figure 3)

$$P_B(t) - 0.125 = (-0.085 \pm 0.01) \exp(-t/15 \pm 3 \text{ days}) \tag{12}$$
$$P_{B1}(t) - 0.07 = (-0.04 \pm 0.01) \exp(-t/15 \pm 3 \text{ days}) \tag{13}$$

and the exchange rate $r = 1$ molecule/3.6 ± 0.5 days.

Again intermolecular proton exchange in the ß-cages, i.e.

$$P_{B1} = P_B^2 N/B = 4.25 \ P_B^2 \quad \text{(broken line in Figure 3) is not found} \tag{14}$$

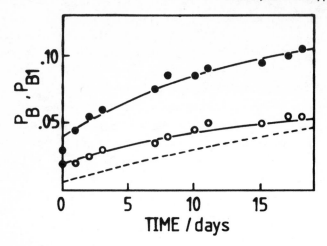

<u>Figure 3:</u> Fractions P_B (●) of all protons and P_{B1} of H_2O-protons (O) in the ß-cages of NaY + 188 mg D_2O + 185 mg H_2O/g as a function of time. Water activity a=0.80. ———after eq.(12)(13), - - - - after (14).

within 3 weeks. The ratio $P_{B1}/P_B=0.5\pm0.2$ shortly after adding H_2O is again in agreement with intermolecular proton exchange equilibrium in the large pores, for which $P_{B1}/P_B=N/A=0.61$ is calculated.

There is evidence that this exchange even takes place in milliseconds or less: In zeolites containing neglectable amounts of Fe^{3+}, the slow relaxation I_A is dominated by intramolecular H-H interaction. Then H_2O and HOD in the large pores should have different relaxation, too, if there were no proton exchange within the relaxation time of several milliseconds. But carefully analysing the relaxation of H_2O/D_2O mixtures we never have found any deviation of I_A from one single exponential.

This fast proton exchange is not surprising, because in the supercages hydrogen bonding takes place like in bulk water, as is known from diffusion and heat capacity measurements [10,11].

Whereas the water activity differed by a factor of 4 in both runs, the same equilibrium exchange rate between the pores has been found. On the other hand, the forward rate of filling the ß-cages is of first order with respect to the water activity and faster by a factor of about 10^3. It is concluded that the backward rate, i.e. passing from the ß-cages back into the supercages, determines the equilibrium exchange rate and suggests the following mechanism: For loadings greater than about 200 mg H_2O/g, i.e. water activity greater than 0.01, almost all ß-cages are filled by 4 water molecules [2,4]. Then one water molecule has to leave its ß-cage (slow backward step) before another molecule can enter into the corresponding ß-cage (fast forward rate). In the same manner it may be explained that the exchange of Sr- and Ba-ions is independent of water activity, too [8].

Exchange in Ca-exchanged Y Zeolites. It has been known from the ion-exchange studies of Sherry [9] that the exchange of Na^+ against Ca^{2+} only is possible in the supercages if performed below about 320K. For complete exchange including ions in the ß-cages, higher temperatures are necessary. Therefore, we studied various faujasites (Linde 13X, 13Y and our NaY) after different ion-exchanges.

All Ca,Na-faujasites, which were exchanged at ambient temperature using ten-fold excess of $CaCl_2$-solutions of different concentrations over 12 hours, showed a fast relaxation I_B identical with the fast relaxation of the mother zeolites. After storing these partly exchanged Ca,Na-zeolites at 350K for several hours, the Ca-type fast relaxation (see below) began to appear.

Using completely (> 98%) exchanged CaY, a quite different fast relaxation has been observed (FID 1 in Figure 4). It is the relaxation of a fixed water molecule as can be seen easily by comparing it with the signal of water in gypsum [6]. As the water molecules remain intact even in the stronger fields of Ca^{2+}, no dissociation and formation of OH-groups is to be expected in Na-faujasites.

This is another proof that the fast relaxation originates from water molecules in the interior of the ß-cages.

Considering the backward rate, the water molecules first have to leave the site of sorption at the cations and then pass through the six-membered rings. The first step will be slower in CaY than in NaY, because of the double charge of Ca^{2+} (the stronger interaction in CaY compared with NaY is clearly reflected by the faster I_B in CaY) and is naturally slower than the equilibrium exchange rate. Adding first 81 mg H_2O/g and then 235 mg D_2O/g to CaY, exchange equilibrium has been reached

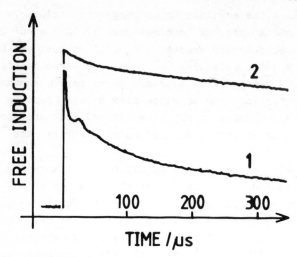

Figure 4: Beginning of proton free induction decay of 81 mg H_2O/g CaY (FID 1) and 15 minutes after additional adding of 235 mg D_2O /g (FID 2).

within 15 minutes: The relaxation (FID 2 in Figure 4) showed a fast component with $T_{2B}=60\pm10$ μs. For fixed HOD $T_2=67$ μs is calculated.

Therefore, the first step must be faster than 15 minutes in NaY, too, and the equilibrium exchange time of several days is given by the second step of passing through the six-membered rings, which are normally occupied by Na-ions. Then removing these ions determines the equilibrium exchange rate.

References

[1] R.M. Barrer and W.M. Meier, *Trans. Faraday Soc.*, 1958, 54, 1074
[2] H. Pfeifer, *Surface Sci.*, 1975, 52, 434
[3] J.S. Murdy, R.L. Patterson, H.A. Resing, J.K. Thompson and N.H. Turner, *J. Phys. Chem.*, 1975, 79, 2674
[4] W.D. Basler, H. Lechert and H. Kacirek, *Ber. Bunsenges. Phys. Chem.*, 1976, 80, 451
[5] W.D. Basler in "Magnetic Resonance in Colloid and Interface Science", ACS Symp. Series 34, H.A. Resing and Ch.G. Wade, Eds. Washington, 1976, p.291
[6] W.D. Basler in "Molecular Sieves II", ACS Symp. Series 40, J.R. Katzer, Ed., Washington, 1977, p.335
[7] J.R. Zimmerman and W.E. Britten, *J. Phys. Chem.*, 1957, 61, 1328
[8] K. Frohnecke and H. Fischbach, *Z. Naturf.*, 1975, 30a, 951
[9] H.S. Sherry, *J. Phys. Chem.*, 1968, 72, 4086
[10] J. Kärger, *Z. Phys. Chem.(Leipzig)*, 1971, 248, 27
[11] W.D. Basler and H. Lechert, *Ber. Bunsenges. Phys. Chem.*, 1972, 76, 1234

W. O. Daly* and W. H. Granville

University of Bradford

INTRODUCTION

The use of zeolitic[1] and other types[2] of molecular sieving adsorbents as a means of separating and purifying gas mixtures is a growing unit operation in chemical engineering. Over the past decade applications of pressure swing adsorption (PSA) have illustrated the potentials of this form of adsorption process. Adsorption process design, particularly of PSA processes, is often an ad hoc affair in comparison with other unit operations such as heat transfer or fluid mechanics. One of the reasons for this situation is the complex nature of the adsorption process.

Adsorption processes are operated in a transient state and the mathematical models involve partial differential equations. The solution of these equations is often prohibitively difficult and simplifying assumptions are necessary to obtain solutions. In new applications of PSA, extensive pilot plant work is often necessary to optimise processes[3]. In fact, there may be many potential applications of PSA separation processes which have been neglected by design engineers because of the lack of relevant adsorption data.

In designing and optimising PSA processes it has been found that the rates of adsorption/desorption of the gas or gases by the zeolite pellets may be an important factor. An example of this is the production of oxygen rich gas using 5A sieves. Consider a bed filled with 5A beads in a simple 2-bed PSA unit being re-pressurised to the adsorption pressure from the revert pressure. There are three distinct flows into the bed, namely the product flow through the system, the flow associated with the pressurisation of the voids and the gas required to bring the adsorbent into equilibrium with the adsorption pressure in the bed. Pilot plant work has shown that the voidage and adsorption surge flows comprise approximately 80% of the flow in the bed. Thus a large portion of the mass transfer takes place under surge conditions. In conventional fixed bed adsorption, the rates of adsorption are also important. However, the surge flows are less important since they are only a small fraction of the total feed into the bed.

A research programme has been initiated to measure adsorption rate and equilibria data pertinent to the operation of PSA processes. The equipment in use comprises two microbalances, which cover both vacuum and high pressure measurements, i.e. 10^{-5} bar to 150 bar, a stainless steel apparatus for measuring multicomponent equilibria data at pressures up to 10 bar, and an apparatus for measuring multicomponent rates of adsorption using the single particle approach.

EXPERIMENTAL

The rate data were measured on a CI Mark 2C vacuum microforce balance[4]. The apparatus is shown schematically in Figure 1. Samples of activated

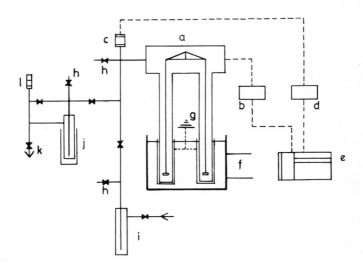

a	glass vacuum container and microbalance head
b	microbalance control unit
c	pressure transducers
d	bridge supply and balance unit for transducer
e	two-pen recorder
f	thermostatically controlled bath
g	earthed anti-static mesh
h	vent valves
i	molecular sieve gas drier, filled with silica gel
j	liquid N_2 cold trap, filled with 13X pellets
k	vacuum pumps
l	vacuum gauge head

Figure 1: Schematic diagram of apparatus

zeolite beads and crystals were subjected to small step changes in pressure. The resulting uptake-time curves were measured along with the gas pressure using a 2-pen recorder. Isotherm data were obtained from the equilibrium uptakes. The range of conditions studied are given below.

Gas	Adsorbent	Temperatures
O_2	4A and 5A Beads	-78.5 and 0 C
N_2	4A Crystals	-78.5, 0 and 30 C
N_2	4A Beads	0, 30 and 50 C
N_2	5A Crystals and Beads	-78.5 and 0 C
CO_2	4A and 5A Crystals and Beads	0, 30 and 50 C

BASIC EXPERIMENTAL RESULTS

Some 300 uptake-time curves were measured and typical examples of these curves are presented in Figures 2 to 6. The fraction uptake is defined as the transient increase in weight divided by the equilibrium change in weight.

Initial inspection of these curves shows that O_2 is rapidly adsorbed on 4A and 5A at ambient temperatures. At -78 C, O_2 on 4A and 5A beads reached 75% fraction uptake after about 80 s. The N_2 data on 5A sieves indicate that the rates on the crystals and beads are approximately the same. For N_2 on 4A sieves, however, there is a large difference in the crystal and bead rates. At 0 C the N_2-4A crystal system reached 75% uptake after about 30 s, whereas the N_2-4A bead system took about 300 s.

For CO_2 the situation is not so clear. All the rate data were found to vary with concentration. In general the crystal rates were faster than the bead rates. Due to the effects of concentration and temperature on the rate data, further analyses are required before more conclusions could be drawn.

ANALYSIS OF RESULTS

Several analytical techniques are available for analysing particle adsorption rate data. Factors such as zeolite crystal size distribution, the bidisperse nature of commercial beads and the effects of concentration on the intracrystalline diffusivity may be taken into account[5-7]. However, many of the models proposed for such analyses are relatively complex. In practical models for adsorption processes the rate data are usually represented by a mass transfer coefficient[8] or a simple Fickian diffusion coefficient with a concentration correction term[9].

The rate data in this paper were analysed using the solution to Fick's law for a step change in concentration for a spherical particle[4] i.e.

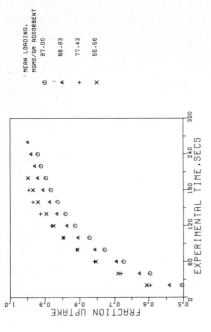

FIGURE 3(A). DESORPTION RATE DATA FOR N2 ON 5A CRYSTALS AT -78 C.

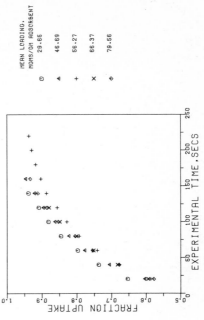

FIGURE 3(B). ADSORPTION RATE DATA FOR N2 ON 5A CRYSTALS AT -78 C.

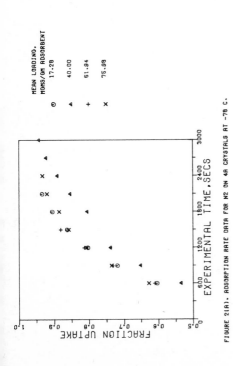

FIGURE 2(A). ADSORPTION RATE DATA FOR N2 ON 4A CRYSTALS AT -78 C.

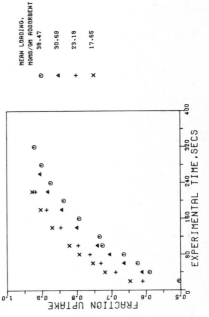

FIGURE 2(B). DESORPTION RATE DATA FOR O2 ON 4A BEADS AT -78 C.

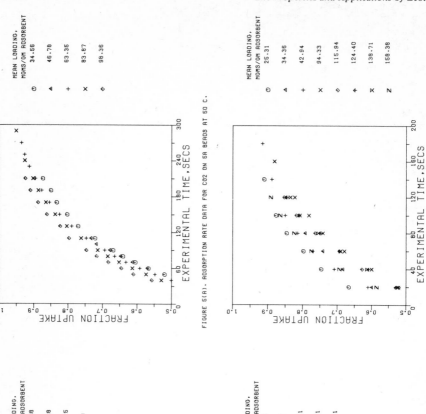

FIGURE 5(A). ADSORPTION RATE DATA FOR CO2 ON 5A BEADS AT 50 C.

FIGURE 5(B). ADSORPTION RATE DATA FOR CO2 ON 5A CRYSTALS AT 50 C.

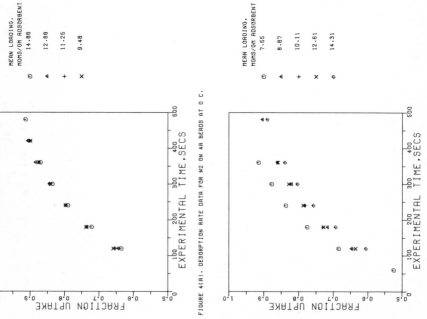

FIGURE 4(A). DESORPTION RATE DATA FOR N2 ON 4A BEADS AT 0 C.

FIGURE 4(B). ADSORPTION RATE DATA FOR N2 ON 4A BEADS AT 0 C.

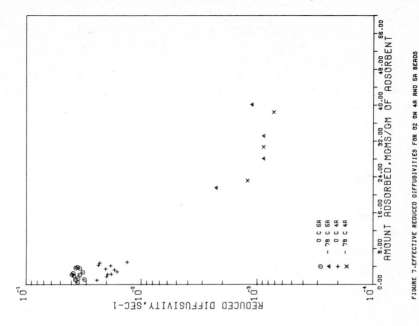

FIGURE 7.-EFFECTIVE REDUCED DIFFUSIVITIES FOR O2 ON 4A AND 5A BEADS

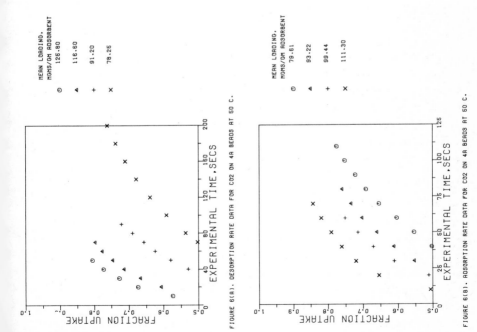

FIGURE 6(A). DESORPTION RATE DATA FOR CO2 ON 4A BEADS AT 50 C.

FIGURE 6(B). ADSORPTION RATE DATA FOR CO2 ON 4A BEADS AT 50 C.

$$\frac{\Delta W_t}{\Delta W_\infty} = 1 - \frac{6}{\pi^2} \sum \frac{1}{m^2} \exp\left(-m^2\pi \frac{D}{R^2}\right)$$

where ΔW_t is the transient increase in weight, ΔW_∞ is the equilibrium increase in weight, D is the effective diffusivity, R is the radius of the adsorbent particle and t is the experimental time.

There are many assumptions implicit in using this equation[4] and it is probable that only a few of them are valid for the system studied in this paper. Thus the analysis presented below is essentially qualitative in nature. However, due to variety in the properties of molecular sieves and their binders[10], particularly for the same sieve from different manufacturers, and the sensitivity of zeolite diffusion properties to the temperature of activation, there must be some doubt concerning the objectiveness of any zeolite diffusion data.

To overcome the problem of zeolite crystal size distribution in obtaining a constant value for the reduced diffusivity[11], the values presented below were calculated from the 75% fraction uptake values. The reduced diffusivities were calculated from the following expression:

$$\frac{D}{R^2}\bigg|_{75\%} = \frac{\tau_{75\%}}{t} = \frac{.0917}{t} \quad s^{-1}.$$

The results for O_2 and N_2 are presented in Figures 7 to 9, whilst Figures 10 to 13 contain results for CO_2 on 4A and 5A beads and crystals. A further set of experiments were performed using CO_2 at 30 C under more controlled conditions to reduce experimental error; these results are presented in Figures 14 and 15. In all the graphs the presence of the binder is not corrected for, i.e. concentrations are expressed in mgms per gm of crystal or mgms per gm of bead.

4A RESULTS

In many cases of adsorption on commercial zeolite beads the intra-crystalline diffusion is thought to be the dominant rate mechanism. Thus it is often possible to measure diffusion data on well characterised zeolite crystal samples and predict the diffusional resistance of commercial pellets or beads from the crystal data. Unfortunately, in the case of 4A zeolites there appears to be a change in the zeolite structure during the pelletisation process, probably due to hydrothermal pore closure. In extreme cases CO_2 molecules can be essentially excluded from the 4A zeolite cage at ambient temperatures[10]. The data for CO_2 and N_2 presented in this paper illustrate

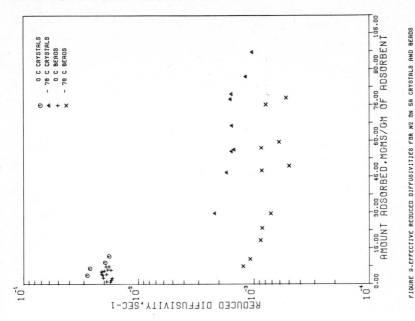

FIGURE 9.EFFECTIVE REDUCED DIFFUSIVITIES FOR N2 ON 5A CRYSTALS AND BEADS

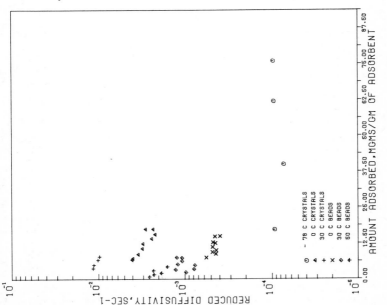

FIGURE 8.EFFECTIVE REDUCED DIFFUSIVITIES FOR N2 ON 4A CRYSTALS AND BEADS

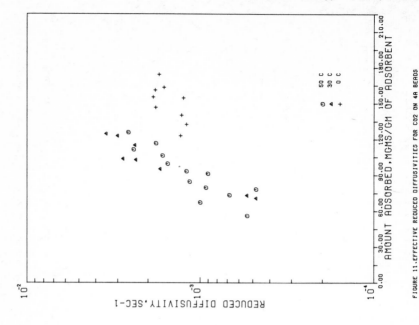

FIGURE 11.-EFFECTIVE REDUCED DIFFUSIVITIES FOR CO2 ON 4A BEADS

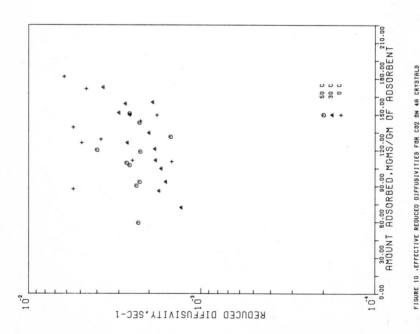

FIGURE 10 .-EFFECTIVE REDUCED DIFFUSIVITIES FOR CO2 ON 4A CRYSTALS

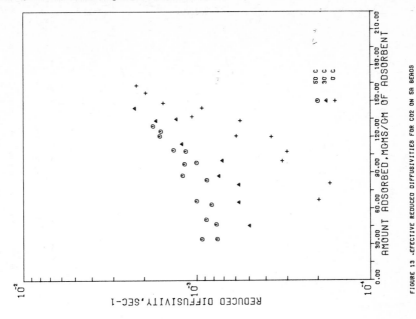

FIGURE 12 .EFFECTIVE REDUCED DIFFUSIVITIES FOR CO2 ON 5A CRYSTALS

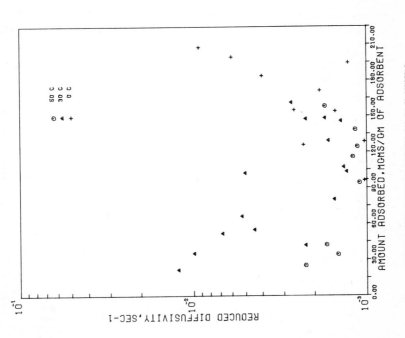

FIGURE 13 .EFFECTIVE REDUCED DIFFUSIVITIES FOR CO2 ON 5A BEADS

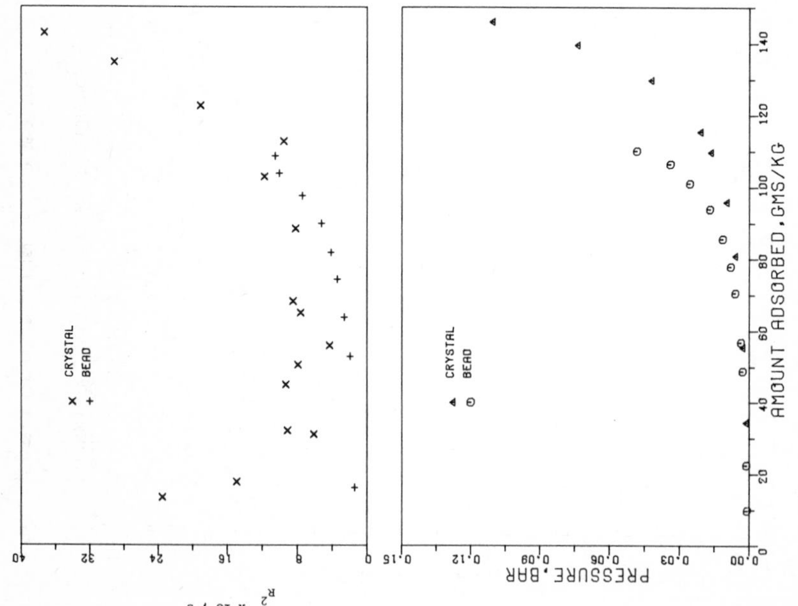

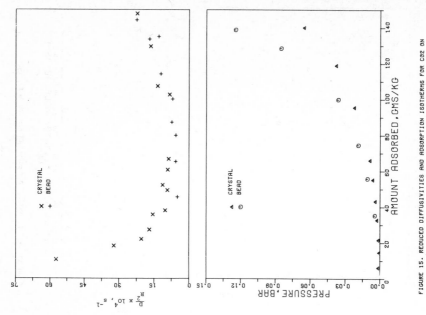

FIGURE 14. REDUCED DIFFUSIVITIES AND ADSORPTION ISOTHERMS FOR CO_2
ON 4A CRYSTALS AND COMMERCIAL BEADS AT 30 C.

FIGURE 15. REDUCED DIFFUSIVITIES AND ADSORPTION ISOTHERMS FOR CO_2 ON
5A CRYSTALS AND COMMERCIAL BEADS AT 30 C.

some effects of this phenomenon on diffusion rates.

Further analysis of the N_2-4A diffusion data[4], see Figure 8, indicates that the crystal diffusion rates were approximately six times greater than those for the beads. However, both the crystal and bead reduced diffusivities had the same activation energies, 21 kJ/mol. Thus it would appear that the N_2 diffusion mechanisms in the crystals and beads are similar except that the pore closure apparently reduces the N_2 mobility by a factor of six. Similar results have been observed for N_2 and CH_4 on 4A crystals and pellets[12]. The crystal diffusivities for both N_2 and CH_4 were approximately 36 times greater than those for the pellets.

In another study[13], it was found that the diffusional resistance of 4A pellets to C_2H_6 was six times greater than for the loose crystals.

The three gases discussed above have the following approximate kinetic diameters, 3.64 Å for N_2, and 3.8 Å for CH_4 and C_2H_6. For CO_2, which has a kinetic diameter of 3.3 Å, the diffusion in 4A zeolites appears to be a more complex process than the three gases discussed above. The data presented in Figure 14 indicate that at high loadings there is apparently no difference between the crystal and bead diffusivities, whereas at lower loadings the crystal diffusivities appear to increase as the concentration decreases. This behaviour could be explained in terms of two different transport mechanisms for the 4A crystals and beads. However, due to the large heat of adsorption of CO_2 at low coverages it may be that the diffusion data were not measured under isothermal conditions.

The only pertinent data in the literature to date for CO_2 are on 4A pellets[15]. The data were measured at 25 C under isothermal conditions using an isotropic exchange technique and were analysed using a bidisperse diffusion model. The intracrystalline diffusion was found to be the dominant rate mechanism; it was also found to be inversely proportional to concentration.

The CO_2-4A data presented in this paper do not exhibit as large a difference between the crystal and bead diffusion rates as for the three gases discussed above, i.e. N_2, CH_4 and C_2H_6. One possible explanation could be that, since the CO_2 molecule is relatively small, 3.3 Å, in comparison with the 4A aperture, the change in the pore size due to the pelletisation process may not affect the CO_2 mobility as much as the effect on the larger N_2, CH_4 and C_2H_6 molecules. Whatever the cause, a more detailed study of this phenomenon is required before the full effect of 4A pelletisation on intracrystalline diffusivities can be understood.

5A RESULTS

Unlike 4A zeolite, the 5A zeolite structure is apparently more stable to the effects of high temperature steaming due to the presence of the divalent calcium cation[10]. The results for N_2 (see Figure 9) appear to substantiate this theory. The slightly lower values for the beads at -78 C may be due to a small macropore resistance. The inverse concentration dependence of the low temperature data is also consistent with other data in the literature[14].

The CO_2-5A diffusion data, see Figures 12, 13 and 15, cannot be readily explained. The crystals diffusivities apparently increase rapidly at low concentrations whereas the bead diffusivities decrease monotonously with concentration. At higher loadings the crystal and bead diffusivities are approximately the same.

To determine whether the bead diffusion rates would increase at very low loadings, an experiment was performed using a single 5A bead from a different batch at 30 C. Again the diffusivities decreased steadily with concentration at loadings down to 10 mgm/gm bead. The temperature dependence of the 5A bead results, see Figure 13, is predictable, unlike the crystal results, which appear to be independent of temperature over the temperature range studied.

One possible explanation for this observation is the effect of the heat of adsorption on the transient change of the sample temperatures. Due to the necessary differences in the bead and crystal sample containers any heat of adsorption liberated could be more readily dissipated by the bead samples. Thus it is possible that the bead adsorption rates were measured under more isothermal conditions than the crystal rates. Whatever the cause, however, it is clear that considerably more research is required to understand fully the diffusion of CO_2 on type A zeolite crystals, and, more importantly, commercial beads.

COMPARISON OF RESULTS

The approximate ratios of N_2 and O_2 diffusivities at ambient temperatures and at atmospheric pressures are presented below.

5A Beads

$$D_{O_2} \simeq 2\, D_{N_2}$$

Oxygen

$$\frac{D}{R^2}\bigg|_{5A} \simeq 2\, \frac{D}{R^2}\bigg|_{4A}$$

4A Beads

$$D_{O_2} \simeq 40\, D_{N_2}$$

Nitrogen

$$\frac{D}{R^2}\bigg|_{4A} \simeq 40\, \frac{D}{R^2}\bigg|_{5A}$$

Also, at pressures up to about 10 bar at ambient temperatures 4A and 5A zeolites adsorb at equilibrium about three times more nitrogen than oxygen for equal partial pressures[4].

The above results illustrate why 5A sieves are chosen in preference to 4A for producing O_2 rich gas from dry CO_2-free air. However, a similar PSA unit filled with 4A sieves and operated correctly could produce N_2 rich air. Such a process would rely on the slower diffusive velocity of N_2 compared with that of O_2, whereas the 5A process relies on the equilibrium separation. Similar comparisons of rate and equilibria data for other gas mixtures could lead to some interesting developments of PSA technology[16]. Also the use of different adsorbents in the same PSA process could provide a useful tool for gas separation.

ACKNOWLEDGEMENTS

Acknowledgements are made to the Science Research Council for a research grant and to Laporte Industries for supplying the zeolite samples.

REFERENCES

1.   D. W. Breck, 'Zeolite Molecular Sieves', Wiley & Sons, New York, 1973.

2.   H. Juntgen, Carbon, 1977, 15, 273.

3.   R. A. Anderson, 'Molecular Sieves II'. ACS, Washington, 1977, 637.

4.   W. O. Daly, PhD Thesis, University of Bradford, 1977.

5.   S. M. Ruthven and K. F. Loughlin, Chem.Eng.Sci., 1971, 26, 1145.

6.   E. Ruckenstein et al, Chem.Eng.Sci., 1971, 26, 1305.

7.   D. M. Ruthven and K. F. Loughlin, Trans.Farad.Soc., 1971, 67, 1661.

8.   I. Zwiebel, C. K. Kralik and J. J. Schnitzer, A.I.Ch.E. Journal, 1974, 20, No. 5, 915.

9.   D. R. Garg and D. M. Ruthven, Chem.Eng.Sci., 1973, 28, 791.

10. C. W. Roberts, Laporte Industries, Personal Communication, 1979.

11. D. M. Ruthven and K. F. Loughlin, A.I.Chem.E. Sym. Series, 1971, 67, No. 117, 35.

12. H. W. Habgood, Can.J.Chem., 1958, 36, 1384.

13. K. F. Loughlin and D. M. Ruthven, Chem.Eng.Sci., 1972, 27, 1401.

14. D. M. Ruthven and R. I. Derrah, J.Chem.Soc.Faraday Trans. 1, 1975, 71, 2031

15. Y. Takenchi and K. Kawazoe, J.Chem.Eng. of Japan, 1976, 9, 46.

16. U.S. Patent No. 2,918,140, 1959.

Heats of Immersion of Co-Substituted
X and Y Zeolites

R.V. Hercigönja, B.B. Radak and V.M. Radak

Faculty of Sciences, University of Belgrade,
and the Boris Kidric Institute of Nuclear Sciences,
Belgrade, Yugoslavia.

Introduction

Heats of immersion in water of A, X and Y zeolites were first studied by Barrer and Cram[1] on samples in the alkali and alkaline earth metal forms. The influence of exchange cations, and of different types of zeolite framework on the magnitudes of the heats of immersion were evaluated in this work. This type of investigation has been extended by Radak et al[2,3] using the same zeolites, but introducing the transition metals as the exchange cations. Based on the heats of immersion of these zeolites in water the electrostatic binding energies of the transition metal ions to the zeolite frameworks were derived. A further extension was presented in a recent paper of Coughlan and Carrol[4] who used A, X, Y and L zeolites exchanged with different mono-, di- and trivalent cations. The heats of immersion in water were interpreted by these authors in the light of available information on cation locations and with reference to the unit cell compositions.

There are considerably less data on the heats of immersion of zeolites in liquids other than water. Tsutsumi and Takahashi[5] measured the heats of immersion of Na-Y and Ca-Y zeolites in different organic solvents, from which they derived the electrostatic field strengths within the zeolite cavities. The electrostatic field has been considered as a source of catalytically active centres in the Ca-substituted zeolite.

In the present work the heats of immersion of Na-X, Na-Y, Co-X and Co-Y zeolites in several organic solvents have been measured. The heat data were then correlated to the dipole moments of the solvents. An attempt was made to use these data for an evaluation of the relative significance of electrostatic fields on the zeolite surface.

Experimental

Zeolite samples: The sodium form of X (Union Carbide) and Y(SK-40) zeolites were prepared for the exchange reaction as described previously[2]. The exchanges of Na^+ ions by divalent Co^{II} ions were performed by equilibrating at ambient temperature, the suspensions being continuously shaken. The exchanged forms were next dried at 373K, and then hydrated by equilibrating them over saturated ammonium chloride solution.

Chemical analyses: The chemical compositions of the zeolites were determined by direct analysis of solid samples (silica gravimetrically, aluminium and cobalt by EDTA titration, sodium by flame photometry and water by the weight loss at 1073K). The results are presented in Table 1.

Table 1. Chemical composition of zeolites

Zeolite	Na^I	Co^{II}	AlO_2^-	SiO_2	H_2O
NaX	11.20	–	29.0	35.66	24.8
CoX	2.47	9.80	26.7	32.66	29.1
NaY	7.45	–	19.4	47.33	26.1
CoY	2.01	6.46	18.5	45.49	27.3

The compositions of the unit cells were calculated using the experimentally determined Si/Al ratio and the fact that the sum of Si + Al atoms must be 192.

Out-gassing: The hydrated samples of known weight were placed in thin-walled ampoules, out-gassed for 48 hours at 573K to a residual pressure of 10^{-3} N m^{-2}, and then sealed in the dry argon atmosphere. The loss of water content of out-gassed samples was controlled by weight-loss determination after out-gassing.

Solvents: The solvents used were of reagent grade (some of them were redistilled) and were stored in flasks over dry zeolite. Solvent water contents were controlled by infrared spectroscopy.

Surface area: Specific surface areas were measured using the BET method by nitrogen adsorption at liquid nitrogen temperatures.

Calorimetry: Heats of immersion were measured in a quasi-adiabatic reaction calorimeter[6] at 298K. The magnitude of the energy produced in the calorimeter was determined by comparing the temperature change measured with that induced by the electrical heater under identical conditions (that is mass, contents, rates of stirring). The heats evolved were determined using the graphical method of Dickinson[7]. All determinations were carried out in duplicate at least, and most

were repeated several times. The precision of the calorimeter amounted to approx-
approximately $\pm1\%$, with a sensitivity of 4 J dm^{-3}.

Results and Discussion

The exchange of cations and their distribution: The compositions of the unit
cells deduced from the chemical analyses were:

$$NaX: \quad Na_{86.2}(AlO_2)_{87.0}(SiO_2)_{105}243.6\,H_2O$$

$$CoX: \quad Co(II)_{32.0}Na_{20.7}(AlO_2)_{87.3}(SiO_2)_{104.7}311.3\,H_2O$$

$$NaY: \quad Na_{55.9}(AlO_2)_{56.8}(SiO_2)_{135.8}250.2\,H_2O$$

$$CoY: \quad Co(II)_{19.6}Na_{15.7}(AlO_2)_{56.3}(SiO_2)_{135.7}271.6\,H_2O$$

These formulae are in agreement with the general formula for an exchanged X
zeolite

$$Co(II)_xNa_{87-2x}(AlO_2)_{87}(SiO_2)_{105} \cdot y\,H_2O$$

and Y zeolite

$$Co(II)_xNa_{57-2x}(AlO_2)_{57}(SiO_2)_{135} \cdot y\,H_2O$$

With the cobalt exchanged samples, the water content per unit cell was
found to be larger in Co-X, in spite of the larger number of cobalt ions in X
form which in turn has essentially the same void volume per unit cell as the Y
zeolite. This can be attributed to the higher negative charge density of the
framework in X which makes possible a more dense packing of both water molecules
and $Co(H_2O)_6^{2+}$ complexes.

The compositions of the unit cells presented above are related to the hydrated forms of investigated zeolites in which the sodium ions from hexagonal
prisms were obviously not exchanged. For the present work, however, data on
the distribution of cations in the dehydrated forms would be of greater interest.
X-ray crystallographic data published to date[8] for Co-Y did not give a precise
analysis of the number of exchanged Co^{2+} ions which may have migrated during
the dehydration process, since the structure determinations were done at two
temperatures only, *i.e.* 473K and 873K. (Our samples were out-gassed at 573K).
However, these results combined with ones obtained from optical spectorscopy[9,10]
enabled the definition of those zeolite regions in which the ions were sited
during the dehydration process. It followed that at the partial dehydration
level a tetrahedral coordination of Co^{II} ion (more stable in X than in Y form)

was prevalent, occupying the S_{II}, S_{II}' and S_I' sites. In the highly dehydrated cobalt forms, octahedral complexes, involving six oxygen atoms of the zeolite framework, would be formed in the S_I sites. This is rather unlikely in our case because of the dehydration conditions that were used. On the other hand, the S_I sites are also not accessible to the solvent molecules which were used in this work. The hexagonal prisms therefore cannot be considered in any further treatment of the present results.

Heats of immersion: The heats of immersion for all four zeolites in various organic solvents were determined   (Table 2). The heats are defined as follows

$$Q_I = \frac{q_{exp}/x}{1-w}$$

Table 2. Heats of immersion at 298K

	$Q_I/J\ g^{-1}$				$Q_I/J\ m^{-2}$			
	n-buta-nol	propionic acid	benzal-dehyde	nitro-benzene	*n*-buta-nol	propionic acid	benzal-dehyde	nitro-benzene
$\mu \times 10^{30}$/C m	5.54	5.84	10.34	14.08	5.54	5.84	10.34	14.08
NaX	285	326	420	508	0.277	0.316	0.408	0.493
CoX	146	200	292	405	0.153	0.209	0.306	0.424
NaY	201	230	271	357	0.223	0.255	0.300	0.395
CoY	239	283	378	455	0.283	0.335	0.448	0.539

$Q_I/J\ g^{-1}$

	n-hexane	n-cyclo-hexane	benzene	carbon tetrachloride
		$\mu = 0$		
NaX	497	410	382	410
CoX	360			
NaY	304			
CoY	444			

Here q_{exp} is the amount of heat evolved on immersing the dehydrated zeolite into the particular solvent. The values were derived directly from the calorimetric measurements assuming that all the water was removed by dehydrating at

573K to a residual pressure of 10^{-3} N m^{-2}. x is the amount of hydrated zeolite taken and w is the mass of water per gram of hydrated zeolite (water content is given in Table 1).

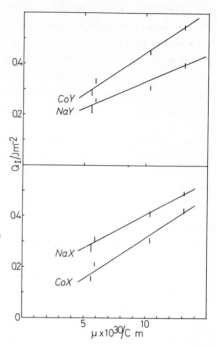

The heats of immersion, Q_I/J g^{-1} of the sodium forms of zeolites in all organic solvents investigated were considerably larger in the X than in the Y form. The influence of charge density in two otherwise identical structures is therefore obvious. Cobalt exchanged Y zeolite showed higher values of heat of immersion than was found with Na-Y samples, in spite of the smaller total number of ions in the former case. This is similar to the results obtained with calcium substituted Y zeolite as compared to the sodium form[5], and also corresponds to the results obtained for the heats of immersion of a number of zeolites in water; higher values for the divalent exchanged form have been found than those observed in

Figure 1. Plots of Q_I Values against Dipole Moment.

the same zeolite containing monovalent ions[1-4]. Such behaviour, however, is not observed with X zeolites when immersed into organic solvents. The heats of immersion of Co-X are lower than the values for Na-X. This might be due to a more stable tetrahedral cobalt configuration in X than in Y zeolite, although this alone could hardly account for such a large difference in behaviour.

In Table 2 the heats of immersion per unit specific surface area are presented. One may notice that the heats of immersion increase with increasing μ of the solvent. For the solvents with $\mu = 0$, however, some very high values for Q_I were obtained. For the other solvents ($5 < \mu \times 10^{30}$/C m < 14) the Q_I values when plotted against the dipole moment (Fig. 1) show a linear dependence. If the slope is taken as an indication for the dipole-framework interaction it could follow that the electrostatic forces are stronger with the cobalt exchanged zeolites.

In order to give a complete analysis of these data it is insufficient merely to consider the dipolar interaction apart from the other forces acting simultaneously in these processes, *viz*, dispersion, repulsion, polarisation, quadrupole interaction. Our further studies and considerations are therefore directed towards getting a more detailed explanation by taking into account all these effects.

The authors thank Professor I. Gal for the available and helpful discussions.

References.

1. R.M. Barrer and P.J. Cram, Advan. Chem. Ser., Mol. Sieve Zeolites, 1971, 102, 105.

2. V.M. Radak, I.J. Gal and B.B. Radak, Thermochim. Acta, 1973, 5, 311.

3. V.M. Radak and I.J. Gal, Proc. 3rd Int. Conf. Mol. Sieves, Leuven University Press, 1973, p. 201.

4. B. Coughlan and W.M. Carrol, J.C.S. Faraday I, 1976, 72, 2016.

5. K. Tsutsumi and H. Takahashi, J. Phys. Chem., 1970, 74, 2710.

6. B.B. Radak, V.J. Matic and V.M. Radak, Croat. Chem. Acta, 1977, 49, 25.

7. H.C. Dickinson, Bull. Nat. Bur. Stand., 1915, 11, 189.

8. P. Gallezot, B. Imelik, J.Chim. Phys., 1974, 71, 155.

9. B. Wichterlova, P. Jiru and A. Curinova, Z. für Phys. Chem. Neue Folge, 1974, 88, 180.

10. H. Hoser, S. Krzyzanowski and F. Trifiro, J. C. S. Faraday I, 1975, 71, 665.

The Influence of Dealumination of Synthetic Y Zeolites on the Equilibrium of Adsorption of C_6-Hydrocarbons

W. Schirmer, H. Thamm, H. Stach and U. Lohse

Central Institute for Physical Chemistry
1199 Berlin, Rudower Chaussee 5, GDR

1. Introduction

The still growing application of zeolite molecular sieves in industrial technologies of separation and catalysis also has stimulated theoretical works on these structurally well-defined microporous adsorbents. Especially, the relations of statistical thermodynamics allowed for the derivation of isotherms, which are able to describe the adsorption equilibrium with high precision [1].

Nevertheless, at present a priori calculations of thermodynamic adsorbate functions are not successful because of the unknown constants of adsorption energy and entropy in these equations. The calculations are complicated by the uncomplete knowledge of the interaction potential as well as by the complicated structure of the molecular sieves [2]. A simplification of the structure of molecular sieves by chemical modification, for example the elimination of the electrostatic field in NaY zeolites by extracting aluminium and sodium cations in such a way that the microporous structure is not altered, should open new ways for theoretical progress.

Dealuminating a commercial NaY zeolite we succeeded in synthesizing an adsorbent which nearly fulfils these demands [3]. This molecular sieve contains less than one Na^+ ion per unit cell. The Si/Al ratio is about 63, and the structure of the crystallites is similar to that of the starting material so that it can be assumed that the adsorption potential is nearly free of electrostatic components. By comparing adsorption data on this adsorbent with those of zeolites containing Na^+ ions essential informations on the influence of the sodium cations should be obtained. In addition to the opportunity for distinguishing different contributions to the interaction energy a careful analysis of the experimental results should yield improved knowledge about the influence of the micropore structure on the level of adsorption energy as well as about the role of adsorbate-adsorbate interaction. Moreover, such a molecular sieve should facilitate the molecular interpretation of adsorption kinetics and dynamics on microporous adsorbents.

We started our investigations on this dealuminated molecular sieve by measuring the adsorption equilibria of hydrocarbons. Beside n-paraffins we studied benzene and cyclohexane. Both molecules are often used in adsorption experiments. Having comparable structures they interact very differently with heterogeneous surfaces. By comparing the thermodynamic adsorbate functions on NaY and its dealuminated form we should be able to find out wether the modification of NaY has led to a homogeneous surface of the adsorbent.

2. Experimental

The starting material for synthesis was crystalline NaY zeolite (type: Zeosorb NaY, producer: VEB CK Bitterfeld, composition of the unit cell: Na_{54} $(AlO_2)_{57}$ $(SiO_2)_{135}$). The process of dealumination was accomplished by cation exchange with ammonium ions, thermal decomposition of the obtained NH_4-forms, thermal treatment with water vapor at high temperatures, and finally by the extraction of the aluminium with diluted mineral acid [4]. The formula of the final product is equal to $Na_{0.7}$ $(AlO_2)_{1-5}$ $(SiO_2)_{191-187}$ (US-Ex, ultrastable-extracted). X-ray investigations showed that the dealuminated zeolite is crystalline with the unit cell dimension a_0 = 2.4256 nm (NaY: a_0 = 2.472 nm) [5]. The occupancy factors are near to one (within the standard deviations). The interatomic distances and bond angles are equal to those which are well-known for pure SiO_2. From porosimetric and adsorption measurements follows that US-Ex contains a secondary pore system beside the faujasite micropore structure. The micropore volume is about 75 % of the NaY pore volume [6]. The adsorbates we have used were spectrally pure. The heats of adsorption were measured by means of a Calvet-type microcalorimeter, the equilibrium pressure by

a capacitance pressure meter. The amount adsorbed was measured
volumetrically using the same pressure meter . While on NaY in
the range of low coverages more than 15 h were necessary for
attaining equilibrium of adsorption, US-Ex equilibrated rapidly
with the vapours applied.

3. Results and Discussion

In figure 1 are plotted isotherms of benzene and cyclohexane on
NaY and US-Ex (amount adsorbed against log p/p_s). In contrast
to NaY, the adsorption isotherms on the dealuminated molecular
sieve show a distinct hysteresis loop that results from the
secondary pore structure of this adsorbent. By comparing the
adsorption of benzene on NaY (curve 1) with that on US-Ex
(curve 3) it can be seen that when in NaY zeolite saturation
capacity is almost approached, in the dealuminated form the
filling of the macropores just starts. The same result one may
see for cyclohexane (curve 2 and 4).

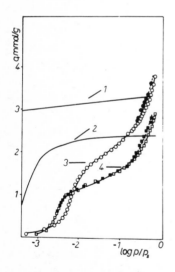

The saturation capacity on NaY
for cyclohexane is about 30 %
less than that for benzene. This
effect might be explained by the
specific interaction between the
π -electrons and the electrostatic field within NaY zeolite. It
is interesting to note that for
log p/p_s-values above -2 this
relation is also valid for US-Ex.
Provided there is no electrostatic field within US-Ex, this
means that the reason mentioned
above can not be the only explanation for the higher adsorption of benzene (compared to
cyclohexane) on NaY. Other causes,
especially steric hindrances,
must be taken into account.

Fig. 1: Adsorption isotherms
of benzene and cyclohexane on
NaY and US-Ex, respectively,
on semilogarithmic plot.
1 - benzene/NaY, 2 - cyclo-
hexane/NaY, 3 - benzene/US-Ex,
4 - cyclohexane/US-Ex.
Black points denote desorption

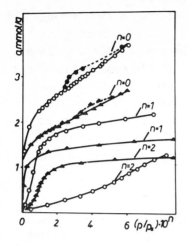

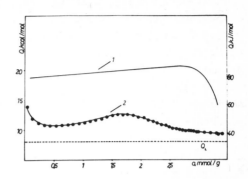

Fig. 2: Adsorption isotherms of benzene and n-hexane on US-Ex. Black points denote desorption.

Fig. 3: Differential heats of adsorption of benzene on NaY and US-Ex, respectively.
1 - benzene/NaY, 2 - benzene/US-Ex
Q_L - here and in the following figures is heat of condensation

The amounts of benzene and n-hexane adsorbed by US-Ex as a function of relative pressure are given in figure 2, where curves corresponding to n = 2 show clearly the behaviour of the isotherm at very low pressures. In this region the isotherms for both adsorbate species are sigmoid in form. Such a behaviour is typical of a strong adsorbate-adsorbate interaction [7]. While in the low pressure range the adsorption of n-hexane exceeds that of benzene, at higher coverages this relationship changes. The higher n-hexane adsorption at a-values below one mmol/g can be attributed to the stronger adsorbate-adsorbate interaction (attraction or repulsion).

The differential heats of adsorption of benzene on NaY and US-Ex are plotted in figure 3. As expected, the heat of adsorption on the aluminium-deficient zeolite is substantially less than on NaY. At a coverage of 0.6 mmol/g (1 molecule per cavity) on zeolite NaY an adsorption heat of 80.0 kJ/mol was measured, while on the US-Ex Q is equal to 45.2 kJ/mol. The difference can be ascribed to the influence of the electrostatic field in NaY. The heats of adsorption also show pronounced differences at larger adsorbate concentration. While on NaY at low coverages

no fall in Q with rising adsorption was found, a distinct decline on US-Ex was measured. At a coverage of 0.3 mmol/g the
curve of the adsorption heat on US-Ex passes through a minimum,
reaches its greatest value at about 1.8 mmol/g, and with further
adsorption gradually approaches a value close to that of heat
of vaporization for the adsorbate. In contrast to US-Ex, the
variation in Q for benzene on NaY is fairly small. Beginning
at low coverages Q rises slightly with increasing adsorption,
reaches its greatest value at a coverage of about 2.7 mmol/g,
and drops rapidly with further adsorption. (Our curve is in
excellent aggreement with that given in [8].)

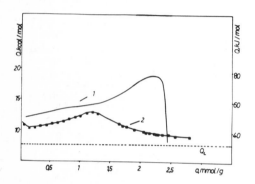

Fig. 4: Differential heats of adsorption of cyclohexane on NaY
and US-Ex, respectively.
1 - cyclohexane/NaY, 2 - cyclo-
hexane/US-Ex

Figure 4 shows the graphs
of the differential heat
of adsorption for cyclo-
hexane on the adsorbents
investigated. Unlike the
curves for benzene on NaY
and US-Ex which signifi-
cantly differ in the magnitude of Q, the heats of
adsorption for cyclohexane
are very near in magnitude.
At an adsorption level of
0.6 mmol/g the heats of
adsorption for cyclohexane
on NaY and US-Ex are 55.7

kJ/mol and 46.9 kJ/mol, respectively. Larger differences occur
as the adsorption is rising. The curve on NaY zeolite exhibits
a distinct maximum as the filling of the zeolite cavities
approaches saturation and then rapidly falls. (These values are
in good accord with those for the cyclohexane/NaY system presented in [9].) The heat of adsorption of cyclohexane on US-Ex
shows a maximum at intermediate a-values and then gradually
falls with increasing adsorption. In the low coverage region
the heat of adsorption for cyclohexane on US-Ex like that for
benzene passes through a minimum.

The calorimetric heats of adsorption for n-hexane on NaY and
US-Ex are plotted in figure 5. As in the case of cyclohexane

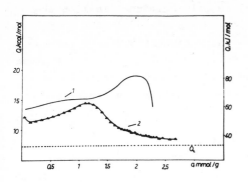

Fig. 5: Differential heat of adsorption of n-hexane on NaY and US-Ex, respectively

the differences in heats of adsorption between NaY and US-Ex are rather small in contrast to benzene. The heat of adsorption for n-hexane on US-Ex follows the same path as that found for the cyclohexane/US-Ex system. The slightly higher Q-values for n-hexane compared to cyclohexane on US-Ex up to a coverage of 1.5 mmol/g can be explained by the stronger dispersion interaction between the chain molecules and the internal surface of the adsorbent. At coverages above 1.5 mmol/g the heat of adsorption for n-hexane is less than that for cyclohexane. As the adsorption isotherms obtained in this coverage range exhibit hysteresis loops which are associated with capillary condensation phenomena the different heats of adsorption can be related to the differences in heats of vaporization. By comparing figures 3, 4 and 5 the following conclusions can be drawn.

1. The heats of adsorption for the adsorbates investigated on NaY clearly demonstrate the specific character of the benzene-adsorbent interaction.
   - At low a-values the heat of adsorption of benzene is substantially higher than those of cyclohexane and n-hexane.
   - With increasing amount adsorbed the heat of adsorption for benzene only slightly rises whereas the curves for cyclohexane and n-hexane show distinct adsorbate-adsorbate maxima.

2. The heats of adsorption for all adsorbates studied on US-Ex follow nearly the same paths.
   - The differences in Q-values between benzene, cyclohexane and n-hexane are small.
   - At the lowest coverage studied the heats of adsorption for all adsorbates drop with increasing adsorption. The rate of decrease is largest for the specific benzene. This be-

haviour suggests that US-Ex still contains a small number
of preferred high energy adsorption sites.
- At high adsorptions (above the maxima in the heat curves)
the heats of adsorption for all molecules are characterized
by a gradual decrease with adsorption. This performance is
consistent with the occurrence of hysteresis loops in the
isotherms and can be attributed to capillary condensation.

From these results we conclude that in a wide range of adsorption (very low and very high coverages excepted) US-Ex represents a microporous and energetic homogeneous adsorbent. Therefore, by comparing heats of adsorption for a given adsorbate on
NaY and US-Ex information on the influence of the electrostatic
field in NaY on the adsorbate-adsorbent interaction can be obtained.

The heats of adsorption of the systems investigated at a coverage
of 0.6 mmol/g (column 2 and 5) and at the maximum of the
heat curve (column 3 and 6) are summarized in table 1. According
to the conclusion above the heats of adsorption on US-Ex can
approximately be regarded as dispersion energies, and the difference between these values and those determined on NaY at equal
amount adsorbed only may arise from the influence of the electrostatic field in NaY.

While the numerical values thus derived for cyclohexane and
n-hexane are mainly due to polarization energy, the corresponding value for benzene includes a specific interaction term.
Assuming, as usual,   that all interaction components are additive and further supposing that the polarization interactions for
cyclohexane and benzene on NaY are near one may estimate the
specific interactions for benzene on NaY to about 26 kJ/mol.
The differences in the heats of adsorptions between a = 0.6
mmol/g and the coverage at which the maximum occurs (column 4
and 7) may be regarded as an expression of the strength of
adsorbate-adsorbate interaction. On both molecular sieves the
corresponding value for n-hexane is larger than that for benzene.
Apparently within the cavities of both adsorbents the interaction of the n-hexane molecules is favoured in comparison to
benzene. Besides, by comparing the ΔQ-values for cyclohexane
and n-hexane on NaY (column 4) and US-Ex (column 7) it can be

__Table 1__ Comparison of heats of adsorption for the systems investigated (Q in kJ/mol)

Adsorbate	NaY			US-Ex			NaY-US-Ex	
	a	b	c	a	b	c	d	e
Benzene	19.1	20.8	1.7	10.8	12.8	2.0	8.3	8.0
Cyclohexane	13.3	18.5	5.2	11.2	13.0	1.8	2.1	5.5
n-Hexane	14.8	19.2	4.4	12.5	14.7	2.2	2.3	4.5

a - $Q_{0.6}$ d - $Q_{0.6}-Q_{0.6}$

b - Q_{max} e - $Q_{max}-Q_{max}$

c - $Q_{0.6}-Q_{max}$

deduced that the higher adsorption potential in NaY also exercises its influence on the adsorbate-adsorbate interaction. Moreover, while on US-Ex the adsorbate-adsorbate interaction appears in pure form (attraction or repulsion), for the analysis of ΔQ-values on NaY it must be taken into account that a restriction of cation movement at high adsorbate concentrations (i. e. a decrease in entropy of the whole adsorbate-adsorbent system) may also yield a contribution to the heat of adsorption at high coverages. Another interesting feature of the benzene adsorption is that on both molecular sieves were found nearly the same values of ΔQ although the numbers of the molecules per cavity (at the maximum of the heat curve) differ by about 50 %. The increase in the heat of adsorption on US-Ex with coverage may be accounted for by the microporous structure of the adsorbent.[*]

The interpretation of the numerical values for ΔQ on zeolite NaY ist more complex. Analysing the data at hand one can not decide whether the benzene molecules are adsorbed localized on

---

[*] The heats of adsorption measured on flat homogeneous surface of graphitized carbon black do not exhibit an increase in Q with amount adsorbed [10]. Apparently the planar orientation of the benzene molecules in monolayer on the carbon black surface does not give rise to a noticeable adsorbate-adsorbate interaction.

energetically high adsorption sites of the same kind only weakly
interacting with each other or if nonlocalized adsorption in a
heterogeneous adsorption potential with strong adsorbate-adsorbate interaction takes place.

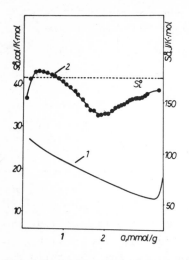

Fig. 6: Differential standard
entropy of benzene adsorbed
on NaY and US-Ex, respectively.
1 - benzene/NaY, 2 - benzene/
US-Ex. S_L^o - here and in fig. 7
is standard entropy of the
liquid adsorbate.

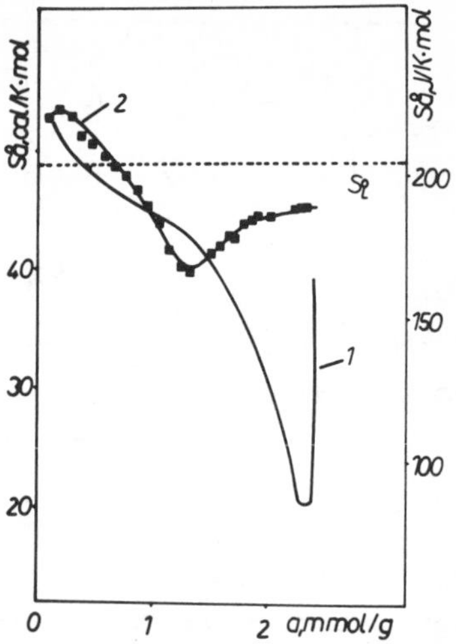

Fig. 7: Differential standard
entropy of cyclohexane adsorbed
on NaY and US-Ex, respectively.
1 - cyclohexane/NaY,
2 - cyclohexane/US-Ex.

The entropies of the adsorbed benzene and cyclohexane on the
molecular sieves investigated (as usual, related to a standard
gaseous state of one atmosphere) are shown in figures 6 and 7,
respectively. As expected the energetics of the adsorption also
strongly influence the entropy curves. While the entropy of the
benzene adsorbed on NaY in the whole range of adsorption is lower
than that on the dealuminated molecular sieve, and monotonically
falls with amount adsorbed, cyclohexane adsorbed on NaY shows
considerable restrictions in mobility at high coverages compared
to low adsorbed amounts and also relative to US-Ex. In aggreement with the discussion above on heats of adsorption, the
entropy data of the adsorbates investigated on US-Ex in contrast
to NaY show no significant differences corresponding to specifi-
ty or nonspecifity of the molecules adsorbed.

References

1 W. Schirmer, K. Fiedler and H. Stach, in "Molecular Sieves II",
   ACS Symposium Series 40, Washington D. C., 1977, p. 305

2 H. Thamm, Ch. Morgeneyer and W. Schirmer, Z. phys. Chemie
   (Leipzig), in press

3 V. Bosaček, D. Freude, R.-G. Kretschmer, U. Lohse, V. Patze-
   lová, W. Schirmer, H. Stach, H. Thamm and Z. Tvarůzková,
   in "Adsorbents, their Production, Properties and Application"
   (in Russian), Publishing House "Science", Leningrad, 1978,
   p. 35

4 U. Lohse, E. Alsdorf and H. Stach, Z. anorg. allg. Chemie,
   1978, 64, 447

5 H. Fichtner-Schmittler, U. Lohse and J. Richter-Mendau,
   J. Catalysis, in press

6 U. Lohse, H. Stach, H. Thamm, W. Schirmer, A. A. Isirikjan,
   N. I. Regent, M. M. Dubinin, Z. anorg. allg. Chemie, in
   preparation

7 H. Thamm, lecture, Symposium on Physical Adsorption, GDR -
   Reinhardsbrunn, April 1978, to be published

8 A. L. Klačko-Gurvič, A. T. Chudiev, Ja. I. Isakov and A. M.
   Rubinštejn, Izv. AN SSSR, Ser. chim., 1967, 1355

9 A. L. Klačko-Gurvič, A. T. Chudiev and A. M. Rubinštejn,
   Izv. AN SSSR, Ser. chim., 1967, 687

10 A. A. Isirikjan, A. V. Kiselev, J. Phys. Chem., 1961, 65, 601

Discussion on Session 3: Sorption Processes.

Chairman: Dr. A.E. Comyns (Laporte Industries Limited, U.K.)

(Throughout this section reference numbers correspond to those in the particular paper under discussion).

Paper 9A (G. Kokotailo and W.H. Meier).

Dr. Vaughan: You have described the Pentasil zeolites as a family of different structures having almost identical X-ray diffraction patterns, compositional ranges and molecular sieving characteristics (i.e. all 10 rings). How do you propose to differentiate new zeolites within this family for patent purposes?

Dr. Kokotailo: Pentasil porotektosilicates exhibit common X-ray diffraction patterns with respect to significant lines indicating common structural characteristics. These common characteristics define the family, individual members of which may vary in composition and minor structural elements. Distinction for patent purposes is a legal matter to be handled by our Counsel.

Dr. Breck: The invention of new words to designate a new zeolite-type structure such as "Pentasil" should be done with caution. For example, "SASIL" is a trade-mark term for a commercial zeolite product. How do you intend to treat Pentasil type structures in the "Zeolite Structures Atlas"?

Dr. Kokotailo: The term "Pentasil" is not intended to be a trade mark. It is intended to describe a particular family of porotektosilicate structures as designated in the text.

Prof. Meier: I agree that new names should not be introduced unless this becomes necessary. The designation "Pentasil" refers to a remarkable and apparently continuous **series** of structures as described in our paper. Structure **types** are properly defined end members of such series. Only these can reasonably be considered for inclusion in the "Atlas of Zeolite Structure Types" (which is also in line with the title of this compilation). Examples of zeolites which for this reason cannot be included in the Atlas are Linde Type T and ZSM-5.

Paper 10 (H. Lechert, N. Kacirek and K.P. Wittern)

Prof. Ruthven: Comparing the absolute value for the self-diffusivity derived from pulsed field gradient measurements with the value calculated from the correlation time (τ) via the Einstein equation ($D = \lambda^2/6\tau$) gives, in principal, a method of calculating the mean square jump distance. Did you make such a calculation and, if so, what value did you find for the mean jump distance?

Prof. Lechert: Calculating the mean jump length from the Einstein equation for the temperature of the T_1 minimum (about $80°C$) for 3.7 benzene molecules in a large cavity of NaX, a value of 0.36 nm is obtained. Analysing the temperature function of T_1 by the translation-relaxation model of Kruger (Z. Naturf. 1969, 24a, 560) values of 0.45-0.49 nm are found. Bearing in mind the approximate nature of these calculations, both values are reasonable if one assumes firstly that the jumps occur between the cations in the faujasite structure, and secondly that they are possibly accompanied by rotational re-orientations (which are indicated from the relaxation analysis with deuterated molecules).

Paper 13 (V.M. Radak, R. Radulovic-Hercigonja, I.J. Gal and B.B. Radak)

Dr. Coughlan: In this paper, all your calorimetric data are represented by integral heats. Now you report integral heats of immersion for molecules of widely differing size and dipole moments *viz:* n-butane, propionic acid, benzaldehyde and nitrobenzene. Precisely because of these properties, the differential heats and packing of the molecules will vary with coverage. Thus when one compares the integral heats of immersion for these molecules, one is not comparing like with like: a large amount of energetic heterogeneity in the differential heat curves versus uptake will affect the integral heat, which is an average value. It is the differential heats for a given uptake, in terms of molecules per unit cell, which should be compared. I should also like to point out that the work of Tsutsumi and Takahashi (J. Phys. Chem. 1970, 74, 2710) is also at fault on a similar point. One cannot take the difference in integral heats of immersion between polar and non-polar molecules of widely different structures and geometries, and then use this difference to give the intrazeolitic electrostatic field directly.

Bearing in mind the above comments,

(i) do you believe that the differences in integral heats that you obtained for Co-X and Co-Y have a realistic interpretation?

(ii) Does the observed linear dependence of the integral heats on μ (reported in the paper) have an absolute physical significance?

(iii) how do you propose to calculate the intracrystalline electrostatic fields from the heats of immersion?

Dr. Radak: Since they are average values the integral heats of immersion cannot be taken as a direct measure of the intrazeolitic electrostatic field. We therefore agree with you that the approach given in the work of Tsutsumi and Takahashi can be misleading. Turning to your direct questions, we would like to point out the following:

(i) and (ii) The differences in integral heats for Co-X and Co-Y, and particularly their linear dependence on μ, cannot have an absolute physical significance,
although they might point to the predominance of electrostatic interactions
particularly when a similar behaviour is noticed in a large number of instances.
(iii) To obtain the intracrystalline electrostatic fields, the determination of
differential heats of immersion is the only experimental approach recognised
so far.

Prof. Sing: Following the comments of Dr. Coughlan, I suggest that the technique
of presorption and heat of immersion should be employed to obtain the differential enthalpies of adsorption for the different sorbate molecules. This approach
should provide an indication of the degree of energetic heterogeneity associated
with each sorption system, and also a test for the soundness of the correlation
between the heat of immersion and dipole moment.

Dr. Radak: Yes, this could be a possible conclusion.

Dr. Rees: In one of our papers (J.C.S. Faraday I 1976, 72, 1840) we showed that
the water content of both zeolites X and Y should be ∿310 molecules p.u.c. at
zero cation volume. It is interesting to note that your Co-X and our Co-X and
Co-Y samples have water contents approaching, or equal to, this above figure.
This suggests that the density of the zeolitic water in these cobalt forms is
much greater than that of liquid water, and suggests that the zeolitic water is
structured differently from liquid water.

Dr. Radak: I remember this paper of yours, and agree with your comments. As you
probably noticed, we have also supposed the water to be structured.

Paper 14 (W. Schirmer, H. Stach and H. Thamm)

Dr. Maiwald: It is customary to check the integrity of ultra-stable zeolites
by X-ray analytical comparison with the mother substance. If you have done this,
how great was the shift of the peaks when going from Na-Y to US-ex-Y? Was any
contribution of the macropores you mentioned indicated by, for example, amorphous
background?

Dr. Stach: An X-ray investigation of the molecular sieve US-ex-Y was performed.
The results have been submitted for publication (H. Fichtner-Schmittler, U. Lohse
and J. Richter-Mendau, J. Catalysis). The structure analysis has not revealed
significant damage of the Y-type zeolite framework. The unit cell parameter
was obtained as a = 2.4256 nm. The line widths of US-ex-Y compared to those of
the mother substance NaY show no significant line-broadening. Therefore, we
concluded that the lattice defects created in the first step of the dealumination process had healed.

The occurrence of secondary pores is not necessarily associated with the presence of amorphous material. As can be seen from electron micrographs holes and cracks penetrate the crystals of US-ex-Y.

Prof. Sing: In the analysis of your isotherm and heat of adsorption data, have you taken into account the adsorption which takes place on the walls of the mesopore at low values of relative pressure (i.e. in the range of micropore filling)?

Dr. Stach: The results of our investigations in the low pressure region (for example the reversibility of the isotherms at low relative pressures) led us to the conclusion that adsorption on the walls of secondary pores may be neglected. By analysis of the adsorption isotherms (by means of t -, t/F - and BET-methods) we determined the micropore volume, the secondary pore volume and the surfaces of the micropore and the secondary pore system. We found the volume of the secondary pores to amount to 40% of the overall pore volume and the corresponding surface of the secondary pore structure to be about 25% of the total US-ex-Y surface. These values are too low to give rise to a considerable contribution to the energetically preferred adsorption in the micropores.

Nevertheless, at very low coverages (0.3 mmol g^{-1}) a noticeable adsorption on defect sites in the secondary pores cannot be excluded.

Binary and Ternary Ion Exchange in Zeolite A

L.V.C. Rees
Physical Chemistry Laboratories, Imperial College of Science
and Technology, London SW7 2AY.

Fundamental studies of cation exchange in zeolites have been carried out mainly because of the crystalline and, thus, well-defined nature of their anionic frameworks. However, the commercial applications of zeolites as cation exchangers have been very limited until recently when zeolite A, in its sodium form, was found to be a suitable replacement of triphosphates as a builder in detergents. Because of the immense financial implications behind this discovery much renewed interest has been shown in cation exchange in zeolite A. This paper will be limited, therefore, to these systems with particular emphasis being placed on the exchanges which are of prime importance in detergency.

Thermodynamics of Ion Exchange

Uni-univalent exchange.

From the isotherm for the exchange

$$A^+ + B^+(Z) \rightleftharpoons B^+ + A^+(Z),$$

where A and B are the two univalent ions and (Z) represents one equivalent of zeolite framework, we obtain the <u>Selectivity Coefficient</u>, K, where

$$K = \frac{A_z}{B_z} \cdot \frac{B_s}{A_s} = \frac{A_z}{B_z} \cdot \frac{m_B}{m_A} \tag{1}$$

A_s and A_z represent the equivalent cation fractions of the ion A^+ in the solution and zeolite phase respectively, B_s and B_z the corresponding equivalent cation fractions of the ion B^+ and m_A and m_B the molalities of the ions A and B in the solution phase.

Any value of K that is different from 1.0 measures the relative difference in preference of the solution and zeolite phases for the two competing ions. In the solution phase it is the activity of the ions that measures their relative preference for the solution phase.

The lower the activity the stronger is the interaction between the ion and the solution phase and the greater the preference of the solution phase for that ion. Thus, the preference of the solution phase for the two counter cations A and B is in the inverse ratio of their activities a_A/a_B.

We can now define a coefficient, K_c, which represents the preference of the exchanger phase <u>only</u> for the two cations relative to the ideal solution state where no preference exists. This ideal solution phase is one of infinite dilution. K_c may, therefore, be called the <u>Corrected Selectivity Coefficient</u>.

If we assume that the exchanger has no preference for the ion A over the ion B, i.e. $K_c = 1.0$, then, if $a_A > a_B$ the exchanger will contain more of ion A than of ion B solely because of the greater tendency of ion A to escape from the solution phase. The ratio A_z/B_z will be equal to a_A/a_B and, therefore,

$$K = \frac{A_z}{B_z} \cdot \frac{B_s}{A_s} = \frac{a_A}{a_B} \cdot \frac{B_s}{A_s} = \frac{\gamma_A}{\gamma_B} \tag{2}$$

Thus, K will be equal to γ_A/γ_B, the ratio of the activity coefficients of the two ions in the solution phase. Since $K_c = 1.0$ then

$$K = K_c \frac{\gamma_A}{\gamma_B} \tag{3}$$

and, therefore, $K_c = K \dfrac{\gamma_B}{\gamma_A}$ (4)

When K_c in the general case is not equal to 1.0 then we obtain the equation for the corrected selectivity coefficient

$$K_c = \frac{A_z}{B_z} \cdot \frac{B_s}{A_s} \cdot \frac{\gamma_B}{\gamma_A} \tag{5}$$

or $\quad K_c = \dfrac{A_z}{B_z} \cdot \dfrac{B_s}{A_s} \cdot \dfrac{\gamma_B}{\gamma_A} \cdot \dfrac{\gamma_{Cl}}{\gamma_{Cl}} = \dfrac{A_z}{B_z} \cdot \dfrac{B_s}{A_s} \cdot \dfrac{\gamma^2_{\pm\,BCl}}{\gamma^2_{\pm\,ACl}}$ (6)

where γ_{+ACl}, γ_{+BCl} are the activity coefficients of the salts ACl and BCl in the mixed salt solution phase.

In dilute solutions of univalent cations (< 0.02 molal), although γ_{+ACl} and γ_{+BCl} may not be unity, their ratio $\gamma^2_{+BCl}/\gamma^2_{+ACl} \simeq 1.0$ within the experimental error. Thus for dilute solutions $K_c \simeq K$.

Can we now "correct" for the interactions in the exchanger phase? Clearly we could introduce the corresponding factor f_A/f_B, where f_A and f_B are the activity coefficients of the ions A and B respectively in the exchanger phase. If we defined them in an exactly analogous manner to γ_A and γ_B we could eliminate all selectivity; i.e. we would have

$$\left[\frac{A_z}{B_z} \cdot \frac{f_A}{f_B} \right] \cdot \left[\frac{B_s}{A_s} \cdot \frac{\gamma_B}{\gamma_A} \right] = 1.0 \tag{7}$$

Thus, $K_c = f_{B/f_A}$, would be nothing more than the ratio of the activity coefficients of the ions in the exchanger phase provided that these activity coefficients are defined in an exactly analogous manner to that for the ions in the solution phase. Although eq. 7 represents a perfectly sound way of interpreting K_c, it cannot be tested experimentally because there is no independent way of evaluating f_{B/f_A} .

K_a, however, the "Thermodynamic Equilibrium Constant" can be defined in this way, i.e.

$$K_a = \frac{A_z f_A}{B_z f_B} \cdot \frac{B_s \gamma_B}{A_s \gamma_A} \tag{8}$$

The standard state chosen is such that f_A and f_B are unity when the exchanger is, respectively, in its pure A and pure B form. K_a is then a true constant and the variation of K_c with A_z is contained within the variation of f_{B/f_A} with A_z.

With these conventions it is possible to show by means of the Gibbs–Duhem equation that

$$\ln K_a = \int_0^1 \ln K_c \cdot dA_z \tag{9}$$

and that
$$\ln f_A = -\int_{K_c^0}^{K_c} B_z \cdot d(\ln K_c) \tag{10}$$

and
$$\ln f_B = +\int_{K_c^1}^{K_c} A_z \cdot d(\ln K_c) \tag{11}$$

where $K_c = K_c^\ominus$ when $A_z = 1.0$ and

$K_c = K_c^1$ when $B_z = 1.0$.

In eqs 9 to 11 any change in the activity of water in the zeolite phase as A ions replace B ions is neglected. f_A and f_B are evaluated from the same graph of $\ln K_c$ vs A_z as is used to evaluate K_a.

K_a is a kind of <u>average selectivity</u> over all sites.

Finally, $\Delta G^\ominus$, the <u>Standard Free Energy</u> change for the transfer of 1 equivalent of ion A from an infinitely dilute solution phase to an exchanger in its pure B form containing 1 equivalent of B ions and the reverse transfer back to the solution phase of 1 equivalent of ion B at infinite dilution is given by

$$\Delta G^\ominus = -RT \ln K_a \tag{12}$$

$\Delta G^\ominus$, as defined above, has been determined for the exchange of Na^+ ions by Li^+, K^+, Rb^+, Cs^+, Ag^+ and Tl^+ in zeolite A and the values obtained are given in Table 1. $\Delta G^\ominus$ is positive when Na^+ ions are replaced by the alkali metal cations. At low A_z values, the selectivity coefficient, K, is greater than 1.0 for exchange of Na^+ by K^+, Rb^+ and Cs^+ e.g. at $A_z = 0.1$, $K = 1.20$ for K^+; 1.8 for Rb^+ and 2.0 for Cs^+. However, overall Na^+ is the preferred ion. $\Delta G^\ominus$ is comparatively large and negative for the exchange $Na^+ \rightarrow Ag^+$ and $Na^+ \rightarrow Tl^+$. The isotherms show high selectivity by the zeolite for these highly polarizable ions at all loadings.

Table 1

Uni-univalent Exchange in Zeolite A

Exchange	Maximum A_z value at 25°C.	Temp K	$\Delta G^{\ominus}$	$\Delta H^{\ominus}$	$T\Delta S^{\ominus}$	Kielland Coeff. C	Reference
			kJ (g. equiv.)$^{-1}$				
Na → Li	1.0	298	+5.43	+9.45	+4.01	−0.21	(1)
Na → K	1.0	298	+0.59	−9.99	−10.58	−0.21	(1)
Na → Rb	0.93	298	+2.84	−10.66	−13.50	−0.70	(1)
Na → Cs	0.63	298	+8.28	−15.88	−24.16	−2.30	(1)
Na → Ag	1.0	278	−16.18				(2)
		298	−16.43	−11.60	+4.83		(2)
		350	−17.56				(2)
Na → Tl	1.0	298	−9.70				(2)

Within the experimental error many of these uni-univalent exchanges show linear $\log_{10} K_c$ vs A_z plots. From these straight line plots we can derive the equation

$$\log_{10} K_c = 2C\, A_z + \log_{10} K_c^1 \tag{13}$$

where $\log_{10} K_c = \log_{10} K_c^1$ where $B_z = 1.0$.

The slope of these graphs is $2C$ where C is the Kielland coefficient[3]. The values of C found for the alkali metal cations are listed in Table 1. The thermodynamic equilibrium constant, K_a, is obtained from the value of K_c at $A_z = 0.50$. The activity coefficients f_A and f_B are obtained from the equations

$$\log_{10} f_A = C\, B_z^2 \text{ and } \log_{10} f_B = C\, A_z^2 \tag{14}$$

In these experiments the B cation is always Na^+. The variations of f_A and f_{Na} with A_z are shown in figure 1.

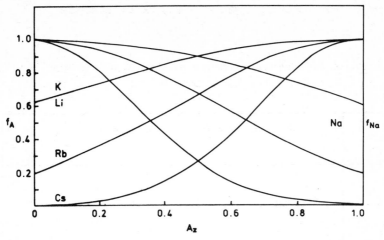

Figure 1. Variation of activity coefficients, f_A and f_{Na} in zeolite phase as a function of A_z where A represents the ions Li, K, Rb, and Cs.

In order to obtain K_c as defined in eq 6 we have to determine the ratio $\gamma^2_{\pm BCl} / \gamma^2_{\pm ACl}$ where $\gamma_{\pm BCl}$ is the mean activity coefficient of the BCl salt solution in the presence of ACl salt, at the relative concentrations given by B_s/A_s. Barrer, Rees and Ward[1] assumed this ratio to be constant at any given molality as given by Glueckauf. According to Glueckauf[4],

$$\log_{10} \frac{\gamma_{+BY}}{\gamma_{+AY}} = \frac{1}{2} \cdot \log_{10} \frac{\overline{\gamma}_{+BY}}{\overline{\gamma}_{+AY}} \tag{15}$$

where $\overline{\gamma}_{+AY}$ is the activity coefficient of the pure AY electrolyte solution at the same ionic strength, I, as that of the mixed electrolyte solution.

Barrer et al calculated, for the $Na^+ \rightarrow K^+$ exchange in zeolite A, that $\dfrac{\gamma^2_{+NaCl}}{\gamma^2_{+KCl}} = 1.0104$ and 1.0877 for I of 0.1M and 1.0M respectively. These activity coefficients have now been determined experimentally[5]. When the molality is $\sim 0.1M$ Glueckauf's value for the activity coefficient ratio is acceptable but when molality is ~ 1.0 the ratio is not constant at 1.0877

but decreases from a value of 1.1321 when $Na_s \to 1.0$ to 1.0642 when
$Na_s \to 0$ (see Table 2).

Table 2. Experimental Activity Coefficient Ratios for Mixed
NaCl/KCl Solutions of Total Molality equal to 1.0M.

	Equivalent Cation Fraction of Na^+ (i.e. Na_s)					
	1.0	0.8	0.6	0.4	0.2	0
$\dfrac{\gamma^2_{\pm\,NaCl}}{\gamma^2_{\pm\,KCl}}$	1.1321	1.1172	1.1046	1.0899	1.0772	1.0642

Thus, small but significant errors in K_c can be introduced, especially
for more concentrated solution phases, if the activity coefficient ratios
are not checked experimentally. In any future accurate study of these
uni-univalent exchanges the activity coefficient ratios of the mixed
solution phase must be ascertained experimentally at each point in the
isotherm.

Barrer et al used solution phases of total molality in the final
points of their exchanges as high as 3.28M for Na - Li; 1.60M for Na - Cs
and 1.0M for Na - K and Na - Rb. Because they assume $\dfrac{\gamma^2_{\pm ACl}}{\gamma^2_{\pm BCl}}$ values as

given by Glueckauf, the final points in their isotherms will lead to values
of K_c at these high A_z values which could be in error. Thus the slope of
their $\log_{10} K_c$ vs A_z plots may be in slight error and the value of K_c at
$0.5\,A_z$ from which K_a is obtained may also be in error to a small extent.
It would be interesting to redetermine these K_c values containing
experimental $\dfrac{\gamma^2_{\pm ACl}}{\gamma^2_{\pm BCl}}$ to ascertain how significant are the original errors.

Sherry and Walton [2] used solution phases which were always 0.1M. Although they used constant values for $\dfrac{\gamma^2_{+ACl}}{\gamma^2_{+BCl}}$ in their analyses, their results should not be in serious error.

Wolf and Furtig [6] did not correct their selectivity coefficients, K, for the activity coefficient ratio factor. Thus, they used K values instead of K_c values in the determination of K_a and $\Delta G^{\ominus}$ and these latter two derived terms are in error because of this. However, as they used $\sim 0.1M$ solutions in the determination of their isotherms the error will not be too large.

Uni-Divalent Exchange. When A is changed from a univalent to a divalent ion then the equations derived for uni-univalent exchange require modification. Let us consider the following interesting specific example,

$$Ca^{2+} + 2\overline{Na^+} \rightleftharpoons \overline{Ca^{2+}} + 2Na^+$$

where the bar above the ion indicates the ion in the zeolite phase. The selectivity coefficient, K, is now defined for this system as

$$K = \frac{Ca_z}{(Na_z)^2} \cdot \frac{m^2_{NaCl}}{m_{CaCl_2}} \tag{16}$$

the corrected selectivity coefficient, K_c, becomes

$$K_c = \frac{Ca_z}{(Na_z)^2} \cdot \frac{m^2_{NaCl}}{m_{CaCl_2}} \cdot \frac{\gamma^4_{\pm\ NaCl}}{\gamma^3_{\pm\ CaCl_2}} \tag{17}$$

and the thermodynamic equilibrium constant, K_a, is given by

$$K_a = \frac{Ca_z}{(Na_z)^2} \cdot \frac{f_{Ca}}{f^2_{Na}} \cdot \frac{m^2_{NaCl}}{m_{CaCl_2}} \cdot \frac{\gamma^4_{\pm\ NaCl}}{\gamma^3_{\pm\ CaCl_2}} \tag{18}$$

To obtain K_a from K_c we now have to use the equation

$$\ln K_a = -1 + \int_0^1 \ln K_c \cdot d\,Ca_z \tag{19}$$

and, finally, $\Delta G^{\ominus}$, the standard free energy per equivalent of exchange,

is obtained from

$$\Delta G^{\ominus} = -\frac{RT}{2} \ln K_a \qquad (20)$$

We can now demonstrate one of the major differences which arises in

uni-divalent exchange which is not present in uni-univalent exchange.

If X_{Na} and X_{Ca} are the respective mole fractions of Na^+ and Ca^{2+} in

solution phase when we have,

$$X_{Na} = \frac{m_{Na}}{m_{Na} + m_{Ca}} = \frac{m_{Na}}{m} \qquad \text{and}$$

$$X_{Ca} = \frac{m_{Ca}}{m_{Na} + m_{Ca}} = \frac{m_{Ca}}{m}$$

where $m = m_{Na} + m_{Ca}$ is the total molality of the equilibrium solution phase.

$$\therefore \quad \frac{(m_{Na})^2}{m_{Ca}} = \frac{(m \, X_{Na})^2}{m \, X_{Ca}} = \frac{m(X_{Na})^2}{X_{Ca}}$$

and K_c as defined in eq 17 can now be written as

$$K_c = \frac{Ca_z}{(Na_z)^2} \cdot \frac{m(X_{Na})^2}{X_{Ca}} \cdot \frac{\gamma^4_{\pm \, NaCl}}{\gamma^3_{\pm \, CaCl_2}} \qquad (21)$$

Thus the relative amounts of Ca^{2+} and Na^+ in the exchanger phase now depend

on both the total concentration of Ca^{2+} and Na^+ as well as their relative

concentrations in the solution phase.

If the ionic composition of the exchanger phase is kept constant then

K_c will remain constant. The composition of the solution phase will vary,

therefore, with m, the total molality of the solution phase. We can see

from eq. 21 that as $m \rightarrow 0$ then the system becomes more selective towards

the divalent ion. This is known as the concentration-valency effect.

The equation for K_c can also be expressed in terms of equivalent

cation fractions in the solution phase to be similar in form to eq. 7.

Secondly, isotherms are always shown in terms of equivalent cation fractions in zeolite phase against equivalent cation fractions in solution phase. Since

$$Na_s = \frac{m_{Na}}{m_{Na} + 2m_{Ca}} \quad \text{and} \quad Ca_s = \frac{2m_{Ca}}{m_{Na} + 2m_{Ca}}$$

we can write

$$K_c = \frac{Ca_z}{(Na_z)^2} \cdot \frac{\left[Na_s(m_{Na} + 2m_{Ca})\right]^2}{\frac{1}{2}Ca_s(m_{Na} + 2m_{Ca})} \cdot \frac{\gamma^4_{\pm \; NaCl}}{\gamma^3_{\pm \; CaCl_2}} \tag{22}$$

$$= \frac{Ca_z}{(Na_z)^2} \cdot \frac{(Na_s)^2}{Ca_s} \cdot 2(m_{Na} + 2m_{Ca}) \cdot \frac{\gamma^4_{\pm \; NaCl}}{\gamma^3_{\pm \; CaCl_2}} \tag{23}$$

$$= \frac{Ca_z}{(Na_z)^2} \cdot \frac{(Na_s)^2}{Ca_s} \cdot 2N \cdot \frac{\gamma^4_{\pm \; NaCl}}{\gamma^3_{\pm \; CaCl_2}} \tag{24}$$

where N = total normality of equilibrium solution phase. In an ion exchange reaction N remains constant whereas m changes as the exchange proceeds.

One could plot $\dfrac{Ca_z}{(Na_z)^2} \cdot \dfrac{(Na_s)^2}{Ca_s}$ against Ca_z as a form of

selectivity curve instead of K against Ca_z but the curve would depend on the total mornality of the solution phase.

Let us look at the $\overline{2Na^+} \rightarrow \overline{Ca^{2+}}$ exchange in zeolite A, as reported by (i) Barrer, Rees and Ward [1]; (ii) Sherry and Walton [2]; (iii) Wolf and Furtig[6].

Barrer et al were simultaneously measuring the heats of ion exchange in a calorimeter as well as determining the exchange equilibrium at the completion of the heat measurement. Their experimental technique required them to increase the initial solution phase molality from 0.005 to 0.54 to allow them to scan the whole range of Ca_z. Their isotherm should, therefore, be corrected for the concentration-valency factor.

For example in the first point in their isotherm the initial solution was 0.005M $CaCl_2$ and in the resulting equilibrium solution Na_s = 0.998 and Ca_s = 0.002. Using Glueckauf's theory [4] the activity coefficient ratio was found to be 1.22 for this equilibrium solution and 1.58 when this solution phase was adjusted to 0.1M. Assuming K_c in eq. 24 to be constant Na_s and Ca_s become 0.975 and 0.025 respectively at 0.1M. Thus Na_s^2/Ca_s decreases from 498 to 38.5 on adjusting the solution phase concentration

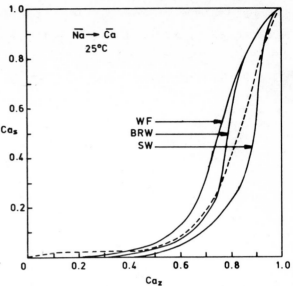

Figure 2.   Na/Ca exchange isotherms. BRW (Barrer, Rees, Ward); WF (Wolf, Furtig); SW (Sherry, Walton). -------BRW isotherm adjusted for a constant 0.1M solution phase.

to 0.1M. The remainder of the isotherm was similarly adjusted and may be compared in figure 2 with the original and with the isotherms of Sherry and Walton, and Wolf and Furtig which were determined at a constant solution phase normality of 0.1N.

However, in these latter two isotherms, when Ca_z > 0.5, the solution phase contains increasing amounts of Ca^{2+} and the molality decreases until at Ca_z = 1.0 the solution phase will be 0.05 M $CaCl_2$. In a strict comparison, therefore, these latter two isotherms should also be corrected for the decreasing molality of the solution phase. Such an adjustment would tend to bring the adjusted Barrer et al isotherm into reasonable agreement with the

Sherry and Walton isotherm at $Ca_z > 0.5$, but would have the opposite effect on the Wolf and Furtig isotherm. The adjusted isotherm derived from the Barrer et al data seems to have too high values for Ca_s at $Ca_z < 0.5$ when compared with the isotherms of Sherry and Walton and Wolf and Furtig in this region. The adjusted values from Barrer, Rees and Ward's data, therefore, must be in error, not because of any serious error in the above calculations, but because of too high Ca_s values in the original data.

Activity Coefficient Ratio

Factor. Let us now look at the $\gamma^4_{\pm NaCl} / \gamma^3_{\pm CaCl_2}$ factor, Γ. In figure 3 a comparison can be made between this ratio calculated according to Glueckauf's theory[4] and experimental values obtained by Moore and Ross[7] for a solution of constant ionic strength of 0.1M. It is obvious from this figure that the Glueckauf theory

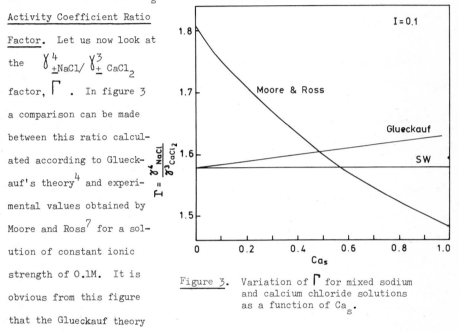

Figure 3. Variation of Γ for mixed sodium and calcium chloride solutions as a function of Ca_s.

predicts ratios which are in serious error from those found experimentally.

Sherry and Walton used a constant value for the activity coefficient ratio in their derivation of K_c. Although they did not state what this constant value was, it probably was 1.58. This value for the ratio has been inserted into figure 3. It is obvious that the K_c data used by Sherry and Walton are, therefore, in error if the experimental activity coefficient ratios found by Moore and Ross are correct. The $\Delta G^{\ominus}$ figure of -3.06 kJ per g. equivalent quoted by Sherry and Walton will be in error, but the error will be less than their experimental error of ± 1.25 kJ.

Wolf and Furtig used the ratio $\dfrac{Ca_z}{\overline{Na}_z} \cdot \dfrac{Na_s}{\overline{Ca}_s}$ in place of K_c to obtain K_a.

Having obtained this incorrect value for K_a they derived $\Delta G^\ominus$ and from the variation in $\Delta G^\ominus$ with temperature they obtained $\Delta H^\ominus$. Because of all the errors involved, their $\Delta G^\ominus$ values must be in very serious doubt and their $\Delta H^\ominus$ value somewhat meaningless.

If $\Delta H^\ominus$ is required it is best to use the calorimetric technique employed by Barrer et al. If use of the variation of $\Delta G^\ominus/_T$ with temperature is made, then the $\Delta G^\ominus$ data must be very accurate. It would be imperative therefore to use experimentally determined Γ values. However, at present these values have not been measured at different temperatures.

Comparison of $\log K_c$ vs

Ca_z Plots. In figure 4 a
comparison can be made of
the variation in $\log_{10}K_c$
vs Ca_z as given by Barrer
et al, Sherry and Walton
and Wolf and Furtig.
In the latter case $\log K_c$
was calculated from their
selectivity coefficient, S,
where $S = \dfrac{Ca_z}{\overline{Na}_z} \cdot \dfrac{Na_s}{\overline{Ca}_s}$. and a
constant value of 1.58 was
used for the activity co-
efficient ratio in this
calculation as this was
the value which was
assumed to have been used

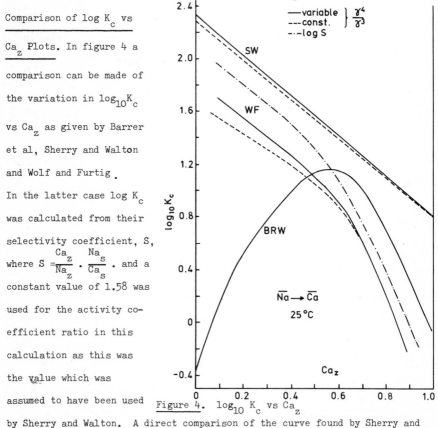

Figure 4. $\log_{10} K_c$ vs Ca_z

by Sherry and Walton. A direct comparison of the curve found by Sherry and Walton and that calculated from the Wolf and Furtig data can be made as their experimental conditions were very similar. The experimental data from which

Barrer et al's plot was derived were not obtained at constant normality of

solution phase and their log K_c plot is not strictly comparable with the

other two plots. However, when the corrections described previously are

applied to one point in this plot to bring the experimental conditions to a

state comparable to the two other studies, it is seen that this correction

has little effect on this one point and it can be assumed that the general

shape of the $\log_{10}K_c$ plot of Barrer et al would not be greatly affected if

corrections for the variations in solution phase normality were made.

All three plots show very large differences in K_c especially at low

Ca_z values. The Sherry and Walton curve shows the highest selectivity for

Ca^{2+} in the zeolite phase at all Ca_z values.

Both the Sherry and Walton and the Wolf and Furtig data have been re-

calculated employing experimental activity coefficient ratios instead of

the constant value of 1.58. In figure 4 these corrected $\log_{10}K_c$ vs Ca_z

plots are also shown and it can be seen that these corrections increase the

selectivity towards Ca^{2+} in the zeolite phase, especially at low Ca_z values.

However, these corrections introduce only a minor change to the $\log_{10}K_c$ plots.

Why does the $\log_{10}K_c$ plot of Barrer et al differ so greatly from the

curve found by Sherry and Walton? Sherry and Walton proposed that true

equilibrium was not established by Barrer et al when they determined the

exchange reaction. However, it is almost certain that equilibrium was estab-

lished at all times by Barrer et al. It seems more likely that their

analysis of Ca_s may have been in error when Ca_s was very small, i.e. at low

Ca_z values. When Ca_z = 0.08 Barrer et al obtained a K value (see eq. 16) of

0.9282 whereas Sherry and Walton found K to be 91.0. What would m_{CaCl_2} in

the equilibrium solution phase have to be in the case of Barrer et al

to change K from 0.9282 to 91.0?

Since $K = \dfrac{Ca_z}{(Na_z)^2} \cdot \dfrac{m^2_{NaCl}}{m_{CaCl_2}}$ we have

$$K = \frac{0.08}{0.92^2} \cdot \frac{(9.84 \times 10^{-3})^2}{m_{CaCl_2}} = 91$$

$$\therefore m_{CaCl_2} = 1.01 \times 10^{-7} \text{ molal.}$$

Barrer et al found experimentally that the equilibrium solution phase was 9.84×10^{-3} M for NaCl and 9.86×10^{-6} for $CaCl_2$. Sherry and Walton's data suggest that this solution phase should have been 1.01×10^{-7} M for $CaCl_2$ assuming that the much higher concentration of NaCl was analysed accurately. It would seem that at these very low $CaCl_2$ concentrations that the accuracy in the analyses carried out by Barrer et al was not good enough. In any future measurement of this and similar exchange systems great care will have to be taken to obtain the true concentrations of the electrolytes in the equilibrium solution phase. Any small contamination of this equilibrium solution phase with Ca^{2+} from the glassware employed, etc. will seriously affect the value of K derived.

The value of K_a is critically dependent on K_c. Thus the differences in $\Delta G^\ominus$ of -0.59 and -3.06 kJ per g equivalent reported by Barrer et al and Sherry and Walton, respectively, are readily explained.

It is interesting to note that Barrer et al found $\Delta H^\ominus$ to be $+8.78$ kJ per g equivalent, while Sherry and Walton reported $\Delta H^\ominus$ to be $+11.29$ kJ per g equivalent, for the $\overline{Na} \rightarrow \overline{Ca}$ exchange. As Barrer et al measured this heat directly in a calorimeter their $\Delta H^\ominus$ value should be fairly accurate. As Sherry and Walton found $\Delta H^\ominus$ from the temperature variation of $\Delta G^\ominus$ and their $\Delta H^\ominus$ is in reasonable agreement with that reported by Barrer et al it would seem that the $\Delta G^\ominus$ values found by Sherry and Walton are reasonably accurate.

Recent Ion Exchange Studies in Zeolite A. Because of the renewed interest in ion exchange in zeolite A and because of the inadequacies of the earlier studies the exchange of Na^+ ions in zeolite A by Ca^{2+} and Mg^{2+} has been redetermined. Particular attention has been placed on those factors which were neglected or which were in possible error in the previous studies.

A sample of zeolite A was used which had been specially prepared by Laporte Industries Ltd., Widnes, Cheshire. The analysis of this material is received, but after equilibration with the water vapour over a saturated solution of calcium nitrate for one week, gave the following unit cell composition:

$Na_{12.65}$ [12.00 AlO_2. 12.02 SiO_2] 26.4 H_2O (Sample 1)

This sample was washed with $\sim$ 5M NaCl solution at $80^{\circ}C$ and then with water until no Cl^- ion could be detected in the wash. The chemical analysis of this material gave the almost ideal unit cell composition of

$Na_{12.05}$[12.00 AlO_2. 12.08 SiO_2] 26.8 H_2O (Sample 2)

Using sample 2 as the starting material isotherms were determined for the following systems at the various normalities, N, and temperatures, T, specified

Na – Ca $\Big\}$ $\Big\{$ N = 0.2, 0.1, 0.05, 0.01 and 0.005 equiv./litre.
Na – Mg   T = 25 and $65^{\circ}C$.

Ca – Mg   N = 0.1 equiv./litre.   T = $65^{\circ}C$.

Na – Mg – Ca N = 0.01 equiv./litre. T = $25^{\circ}C$.

Exchange was followed for at least 5 days at $65^{\circ}C$ and 10 days at $25^{\circ}C$. The resulting isotherms are shown in figures 5 to 7.

The corrected selectivity coefficient, K_c, was calculated from eq. 24 for both the Na – Ca and the Na – Mg exchanges and the resulting log K_c vs Ca_z plots are shown in figures 8 and 9 respectively. The activity coefficient ratios, Γ, were obtained from the experimental data of Moore and Ross and are presented in figure 10. However, the activity coefficients of pure Mg Cl_2 solutions at low molalities are not available in the literature and, because of the lack of a Mg ion selective electrode, no experimental values of Γ for Na/Mg mixtures are available. Consequently, Γ values of 1.9, 1.6, 1.5, 1.25 and 1.2 were chosen for solutions of 0.2, 0.1, 0.05, 0.01 and

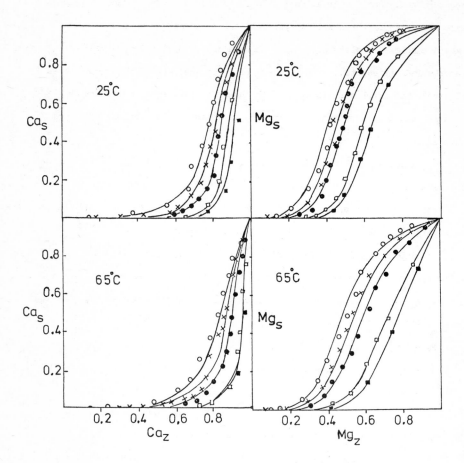

<u>Figure 5.</u> Exchange isotherms at 25 and 65°C for Na/Ca and
Na/Mg exchange.

○ 0.2N;   X 0.1N;   ● 0.05N;   □ 0.01N;   ■ 0.005N

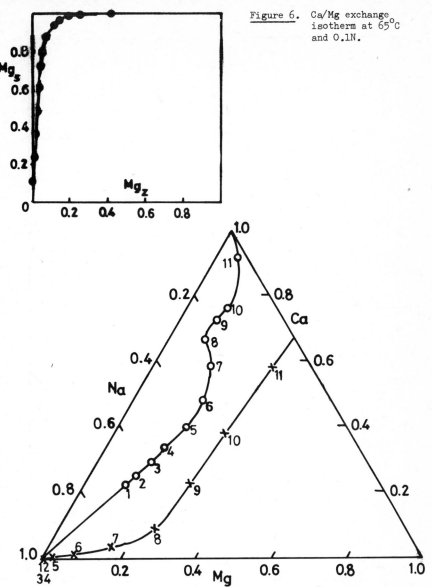

Figure 6. Ca/Mg exchange isotherm at 65°C and O.lN.

Figure 7. Na/Ca/Mg ternary exchange isotherm at 25°C and O.OlN.
Initial solution phase had constant Ca:Mg ratio of 2:1
O zeolite phase;   X solution phase.
Numbers indicate corresponding points in solution and
zeolite phases at equilibrium.

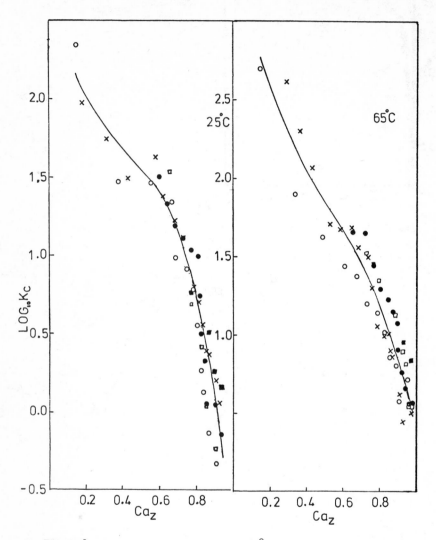

<u>Figure 8</u>. $\log_{10} K_c$ vs Ca_z at 25 and $65°C$.
Symbols as in figure 5.

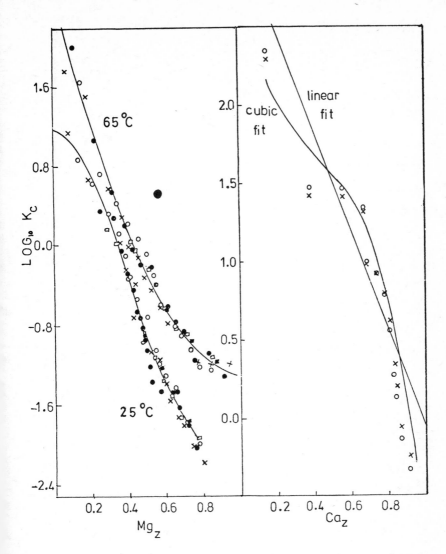

Figure 9. $\log_{10} K_c$ vs Mg_z at 25 and $65^\circ C$

Symbols as in figure 5.

Figure 11. $\log_{10} K_c$ vs Ca_z at $25^\circ C$ and $0.2N$

O   calculated from Moore and Ross data

X   calculated from Glueckauf theory

0.005N respectively. These
values are the values found
by Moore and Ross for the
Na/Ca system at $Ca_s = 0.5$
(see figure 10) and are
believed to be the most
reasonable to use for the
Na/Mg system until
values measured experi-
mentally become available.
Figure 10 also shows the
values of Γ calculated
according to Glueckauf's
theory. Once again, the
serious error in the theory
is demonstrated. This
error becomes more and more
serious as the concentration

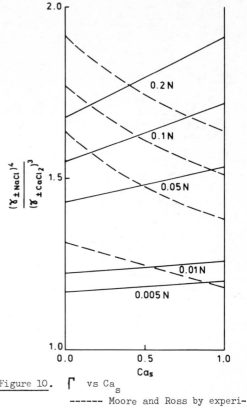

Figure 10. Γ vs Ca_s

------ Moore and Ross by experiment.
——— Glueckauf theory

of the solution phase increases. However, it is interesting to note that
these errors in Γ do not introduce very serious errors in $\log_{10} K_c$ as
figure 11 demonstrates. The differences are not large even in this most
concentrated solution case and would not lead to a great difference in K_a
and thus $\Delta G^{\ominus}$ when one considers the scatter of the data in figure 11.

    Smoothed curves were fitted through the experimental points of the
$\log_{10} K_c$ vs A_z plots in figures 8 and 9 derived from a polynomial equation of
the form

$$\log_{10} K_c = C_o + C_1 A_z + C_2 A_z^2 + C_3 A_z^3 \qquad (25)$$

The polynomials obtained from these best fits are given in Table 3 which also
lists the error of the fit, R, where R is defined as

$$R = \sqrt{\frac{\Sigma (\log K_{c_{(obs)}} - \log K_{c_{(cal)}})^2}{N - M - 1}} \qquad (26)$$

N is the number of pairs of $\log K_c$, A_z values and M is the order of the polynomial.

The smoothed values of K_c calculated from eq. 25 were introduced into eq. 24 and predicted isotherms at different normalities were obtained. These isotherm contours are shown in figure 5 as continuous lines where they may be compared with the experimentally determined points.

The experimental points are in excellent agreement with the predicted isotherms. The concentration-valency effect is clearly demonstrated to be obeyed in these systems. The excellent agreement in the Na-Mg isotherms also indicates that the Γ values chosen cannot be in serious error. One could now predict isotherms for any normality of solution phase with reasonable confidence although the agreement between the experimental and predicted isotherms is at its best at the lower concentrations.

The thermodynamic equilibrium constant, K_a, for the Na/Ca and Na/Mg systems was calculated from eq. 19. The polynomial expressions in Table 3 were used in the integration. Table 3 lists the values of K_a obtained and also the values of $\Delta G^\ominus$ derived from these K_a values according to eq. 20. The standard enthalpies of exchange were calculated from the values of K_a at 25 and 65°C and these are also given in Table 3. The value of 12.2 kJ (g equiv.)$^{-1}$ obtained for $\Delta H^\ominus$ for the Na/Ca exchange may be compared with the calorimetric value of 8.8 kJ (g equiv.)$^{-1}$ determined by Barrer, Rees and Ward [1]. This latter value is likely to be the more accurate value demonstrating that, when dealing with the small $\Delta H^\ominus$ values encountered in ion exchange studies, and even when equilibrium measurements are made with the utmost precision, $\Delta H^\ominus$ values derived from the temperature dependence of K_a do not lead to very accurate $\Delta H^\ominus$ values.

It can be seen from Table 3 that $\Delta G^\ominus$ for the Na/Mg exchange is positive i.e. the exchange shows an overall selectivity towards Na. Inspection of the

Table 3   Na$^+$ → $\frac{1}{2}$Ca^{2+} and Na$^+$ → $\frac{1}{2}$Mg^{2+} Exchange in Zeolite A

Exchange	Temp °C	Polynomial equation for log K_c vs A_z where A = Ca or Mg	R	K_a	$\Delta G^\ominus$	$\Delta H^\ominus$	$T\Delta S^\ominus$	$\Delta S^\ominus$
					kJ (g. equiv)$^{-1}$			J (g.equiv.K)$^{-1}$
Na$^+$ → $\frac{1}{2}$Ca^{2+}	25	$\log_{10}K_c = 2.83 - 6.03Ca_z + 11.5Ca_z^2 - 9.11Ca_z^3$	0.197	8.66	-2.68	12.2	14.9	50
	65	$\log_{10}K_c = 3.41 - 5.28Ca_z + 6.32Ca_z^2 - 4.01Ca_z^3$	0.167	27.8	-4.69		16.9	50
Na$^+$ → $\frac{1}{2}$Mg^{2+}	25	$\log_{10}K_c = 1.46 - 2.66Mg_z - 6.34Mg_z^2 + 5.08Mg_z^3$	0.127	0.0717	3.26	18.6	15.5	52
	65	$\log_{10}K_c = 2.61 - 7.74Mg_z + 3.57Mg_z^2$	0.110	0.427	1.20		17.4	51

isotherms in figure 5 may tend to suggest the opposite. When $Z_A \neq Z_B$
one cannot be certain of the sign of $\Delta G^{\ominus}$ from the shape of the isotherm
and because of the concentration-valency effect this uncertainty increases
with decrease in the normality of the solution phase. Secondly, inspection
of the isotherms in figure 5 shows that the inflexion point tends to
disappear with dilution of the solution phase. It is obvious, therefore,
that one should not attempt to conclude that there is only one set of
exchange sites when an isotherm shows no inflexion point especially when the
solution phase is very dilute.

From the value of $\Delta H^{\ominus}$ calculated above the equilibrium constant, K_a,
could be derived at any temperature T. However, because the log K_c vs A_Z
plots at 25 and $65^{\circ}C$ are not parallel, the activity coefficients of the
two ions in the zeolite phase must change with temperature. The relationship
therefore between log K_c and A_Z cannot be predicted from K_a.

The activity coefficients, f, of the two ions, Na/Ca and Na/Mg in the
zeolite phase can be calculated from eqs. 27 and 28.

$$\ln f_A = - Na_Z - Na_Z \ln K_c + \int_0^{Na_Z} \ln K_c \, d \, Na_Z \qquad (27)$$

$$\text{and} \quad \ln f_{Na}^2 = A_Z + A_Z \ln K_c + \int_0^{A_Z} \ln K_c \, d \, A_Z \qquad (28)$$

where A represents the divalent ion i.e. either Ca or Mg. The resulting
pairs of activity coefficients for the Na/Ca and Na/Mg exchange at 25 and
$65^{\circ}C$ are shown in figure 12.

Figure 6 shows the Ca/Mg isotherm at $65^{\circ}C$ and a solution phase normality
of 0.1 equiv/litre. Because both ions have the same valency the isotherm
should not be greatly dependent on the concentration of the solution phase.
The isotherm demonstrates the very high selectivity for Ca. A solution
containing a 2:1 ratio of Ca:Mg if placed in contact with pure Ca A zeolite
would exchange little Mg into the zeolite phase. When Mg_Z is 0.2 the solution
phase has a Ca_s value of 0.012 while at Mg_Z equal to 0.42 the Ca_s value was

0.0026 which represents only 5.3 p.p.m. of Ca in the solution phase.

The interesting Na/Mg/Ca ternary exchange isotherm at 25°C is shown in figure 7. In all measurements the starting solution contained only Ca/Mg at the constant ratio of 2:1 and had a constant total normality of 0.01 equiv./litre. The isotherm was scanned by varying the weight of Na-A zeolite placed in contact with a constant volume of solution. The isotherm shows initially, at small divalent ion loadings, that little divalent ion remains in the solution phase (see points 1-4). Therefore, a very high

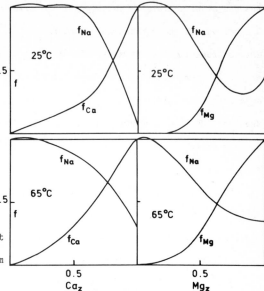

Figure 12. Activity coefficients of Na/Ca and Na/Mg in zeolite phase as a function of Ca_z and Mg_z respectively.

selectivity of Ca/Mg over Na is demonstrated in this region. The very small amount of divalent ion in the solution phase was found to be pure Mg within the limits of detection for Ca. This situation changed little until 58% of the Na had been replaced. After point 5 in the isotherm Ca begins to compete selectively not only for Na but also for the Mg now present in the exchanger. Mg_z attains a maximum value of 0.184 when 8 Na ions per unit cell have been replaced. Mg_z decreases rapidly to ~ 0.1 during exchange of the next Na ion and decreases further to 0.06 when the zeolite contains some 91 equivalent % of Ca. Although not measured it is expected that at even higher Ca loadings the zeolite would give the very high selectivities shown for Ca in the binary Ca/Mg isotherm in figure 6.

It has been found in previous studies that calcium and ferrous ions prefer to be located in the six-membered oxygen windows, S_1, sites to the eight-membered, S_2, sites of zeolite A. Initially each entering Ca ion probably exchanges with one S_1 and one S_2 sited Na ion until all four Na ions located in S_2 sites have been replaced. If Mg behaves similarly to Ca then the ternary isotherm results suggest that Ca ions do not compete with Mg ions until all four S_2 sited Na ions have been replaced.

At the maximum in the Mg loading there are 1 Mg, 3 Ca and 4 Na ions per unit cell. At complete exchange the six divalent ions have eight S_1 sites available yet Ca competes very selectively for these sites at the expense of Mg. These results, along with the difficulty of exchanging more than two Mg ions into Ca-A zeolite, which is demonstrated in figure 6, suggest that Mg ions are sited in S_2 sites in preference to S_1 sites.

Acknowledgement

The studies reported here were carried out by Mr. S. Barri, who was supplied with a grant and supplementary equipment by the Procter and Gamble Company, to whom we are gratefully indebted.

REFERENCES

(1)   R.M. Barrer, L.V.C. Rees and D.J. Ward, Proc. Roy. Soc., 1963, A273, 180.

(2)   H.S. Sherry and H.F. Walton, J. Phys. Chem., 1967, 71, 1457.

(3)   J. Kielland, J. Soc. Chem. Ind., London, 1935, 54, 232.

(4)   E. Glueckauf, Nature, 1949, 163, 414.

(5)   H.S. Harned and R.A. Robinson,"Multicomponent Electrolyte Solutions" Volume 2 Int. Encyclopedia of Physical Chemistry and Chemical Physics, Pergamon Press, Oxford, 1968, p 21.

(6)   F. Wolf and H. Furtig, Kolloid-Z.Z. Polymer, 1965, 206, 48.

(7)   E.W. Moore and J.W. Ross, J. Appl. Physiol., 1965, 20, 1332.

Sodium Aluminium Silicates in the Washing Process
Part VII: Counterion Effects

by M.J. Schwuger and H.G. Smolka
Laboratories of Henkel KGaA, Düsseldorf, Germany

1. Introduction

Since the first positive results were obtained in trying to substitute penta-sodium-triphosphate (STP) by sodium-aluminium-silicates
in detergents, many studies were carried out in the areas of physical chemistry, application, ecology and toxicology [1-12]. This
class of compounds possesses such an outstanding set of favorable
properties that it may be used on a large scale [13]. As shown in
previous papers [6, 11] water-insoluble sodium-aluminium-silicates
have, in addition to ion exchange, new properties that show positive effects in the washing process. Conventional laundry detergents and cleansers generally contain a relatively high ammount of
counterions because most ingredients are present as Na-salts. The
strong counterion effect enhances the adsorption of anionic surfactants at different interfaces as a result of the compression of
the electric double-layer. This effect is, therefore, particularly
important for the detergency [5].

A high degree of substitution of STP in laundry detergents by
SASIL* signifies that the counterion concentration of the newly
formulated products will be strongly reduced. The manifold counterion influences between the components of a laundry detergent and
the additional counterion effect of magnesium and calcium on application in hard water were the starting point for a comprehensive
investigation of the counterion effects with regard to the use of
SASIL in laundry detergents.

2. Results and Discussion

With respect to the counterion influences three areas must principally be distinguished:
a) Influence of counterions on SASIL, particularly with respect to
   ion exchange and detergency.

---

*SASIL (Sodium-Aluminium-SILicate) is a trademark of Henkel KGaA.
In the present study SASIL is a synthetic crystalline Zeolite A
with an average particle diameter of 5.6 µ.

b) Influence of SASIL on surfactants and counterion effects in surfactant/SASIL-combinations.

c) Counterion effects by multivalent ions, particularly magnesium.

2.1 Counterion Influences on SASIL

Investigations on the mechanism of action of surfactants and complexing agents in the washing process have shown that additional
water softening is advantageous apart from the specific action that
both classes of substances show at interfaces.[15]. The most substantial principles that necessitate the elimination of calcium
ions can be summarized as follows:

a) Ca-ions can form sparingly soluble salts especially with
   anionic surfactants so that the whole quantity of the surfactants
   is not available for the washing process.

b) Sparingly soluble organic and inorganic calcium salts may depo-
   sit on the fabric during the washing process and may result in
   heavy incrustations after repeated wash cycles. These incrustations may lead to fibre damage.

c).The counterion effect by dissolved Ca-ions in the solution reduces the extraction of multivalent cations from soil and textile fibre. Salts of multivalent cations are almost always found
   in soil and on textile fibres. For the most part they consist
   of inorganic salts. Frequently, however, soil components are
   connected with the fibre via cation bonds, too. These can be
   formed by carboxylic groups that always exist in cotton by oxidation and by active centers of metal oxides or by soaps originating from sebum. The saltlike bonds are disconnected by dissolution of calcium in particular. Then the multivalent cation
   is eliminated from the system by ion exchange with SASIL. As a
   consequence of the ex-
   traction of calcium ions
   defects develop in the
   "soil structure" that
   lead to a loosening of
   the texture and effect
   an easier separation
   from the substrate.
   The correlations between
   water softening and detergency are shown in
   Fig. 1. A reduction of

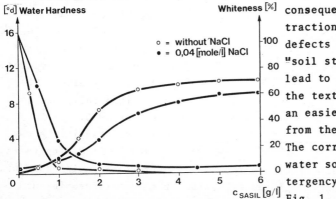

Fig. 1   Influence of NaCl on Water Softening and Detergency of SASIL

the water hardness of initially $16°d$ to approximately $3-4°d$ does not yet effect an appreciable detergency. Only a large elimination of the calcium from the system results in a significant increase in the removal of soil. Because SASIL is an ion exchanger Na-counterions compete with the Ca-ions so that, in spite of a high selectivity, the exchange against calcium is reduced by the excess of Na-ions. The influence of Na-ions on the equilibrium exchange capacity for calcium is shown by Eq. 1 [8, 16].

$$Q_{Ca^{++}} = \frac{Q_m \; c_{Ca^{++}}}{c_{Ca^{++}} + 2 \dfrac{b_2}{b_1} (c'_{Na^+} + 2 Q_{Ca^{++}})} \tag{1}$$

Q_{Ca}^{2+} -   Charge of the ion exchanger with calcium

Q_m -   Maximum ion exchange capacity for calcium

c_{Ca}^{2+} -   Ca-concentration in the solution

c'_{Na}^+ -   Na-concentration in the solution before ion exchange

b_1, b_2 -   Ion specific constants

According to Eq. 1 the amount of calcium remaining in solution is larger in the presence of large Na-counterion concentrations than in counterion-free systems. The example shows that the somewhat smaller water softening also results in a certain reduction of soil detaching. Fig. 2 shows the example of the maximum ion exchange capacity for calcium Q_m and the limits between which a reduction of the exchange capacity can vary in practice. According to the mass action law, however, the counterion influence also takes effect on water-soluble complexing agents. In Fig. 3a and 3b a comparison of STP, Na-citrate and SASIL is given. The counterion influence in the case of STP is less distinct than in the case of SASIL. According to the different binding mechanisms the temperature influence is yet opposite.

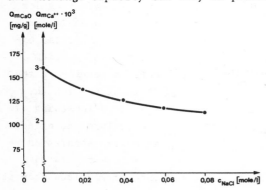

Fig. 2 Counterion Effect on the Maximal Amount of Exchangeable Calcium

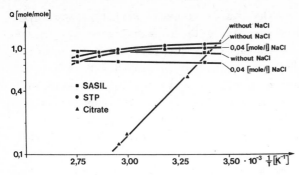

Fig. 3a) Effect of Temperature on Calcium Binding Ability

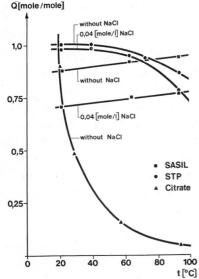

Fig. 3b) Effect of Temperature on Calcium Binding Ability

The base of the calcium binding in detergency is on the one hand the sequestration of Ca-ions by STP, and on the other hand the ion exchange by SASIL. With STP and citrate it is a homogeneous equilibrium, whereas with SASIL a heterogeneous one exists. The description of the formation constants may occur according to the general formulation for chemical equilibria.

$$\ln k = -\frac{\Delta G}{RT} = -\frac{\Delta H}{RT} + \frac{\Delta S}{R} \qquad (2)$$

In first approximation, with the help of Eq. 2 a study of the temperature dependence of the calcium bond may be carried out. As expected, the equilibria for STP and citrate show a negative temperature coefficient. The citrate equilibrium proceeds according to Eq. 2. On the contrary, the temperature dependence for the STP equilibrium is more complex. At low temperatures it shows only a small temperature coefficient that further diminishes with increasing temperature. This indicates that the description by Eq. 2 is no more sufficient and the enthalpy of complex formation is no more temperature independent in the temperature range considered.

There exists another principal course for the equilibrium calcium binding of SASIL. A slightly positive temperature coefficient can be observed. The thermodynamic explanation of this result by Eq. 2 can only be conducted by the hypothesis of a positive reaction ent-

halpy ΔH. Caloric data for the ion exchange by zeolite X showed a
positive value for ΔH with respect to the Ca-exchange [16]. Therefore, it seems justifiable to suppose a positive reaction enthalpy
for SASIL as well. This positive enthalpy is probably caused by
solvation and resolvation in connection with the incorporation of
Ca-ions and the release of Na-ions. Thus it may be supposed that
the hydrate shell of the Ca-ion will be partly stripped off on
entering into the zeolite and that the incorporated Ca-ion stays
yet partly hydrated. Since the Na-ions can leave SASIL only in a
hydrated form, "free" water molecules must still be in the crystal
for the build-up of a hydrate shell. Such processes certainly
effect a net positive formation enthalpy that additionally will be
temperature dependent itself like the underlying solvation reactions.

2.2 Counterion Effects in Mixtures of Surfactants and SASIL

The most important laundry detergent ingredients for which counterion effects may be strongly pronounced are anionic surfactants.
Their mechanism of action is directly connected with the adsorption
ability at all interfaces involved. Since in alkaline aqueous
environment all solid and liquid interfaces are negatively charged,
a mutual repulsion occurs between the interfaces and the hydrophilic groups of anionic surfactants. So far a potential barrier is
to be overcome, in that the hydrophobic interactions between the
hydrophobic molecule part and the interfaces will prevail in comparison to the repulsion and attraction forces by the solvent.
Addition of electrolyte effects the compression of the electric
double-layer and thus the reduction of the mutual repulsion of the
equally charged partners. In virtue of the mass action law the
solubility and the dissociation of the anionic surfactants are also
reduced. The consequence of this electrolyte effect is the enhanced
adsorption of the anionic surfactants and the reduction of the
equilibrium concentration in order to attain the complete occupation of the interface.

Thus, in connection with the compression of the electric double-
layer, a series of characteristic properties of the anionic surfactants are changed. As an example for this a plot of the surface
tension of sodium-n-dodecyl sulfate (SDS) vs. concentration with
or without NaCl or $CaCl_2$ is shown in Fig. 4. At a given
surfactant concentration the NaCl containing solution is more surface active. In the same positive sense for the laundering process

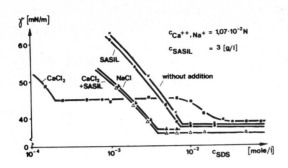

Fig. 4 Counterion Effects on Surface Tension of SDS

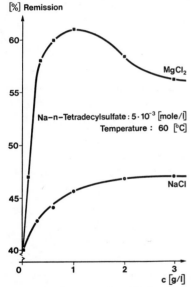

Fig. 5 Counterion Effects on Detergency (Polyester/Cotton)

a series of interfacial and colloidal-chemical characteristics of anionic surfactants are changed. Fig. 5 shows the soil removal by a comparably less suitable surfactant with increasing electrolyte concentration. It is recognized that small additions of electrolyte effect a strong increase in the detergency. With regard to the counterion type Fig. 5 shows that the influences increase with increasing valence according to the Schulze-Hardy Rule, so that $MgCl_2$ has a stronger positive effect than NaCl.

Ca-ions principally have the same properties as Mg-ions. In Fig. 4 this is shown regarding the surface tension/concentration-curve of SDS. The surfactant concentration that is necessary to attain a definite surface tension is significantly lower in the presence of Ca-ions than in the presence of Na-ions. In comparison with the NaCl curve, however, the $CaCl_2$ curve exhibits a plateau at significantly higher surface tensions. Whereas the inflexion point in the surface tension/concentration-curve with and without NaCl addition shows the formation of surfactant micelles, the first inflexion point after addition of $CaCl_2$ indicates the formation of a new solid phase. This may be examined by solubilization of sparingly soluble dyestuffs. In Fig. 6 the solubilization of Sudanred G vs. the surfactant concentration is presented under equal test conditions as in Fig. 4. By addition of NaCl the solubili-

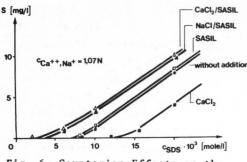

Fig. 6 Counterion Effects on the
Solubilization of Sudanred G by SDS

zation starts at lower concentrations which corresponds to
the discussed behavior of the
$\gamma/\log$ c-curve. In contrast,
the addition of $CaCl_2$ effects
a significant reduction of the
solubilization. Due to the
poor solubility of the Ca-
salts of SDS no micelles are
formed at the first inflexion
point of the $\gamma/\log$ c-curve.
In the solution there are

only those SDS ions present that are in equilibrium with the solid
precipitated Ca-surfactant. This is one of the reasons why calcium
has such a disadvantageous effect in the laundering process in
spite of the strong reduction of the surface tension and, by analogy, the increase of the adsorption.

Considering the system SDS/SASIL, one observes that both partners
do not enter into significant interactions. Due to the high negative
charge of the SASIL surface no surfactant adsorption occurs [8].
For the same reason SASIL will not influence the surface tension.
The Na^+-counterions are located in the pores of the crystal lattice so that they cannot exert an influence on the adsorption and
the formation of micelles of SDS. In the $CaCl_2$/SASIL system the
surface `tension/concentration-curve of SDS is approximately the
same as with NaCl. The Ca-ions are eliminated from the aqueous environment by ion exchange and the Na-ions that are advantageous for
the system are set free. Thus the surface tension and the critical
micelle concentration of SDS are shifted to lower values. The entire conditions are reversed and the solubilization, too, begins
at the corresponding lower surfactant concentration (Fig. 6).

Emulsification with paraffine oil and adsorption on graphite show
that these discussed effects are not limited to the interface water/
air and the formation of micelles. As a measure for the interfacial
tension and the emulsification the number (n) of paraffine oil
droplets formed from 1 ml flowing out of a capillary into
different solutions under comparable conditions is presented in
Fig. 7. The dynamic interfacial tension decreases with increasing
number of droplets formed. At the same time the tendency to
emulsification increases. SASIL enhances the tendency for emulsi-

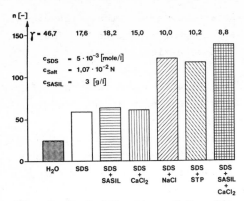

Fig. 7 Emulsification of Paraffin Oil

fication very strongly in hard water, whereas the emulsifying ability of SDS is hardly influenced with SASIL only. It is worth remarking that the increase in the system SDS/ SASIL/CaCl$_2$ is somewhat higher than in the system SDS/NaCl. This may probably be explained by the fact that there is a small residue of Ca-ions which is still in equilibrium with SASIL still having a positive effect in connection with the effects described for the compression of the electric double-layer. It causes the existing small difference due to the stronger effect with bivalent ions. STP shows only a counterion effect that is equivalent to the one of NaCl. Fig. 7 depicts also the static interfacial tensions measured in mN m^{-1} after 30 min. In spite of somewhat more advantageous static interfacial tensions in the system SDS/CaCl$_2$ in comparison to SDS the dynamic behavior is equal since here the kinetics of the dissolution of Ca-n-dodecyl sulfate during the formation of new interfaces is included in the consideration.

The conditions at the solid/liquid interface are completely analogous. Fig. 8 shows that the adsorption of SDS at graphon, a graphitized carbon black, is not influenced by SASIL. In calcium containing water, however, a significant increase can be observed for the same reason as at other interfaces.

2.3 Magnesium Ions in the Laundering Process

The positive counterion effect of Na-ions is masked in the presence of Ca-ions due to the strong decrease of the solubility of the surfactant. Additionally, the electric double-layer is compressed with Ca-ions to such a degree that the electrostatic repulsion between the particulate soil and the fibre is impaired. Mg-ions, however, do decrease the solubility of the usual surfactants but essentially to a lesser degree than Ca-ions. Likewise, the compression of the electric double-layer with Mg-ions is stronger than with Na-ions according to the Schulze-Hardy Rule. The disadvantageous properties of Ca-ions, though, do not exist here.

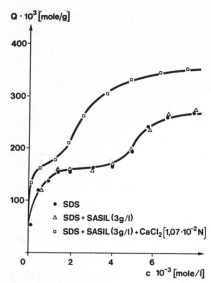

Fig. 8   Adsorption Isotherm
of SDS on Graphon at 25 °C

Therefore, Mg-ions take a particular mid-position among the ions
often occurring in nature.

With respect to the influence of
Na-ions on the interfacial and
surface tension, it was shown that
the origin of the ions is not de-
cisive for the action. For example,
the counterions from NaCl or Na_2SO_4
are equal to those from STP [17].
Fig. 9 depicts, however, that bivalent Mg-ions are able to decrease
the surface tension of SDS significantly more than those with monovalent cations. In the case of a
pure anionic surfactant/STP-system
no specific complexing agent properties exist other than the un-
specific electrolyte
effect [18]. Significant
influences, however, can
additionally be observed
when salts of bivalent
ions are present. With
increasing concentrations
of STP at a constant con-
centration of SDS and
$MgCl_2$ (Fig. 10) the stronger reduction of surface
tension by Mg-ions is
substituted by the weaker
action of the Na-ion, as
Mg-ions are complexed to

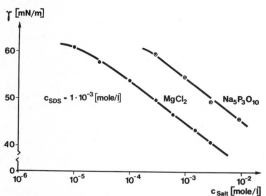

Fig. 9   Influence of Na- and Mg-Salts
on the Surface Tension of SDS

an increasing degree. As soon as the STP concentration is equivalent to the present Mg-ion concentration a maximum in the γ/log c-
curve is observed. Further increase of the Na-ion concentration
manifests itself by another decrease of the surface tension. The
descending curve above the maximum coincides with the STP curve of
Fig. 9 within the tolerance. Thus the complexed Mg-ions exert
no more influence on the surface activity of SDS. This means that
STP undoes the positive action of Mg-ions with respect to the sur-

type A zeolites show a higher selectivity to calcium than to magnesium [20]. Moreover, the Mg-ion exchange kinetics is especially
slower at lower temperatures [6], because the penetration into
the pores is impeded due to their larger hydrate shell. Therefore
SASIL eliminates calcium practically completely from the system,
whereas under practical conditions magnesium would still remain in
the solution to a certain degree. Thus, when using SASIL in laundry detergents, there exists the possibility to take advantage of
the positive counterion effects of magnesium. These effects do not
manifest themselves when phosphate is used due to its high complex
formation constant for magnesium, too.

This effect is illustrated in Fig. 12 by the emulsification of
paraffine oil by SDS. It can be shown that the advantageous properties of $MgCl_2$ are not influenced by SASIL. Using the droplet counting method as a measure for the dynamic interfacial tension and the emulsification tendency of an oil, it is recognized that the largest number of droplets is formed in the SDS/SASIL/$MgCl_2$ system and is signi-

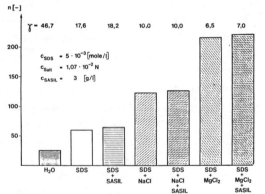

Fig. 12 Emulsification of Paraffin Oil

ficantly higher than in NaCl containing systems. The static interfacial tensions γ indicated in Fig. 12 correspond with the number
of droplets. Equivalent STP concentrations form chelate-complexes
with magnesium and reduce the counterion effect to that of the Na-
ions.

The results indicate that there is probably a significant improvement of the partial processes of the laundering operation that are
connected with the wetting and the roll-up of oily soil as well as
with their elimination via emulsification in the system Mg-ions/
anionic surfactant/SASIL. Model detergency experiments with the
technically widely applied linear alkyl benzene sulfonate (LAS)
and SASIL confirm that the presence of Mg-ions virtually has a
positive effect under the substantially more complex conditions of
the laundering process. The effect of Mg-counterions on LAS is

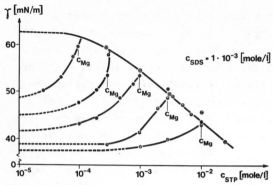

Fig. 10   Influence of STP on the Surface
Tension of SDS at various MgCl$_2$-concns,

face activity. There-
fore the specific stron-
ger counterion action
of magnesium does not
manifest itself in
usual conventional
laundering systems due
to the unselective com-
plexing of Ca- and Mg-
ions by STP.

The particular mid-posi-
tion of magnesium

between calcium and sodium may be illustrated particularly well
by solubilization. Calcium-ions significantly diminish the solubilization ability of anionic surfactants. This deterioration can
be nullified by addition of ion exchangers or complexing agents.
In contrast, the solubilization by SDS in the absence of SASIL
significantly increases with increasing Mg-ion concentration. The
solubilization in Fig. 11 passes through a maximum. This indicates

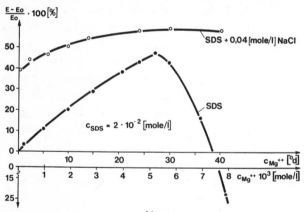

that the formation
of sparingly soluble
Mg-salts, unlike
calcium, occurs at
essentially higher
concentrations. It
also indicates that
small amounts of
magnesium have a
pronounced posi-
tive effect on the
colloid-chemical
properties. With
addition of NaCl
the influence of

Fig. 11   Effect of Mg^{++}-Ions on the Relative
Increase of Solubilization of Orange OT by SDS

Mg-ions is no more that large since the effect is partly masked by
Na-ions, but still observable. In Fig. 5 it was already depicted
that the stronger Mg-counterion effect and, as a consequence there-
of, the more advantageous interfacial and colloid-chemical properties of surfactants also effect the complex laundering process.
Proceeding on the assumption that STP is to be substituted by
SASIL in laundry detergents, it has to be taken into account that

analogous to that on Na-n-tetradecylsulfate. It is remarkable that the positive influence of Mg-ions also remains after addition of Na-ion concentrations relevant to practical conditions. For the actual test conditions the maximum of the action of Mg-ions is in the range of 25 ^{o}d of Mg-hardness. Additions of NaCl weaken the effect. This effect may even disappear at especially high NaCl concentrations. This was observed in the system LAS/SASIL/magnesium after addition of 0.08 mole/1 NaCl. In laundry detergents the counterion concentration varies and is dependent on the special formulation. The value of 0.04 mole/1 Na-counterions supposed in Fig. 13 probably represents an average possible value for a relatively high substitution degree of STP. The location of the maximum is like all complex processes very strongly dependent on the special constitution of the surfactant, on the textile fibre, on the type of soil and on the composition of the laundry detergent. With poorly soluble surfactants it will be shifted to lower Mg-concentrations. In general it can be concluded that there is a positive effect on the net result if a part of the Mg-ions, originating from natural sources, in extremely rare cases up to approximately 5-6 ^{o}d, remains in the washing system. Therefore in the literature washing systems with high amounts of Mg-ions can be found [21-23].

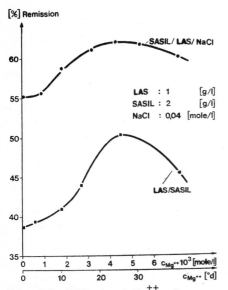

Fig. 13   Effect of Mg^{++} -Ions on Detergency on Polyester/Cotton at 60°C

3. Conclusion

SASIL and surfactants do not enter into substantial interactions in de-ionized aqueous systems. In calcium-containing systems the negative effect of the Ca-ions is inhibited by ion exchange and a positive indirect counterion effect is exerted on the surfactant by the Na-ions set free from SASIL. These positive indirect counterion effects can be observed at all interfaces and in the colloidal state in solution. Mg-ions are more effective than Na-ions due to stronger counterion effects. Since a part of Mg-ions from the

water hardness is not exchanged by SASIL there will result a
positive effect on laundering and cleaning by Mg-ions in SASIL-
containing detergents.

References

[1] DOS 2412837, 1974, Henkel KGaA

[2] DOS 2412838, 1975, Henkel KGaA

[3] DOS 2412836, 1974, Henkel KGaA

[4] DOS 2531342, 1976, Henkel KGaA

[5] P. Berth, G. Jakobi, C. Schmadel, M.J. Schwuger and C.H. Krauch,
Angew. Chem., 1975, 87, 115

[6] M.J. Schwuger and H.G. Smolka, Colloid Polymer Sci., 1976, 254,
1062

[7] H.G. Smolka and M.J. Schwuger, Colloid Polymer Sci., 1978, 256,
270

[8] M.J. Schwuger and H.G. Smolka, Colloid Polymer Sci., 1978, 256,
1014

[9] M.J. Schwuger, H.G. Smolka and C.P. Kurzendörfer, Tenside,
Detergents, 1976, 13, 305

[10] C.P. Kurzendörfer, M.J. Schwuger and H.G. Smolka, Tenside,
Detergents (in print)

[11] H.G. Smolka and M.J. Schwuger, Tenside, 1977, 14, 222

[12] P. Berth, J. Amer. Chem. Soc., 1978, 55, 52

[13]Europa Chemie, 1978, 24, 393

[14]M.J. Schwuger, Ber. Bunsengesell. für Physik. Chemie,(in print)

[15]G. Jakobi and M.J. Schwuger, Chemiker-Z., 1975, 99, 182

[16]F. Wolf, D. Ceacareanu and K. Pilchowski, Z. Phys. Chemie, Leipzig, 1973, 252, 50

[17]H. Lange, in Tenside - Textilhilfsmittel - Waschrohstoffe, edited by K. Lindner, Wissenschaftliche Verlagsgesellschaft mbH, Stuttgart, 1971, p. 2241-2275

[18]H. Lange, Kolloid-Z., Z. Polymere, 1951, 121, 66

[19]H. Lange, unpublished results

[20]F. Wolf and H. Fürtig, Kolloid-Z., Z. Polymere, 1965, 206, 48

[21]US-PS 2766212, 1956, General Aniline & Film Corporation, New York

[22]US-PS 3202613, 1965, Colgate-Palmolive Co., New York

[23]DOS 2628759, 1976, Procter & Gamble Co., Cincinnati, Ohio (V.St.A.)

Natural Chabazite for Iron and Manganese Removal from Water

R.Aiello[o], C.Colella[oo], and A.Nastro[o]

[o]Dipartimento di Chimica, Università della Calabria, Arcavacata di Rende (CS), Italy

[oo]Istituto di Chimica Applicata, Facoltà di Ingegneria, Università di Napoli, Italy

Abstract

The results of an experimental investigation into the possibility of using chabazite tuff, suitably treated, in the process of iron and manganese removal from water are presented.
The remarkable aptitude of the natural material to act as manganese zeolite is emphasized, also on the ground of previous results.
Through the study of the breakthrough curves of model waters, containing each time iron, manganese or both cations, obtained either with manganese zeolite or with tuff beds, and through the evaluation of other chemical changes occurring in the system during the process (pH variation, precipitation, cation concentration in the eluate) a reaction mechanism is proposed, characterized by the concomitant action of ion exchange and oxidation for iron removal and ion exchange and absorption for manganese removal.

Introduction

The use of manganese zeolite (i.e. a natural cation exchanger bearing a manganese oxide layer) in the practice of industrial water conditioning has been reported for a long time[1]. Its aptness to remove iron and manganese from water is very marked, allowing reduction of the concentration of the above species in the influent below the tolerance limits for many industrial applications (often less than 0.1 ppm for each cation[2]).

The natural ion exchanger usually employed is greensand, namely glauconite, which is processed beforehand by treatment with manganous salt and potassium permanganate solutions. For a few years a clinoptilolite tuff has been successfully employed in Japan in place of greensand[3].

In two recent papers[4,5] it has been reported that the Campanian tuff[6], eminently chabazitic, but commonly also containing little amounts of phillipsite, can act as manganese zeolite, showing, also because of its larger ion exchange capacity, an aptitude to iron abatement greater than that of greensand and clinoptilolite tuff, which is moreover maintained even after several regenerations.

The aim of this work is first of all to test the above material for manganese removal from water, comparing the results with those obtained with iron, secondly to extend the experiments to model waters containing both iron and manganese and lastly to investigate the mechanism of the process.

Experimental

The sample of chabazitic Campanian tuff, coming from Mercogliano (Avellino), was intentionally chosen free from X-ray detectable phillipsite, in order to have a model that, even if practically not too different from the common occurrence, could more simply be be investigated. The rock was ground and the grain size fraction between 35 and 50 mesh employed in all the experiments.

The chemical analysis (X-ray fluorescence) of the grains appears in Table 1. Their chabazite content, evaluated on the basis of the water vapour amount adsorbed by the Na-exchanged form, was near 50%.

Table 1

Chemical analysis of chabazite tuff grains

SiO_2	54.53
Al_2O_3	17.32
TiO_2	0.53
Fe_2O_3	4.31
MnO	0.14
P_2O_5	0.07
CaO	4.87
MgO	1.19
Na_2O	0.57
K_2O	5.45
H_2O	10.81
	99.79

The preparation of manganese zeolite was effected as follows. 5-gram beds (depth of filtration layer: about 25 mm) of chabazite tuff, in the original cationic form, or sometimes previously K-exchanged, were prepared, and the grains were accommodated in a glass column (diameter: 20 mm), fitted with a porous sintered glass disk.

The manganese zeolite was obtained by passing through the bed, washed several times with distilled water, 0.7 liters of a 1M solution of $MnSO_4$ and successively, after suitable washing, 0.7 liters of a 5 g/l solution of $KMnO_4$. The flow rate in both cases was 0.4 m/h . The colour of the tuff grains after preparation was dark brown.

Plate 1 shows two scanning electron micrographs (Leitz AMR 1200) referring to the tuff before (A) and after the above treatments (B). By comparing both images it can be noticed in the second one the manganese oxide precipitated on the chabazite particles, according to the reaction:

$$(7-2x)Mn^{++} + (2x-2)MnO_4^- + (8-3x)H_2O \longrightarrow 5MnO_x + (16-6x)H^+$$

where x can vary in a wide range with a maximum value of 1.95^7.

A little oxide specimen collected from the treated grains, tested by X-ray, turned out to be completely amorphous.

The amounts of manganese (evaluated, after a suitable acid attack of the solid, through atomic absorption spectrophotometry) taken by the original tuff sample after ion exchange ($MnSO_4$) and after oxidation ($KMnO_4$) were 8.0 and 11.8 mg/g tuff, respectively, while that released back-exchanging manganese zeolite with KCl was 1.1 mg/g tuff. According to the previous reaction and from the above data, the average oxidation number of manganese (2x) in MnO_x can be calculated from the equation:

$$\frac{7 - 2x}{5} = \frac{8.0 - 1.1}{11.8 - 1.1}$$

which gives 2x = 3.78. Accordingly, the oxidizing capacity of manganese zeolite for Fe^{++}, supposing the following redox reaction:

$$MnO_x + (2x-2)Fe^{++} + (5x-6)H_2O \longrightarrow (2x-2)Fe(OH)_3 + Mn^{++} + (4x-6)H^+$$

will be 13.8 mg Fe/g tuff.

Runs of iron and/or manganese removal from water were effected by percolating through the bed solutions of Fe^{++} and/or Mn^{++}, made from reagent grade sulphates,of total concentration ranging between 1 and 5 ppm with a flow rate from 0.8 to 1.6 m/h . The contact time was accordingly from about 110 to 55 sec. A liquid column of about 30 mm was maintained over the bed and care was taken to avoid the contact of the influent solution with atmospheric oxygen.

Fe^{++} (colorimetric), Mn^{++} and K^+, in the runs with K-exchanged tuff samples (spectrophotometric), and pH were measured in the

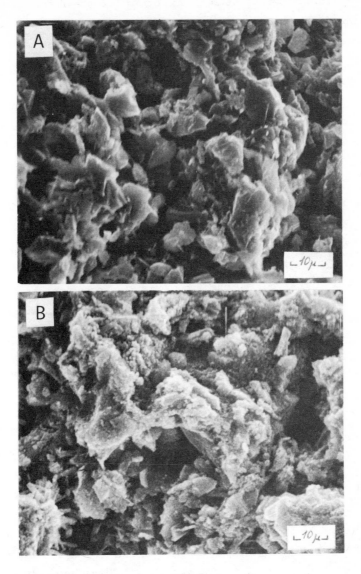

Plate 1 - Scanning electron micrographs
of the original chabazite tuff (A) and of
same material after treatment with manganese sulphate and potassium permanganate
solutions (B).

successive 1-liter fractions of eluate collected.

The breakthrough points at Fe^{++} and/or Mn^{++} concentrations equal to 0.1 ppm were chosen as a measure of the aptness of the ion exchanger to act as manganese zeolite in the iron and/or manganese removal from water.

Ion exchange experiments were arranged in flow with beds and elution conditions analogous to those above described. Back-exchange runs were sometimes performed on finely ground specimens by passing through them a 1M solution of KCl with a flow rate of 0.4 m/h . .

Results and discussion

Figures 1 and 2 show the influence of the initial concentration of the influent and of the flow rate of the percolating solution on the abatement of iron (upper part of the figures) or manganese (lower part) in the water.

To be noticed that during the iron removal process no detectable manganese comes out from the bed before the breakthrough point of iron.

The experimental conditions are: for Fig.1, cation concentration in the influent equal to 5 (a), 2.5 (b) and 1 ppm (c), flow rate constant at 1.2 m/h ; for Fig.2, cation concentration in the influent constant at 2.5 ppm , flow rate equal to 1.6 (a), 1.2 (b) and 0.8 m/h (c).

Similarly to a general ion exchange process in flow, the values of the breakthrough point are a reciprocal function of the influent concentration (Fig.1) and of the flow rate (Fig.2).

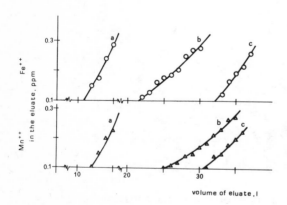

Fig.1 - Breakthrough curves relating to the iron (o) or manganese (Δ) removal by manganese chabazite tuff percolated by 5 (a), 2.5 (b), 1 ppm (c) Fe^{++} or Mn^{++} solutions with a flow rate of 1.2 m/h .

From the data of Fig.1, however, the process appears more effective with higher Fe^{++} or Mn^{++} concentration in the original solution. It must be noticed, moreover, from the breakthrough point of Fe^{++} in Fig.2, that at low flow rate the actual iron abatement from manganese zeolite remarkably exceeds its oxidizing capacity (see experimental) which means that some iron is removed according to another mechanism, plausibly by ion exchange. The oxidizing action of manganese oxide is, on the other hand, proved both by the grain colour getting lighter or by the evidence of the formation of some hydrated ferric oxide, identified,

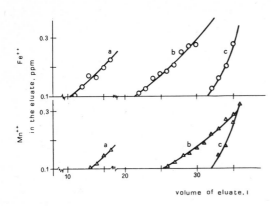

Fig.2 - Breakthrough curves relating to iron (o) or manganese (Δ) removal by manganese chabazite tuff percolated by 2.5 ppm Fe^{++} or Mn^{++} solutions with flow rates of 1.6 (a), 1.2 (b), 0.8 m/h (c).

after dissolution, by means of standard qualitative analytical procedures for Fe^{+++}.

The comparison between the breakthrough curves of iron and of manganese, while varying the elution conditions, points out some discrepancies in the breakthrough point of the cations. This can be justified by considering that manganese, either directly present in the influent or coming from the reduction of MnO_x (iron removal) is kept on the bed, besides by ion exchange, in a rather different way from iron, namely thanks to the absorbing properties of manganese oxide surface towards multivalent cations, especially Mn^{++}[7]. Two more points must be emphasized. First of all, during the process pH remains practically constant in the various fractions collected and nearly unchanged from the initial value (5.5, on average, in the effluent compared with about 6.0 in the influent). This seemingly contrasts with the involved acid-generant reactions: the redox reported in the experimental and the above mentioned absorption process[7], but can be justified by considering that zeolite can partly neutralize H^+, both by ion exchange and by hydrolysis. Secondly, the exhausted manganese chabazite tuff bed can be successfully regenerated by potassium permanganate in the same experimental conditions as in the preparation. For example, three successive regenerations of the bed, passed through by a 2.5 ppm Fe^{++} solution with a flow rate of 1.2 m/h (upper part of Fig.1, curve b), allowed to collect 20, 21 and 18 liters of effluent with a Fe^{++} content lower than 0.1 ppm[4].

The reaction mechanism, which the presented data have already helped to explain, has been further investigated, focusing the attention upon iron removal only, which, as said before, involves also some manganese abatement. The study of this mechanism requires in fact the analysis of two stages: the first one is the interaction between Fe^{++} and the system MnO_x-ion exchanger; the second is that between Mn^{++}, coming from the reduction of MnO_x, and the same system. In order to make the system easier to be studied, it was thought advisable to employ tuff samples previously K-exchanged, even if this inevitably affects the extent of iron removal[5].

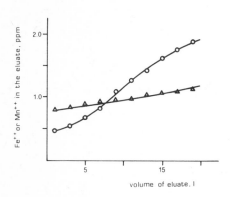

Fig.3 - Elution curves of Fe^{++} (o) and Mn^{++} (Δ) through chabazite tuff. Cation concentration in the influent 2.5 ppm; flow rate of the solution 1.2 m/h.

Figure 3 shows the elution curves of Fe^{++} and Mn^{++}, obtained by percolating through a 5-gram bed of K-exchanged tuff 2.5 ppm solutions with a flow rate of 1.2 m/h. It can be seen that chabazite tuff shows a certain selectivity for the cations, especially for Mn^{++}; nevertheless the cation concentration in the effluent is higher than 0.5 ppm already from the early fractions collected, which means that ion exchange, although playing an important role in the global process, does not succeed, even in absence of competing cations, to reduce the iron and manganese content in the water below the required limits.

In Fig.4 the whole process of iron removal by manganese chabazite tuff (previously K-exchanged) is summarized. The experimental conditions are again: Fe^{++} concentration in the influent 2.5 ppm , flow rate of the solution 1.2 m/h . The dotted line refers to the influent concentration; the other three curves give the concentration trend of K^+, Fe^{++} and Mn^{++} in the successive 1-liter fractions of the eluate.

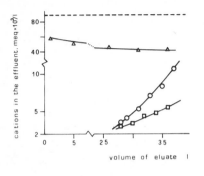

volume of eluate l

Fig.4 - Concentration trend of effluent K^+ (Δ), Fe^{++} (o) and Mn^{++} ($\square$), relating to iron removal by manganese chabazite tuff, beforehand K-exchanged, percolated by a 2.5 ppm Fe^{++} solution with a flow rate of 1.2 m/h. The dotted line refers to the initial Fe^{++} concentration in the influent.

From the data of this figure it can be calculated that the equivalent ratio between effluent K^+ and influent Fe^{++} before breakthrough points of iron and manganese ranges between 0.65 and 0.50, compared with 1 in the hypothesis of complete exchange of iron and with 0.56 in the hypothesis of complete oxidation of Fe^{++}, according with the reaction reported in the experimental, followed by the exchange reaction: $Mn^{++} \rightarrow 2K^+$. Values above 0.56 indicate a concomitance of ion exchange and oxidation, values below 0.56 would mean that some Fe^{++}, or more probably, some Mn^{++} is absorbed on the manganese oxide. The actual extent of the above ratio therefore allows one to deduce that ion exchange for Fe^{++} occurs even when oxidation is possible as antagonist and that Mn^{++} produced by reduction (and probably also that to be removed directly from water in the manganese removal process) is kept on the bed partly by ion exchange and partly by absorption.

The last two figures show the action of manganese chabazite tuff on the concomitant removal of Fe^{++} and Mn^{++} from the influent solution. Figure 5 allows to deduce the influence of the flow rate of the solution (a: 1.6, b: 1.2 and c: 0.8 m/h) on the elution of iron and manganese at constant total concentration (1.25 ppm for each cation).It has to be noticed that the manganese to be actually re-

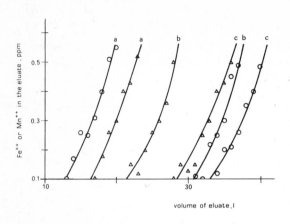

moved includes also that produced by reduction of MnO_x; therefore its total amount is normally larger than that of iron. The breakthrough points of the couples of curves b and c reported in Fig.5 are consistent with the above remark, while those of the curves a are in contrast with it. This discrepancy can be explained supposing that, at higher flow rate

Fig.5 - Breakthrough curves, relating to simultaneous iron (o) and manganese (Δ) removal by manganese chabazite tuff percolated by 1.25 ppm Fe^{++} and 1.25 ppm Mn^{++} solutions with flow rates of 1.6 (a), 1.2 (b), 0.8 m/h (c).

(curves a), ion exchange prevails over both oxidation of iron and absorption of manganese, so that the greater selectivity of chabazite for Mn^{++} than for Fe^{++} allows the latter to precede the former in the eluate. Decreasing the flow rate (curves b and c), first oxidation (notice the remarkable shift of the breakthrough point of iron in comparison with manganese passing from curves a to b) and then absorption take place in larger extent, which justifies the inversion in the breakthrough order of the cations. Figure 6 refers to the influence of the different concentration of iron and manganese

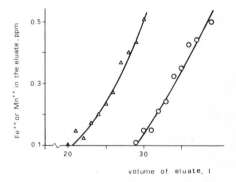

Fig.6 - Breakthrough curves, relating to iron (o) and manganese (Δ) removal by manganese chabazite tuff percolated by 2 ppm Fe^{++} and 0.5 ppm Mn^{++} solutions with a flow rate of 1.2 m/h.

in the influent (Fe^{++} 2 ppm , Mn^{++} 0.5 ppm) on the elution of the cations at constant flow rate (1.2 m/h). The values of the break-through points can be once again justified supposing in these ex-perimental conditions the predominance of oxidation on absorption. Following iron oxidation, more manganese ion is produced and therefore less manganese dioxide, disfavouring consequently the absorption of Mn^{++} on its surface.

Conclusions

The results above discussed confirm the remarkable aptitude of cha-bazite tuff to act, after suitable treatment, as manganese zeolite in the process of iron and manganese removal from model water. The complex mechanism of the process,although not clarified in a quantitative way, is characterized by the concomitant action of ion exchange and oxidation for iron removal and ion exchange and absorption for manganese removal. The experimental data also sug-gest that, with increasing flow rate, ion exchange prevails on oxi-dation and absorption.

The work is now continuing with the aim to test the material with waters containing other competing cations.

Acknowledgment

Thanks are due to M.Bruno for his help in carrying out some of the experimental work.

References

[1] E.Nordell, "Water treatment for industrial and other uses", van Nostrand Reinhold Company, New York, 1961, Chapter 14, p.399.

[2] S.B.Applebaum, "Demineralization by ion exchange", Academic Press, New York, 1968, Chapter 2, p.16

[3] K.Torii, "Natural Zeolites. Occurrence, properties, use", Pergamon Press, Oxford, 1978, p.441.

[4] R.Aiello, A.Nastro and C.Colella, Effl. water treat. jour., 1978, 18, 611.

[5] C.Colella, A.Nastro and R.Aiello, in press.

[6] R.Sersale, "Natural zeolites. Occurrence, properties, use", Pergamon Press, Oxford, 1978, p.285.

[7] J.J.Morgan, W.Stumm, J. Coll. Sci., 1964, 19, 347.

Influence of Outgoing Cation in Ion Exchange. Transition Metal Ion Exchange in K-Y Zeolite

A. Maes, J. Verlinden and A. Cremers

Centrum voor Oppervlaktescheikunde en Colloïdale Scheikunde
Katholieke Universiteit te Leuven
de Croylaan 42, B-3030 Leuven (Heverlee), Belgium

In a recent paper from this laboratory [1] it was hypothesized that one of the reasons for obtaining low maximum exchange levels in CaY zeolite is the existence of a stable Na-hydrate in NaY. This paper examines in some detail the effect of the nature of the outgoing cation on the ion exchange equilibria of the transition metal ions Co, Zn, Ni and Cd in zeolite Y. In particular, some detailed analysis is given of the purely thermodynamic aspects of the ion exchange reaction at very high exchange levels.

Materials and Methods

The NaY and KY zeolites used are those described in earlier papers [2,3]. Ion exchange isotherms are obtained on KY zeolite at 0.01 total normality using membrane dialysis methods and two weeks exchange to ensure equilibrium [2]. The following systems were studied: K-Co and K-Cd (5°, 25°, 45°C); K-Ni (5, 25°C); K-Zn (25°C); in addition, the Na-Cd system was also studied (25°C). The pH of initial and equilibrium systems is 5 to 5.5, which should eliminate interference from hydroxide precipitation. In all systems (except the Na-Cd case in which both ions were monitored), the exchange isotherms were obtained from concentration changes of the transition metal ion in the liquid phase, a procedure which is justified on the basis of the stoichiometric exchange in Y zeolite [2].

Maximum exchange levels are obtained at different temperatures by exhaustive saturation (0.005 M salts) using radiotracer methods described earlier [2,3]. Occasionally, exhaustive saturation methods were used to determine the equilibrium zeolite composition at high equivalent fractions of the transition metal ion in the liquid, i.e. 0.8 to 0.9.

Results

The ion exchange isotherms obtained at different temperatures on KY zeolite
for the pairs K–Co, K–Cd (5, 25, 45°C), K–Ni (5–25°C) and K–Zn (25°C) are
shown in figure 1.   The Na–Cd data at 25°C are also shown in figure 1.

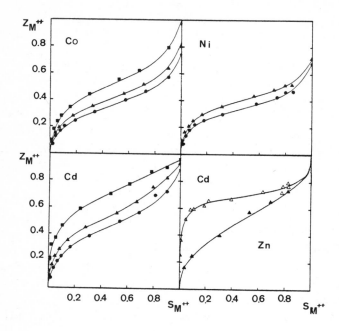

Figure 1   Exchange isotherms for M^{2+}–KY at 5 (●), 25 (▲) and
45°C (■) and for Cd–NaY at 25°C (Δ).

Similar to what one observes in NaY, the exchange reaction is endothermic
and characterized by sigmoid isotherms which, in most cases, terminate at
exchange levels well below the crystal charge.   In all cases, except for Ni,
the maximum values exceed the 70% exchange level, suggesting the participation
of small cage sites.   Temperature effects on maximum exchange levels are
summarized in table 1.   It is seen that the temperature dependence of
maximum exchange levels varies with the nature of the metal ion; throughout
the temperature range studied these levels increase in the order Ni< Co< Zn
< Cd for NaY and KY, a sequence which is at variance with the one reported
on NaY by Radak and Gal [4].

Table 1. Maximum exchange expressed in fractions of the CEC at 5,25 and 45°C on KY and NaY[3] zeolite.

	°C	KY	NaY [3]
Cd	5	0.89	–
	25	0.94	0.98
	45	0.96	–
Zn	5	0.76	0.80
	25	0.89	0.95
	45	–	1.00
Co	5	0.74	0.74
	25	0.81	0.80
	45	0.97	0.98
Ni	5	0.68	0.70
	25	0.72	0.70
	45	–	0.71

In all systems, except again for Ni, complete or nearly complete exchange can be achieved at 45°C on both NaY and KY. The maximum exchange levels, obtained at lower temperatures, on KY are barely different, 5% at most, from those obtained on NaY: they are slightly higher (Cd) or lower (Zn). This behavior contrasts with the one reported for Ca in KY and NaY [1] which would indicate that the presence of a stable Na hydrate in the sodalite cage as an impediment to complete saturation in the case of Ca cannot be generalized to transition metal ions: this supports the view [3] that, among other factors, the ability of transition metal ions to adapt to the coordination requirements in the small cages is of importance.

The exchange isotherms on KY are less steep than those found in NaY[3] under otherwise identical conditions as is to be expected from the higher selectivity for K ions, as compared to Na, in the supercages [5]. It is furthermore apparent that a selectivity reversal occurs around the 60-70 % exchange level, as found in the case of Na, suggesting that the exchange in small cage sites is minimal below the 60% exchange level. In the initial stages of the exchange reactions, and surely at 5 °C, the participation of small cage sites is expected to be minimal, as inferred from temperature induced irreversibility measurements[2,6].

An idea about the selectivity sequence in the big cavities in NaY and KY is obtained by comparing the (un-normalized) ion-exchange isotherms at low metal loading, shown in figure 2.

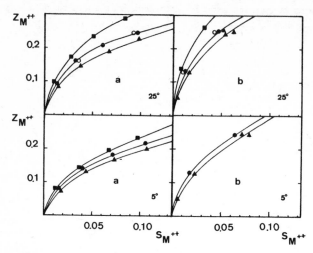

Figure 2. Low loading portion of exchange isotherms of Co(▲),
Ni(●), Cd(■) and Zn (O) at 5 and 25°C in zeolite
KY(a) and NaY(b)

The selectivity sequence (which is in no way affected by the differences in maximum exchange levels which differ at most by 20%) is Cd > Zn ≳ Ni > Co, irrespective of temperature (5 or 25°C) or the nature of the outgoing cation. The ratio of the (un-normalized) selectivity coefficients of Cd, Zn, Ni, Co versus Na and K, as calculated from interpolated values at 15% exchange in figure 2, is about 20 (see table 2) irrespective of the nature of the metal ion and temperature. This observation is consistent with the conclusion that, in all four cases, the same (big cavity) sites are involved. The resulting value of $_{Na}^{K}K_c$ of about 4.5 corresponds with the earlier estimate of ∿5 for the preference of K over Na in the supercages, based on Na-Cs and K-Cs equilibria in the supercages[5].

Table 2. Ratio of un-normalized selectivity coefficients of transition metal
ions versus Na and K: $\frac{M^{++}+K_c}{Na^++K_c} / \frac{M^{++}+K}{K^++K_c}$

Temperature(°C)	Co	Cd	Ni	Zn
5	21.4	–	21.8	–
25	19.1	23.6	19.9	22.0

All selectivity data are summarized in figure 3 in terms of the logarithms of the (normalized) selectivity coefficients versus ion exchanger composition.

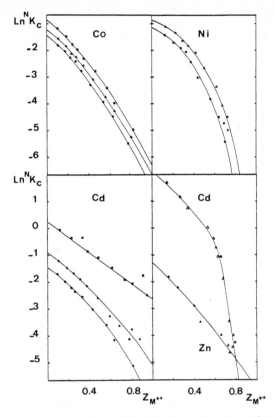

Figure 3. Surface composition dependence of the normalized selectivity coefficients for M^{2+}-KY at 5($\bullet$), 25($\blacktriangle$) and 45°C($\blacksquare$) and for Cd-NaY at 25°C($\triangle$).

The corresponding free energies of exchange, at the various temperatures studied, as obtained by the adapted procedure[7] of Gaines & Thomas[8] are shown in Table 3. The affinity sequence obtained in KY, Cd > Zn ⇋ Ni > Co corresponds with the selectivity sequence at low metal loading, but differs from the one based on $\Delta G°$ values in NaY.

Table 3.   Standard free energies in kJ equiv^{-1} for the exchange of Cd, Zn, Co and Ni on KY zeolite and Cd on NaY.

M^{++}	Exchanger	5°C	25°C	45°C
Cd	KY	5.20	4.60	2.95
	NaY	–	3.25	–
Zn	KY	–	5.54	–
Co	KY	5.62	5.62	5.53
Ni	KY	5.44	5.15	–

Discussion

The values for the free energy of (incomplete) exchange for each transition metal ion versus either Na^+ or K^+ (table 3) at different temperatures are meaningful as such since the standard and reference states are unambiguously defined, i.e. the homoionic Na^+ or K^+ forms, μ^o_{Na} and μ^o_K and the maximum exchanged transition metal ion forms $\mu^o_{M(Na)}$ and $\mu^o_{M(K)}$. However, these standard chemical potentials differ in regard to the nature of the ion, Na^+ or K^+, neutralizing the fraction of exchange sites which appear to be inaccessible to M^{++}. Therefore, even when identical maximum exchange levels are obtained on NaY and KY, quantitative predictions of selectivity based on binary equilibria versus a common third ion are unjustified and, at the very least, hazardous. This was nicely demonstrated for the Na–K–Cs and Na–K–Co Y zeolite in a recent paper[5]. Since furthermore, maximum exchange limits may vary with temperature for a given metal ion, the use of Van 't Hoffs law for obtaining ΔH^o values is not allowed, since the standard chemical potentials for two temperature conditions are different. If, however, maximum exchange levels are identical at different temperatures, as seems to be the case for Ni, meaningful ΔH^o values can be obtained, notwithstanding the fact that the manner of charge neutralization may be different at the two temperatures. By the same token, a comparison of free energies of exchange of two metal ions against a common reference cation, say Na^+, is also unjustified unless identical maximum exchange levels are obtained, as was the case for alkaline earth metal ions in X and Y zeolites[9] and transition metal ions in mordenite[10].

Consequently, the only cases for which a comparison is justified here are the K–CO and the K–Cd systems at 45°C which show clearly that, over the entire composition range (see fig.3), Cd^{++} ions are preferred over Co^{++} ions: the overall effect amounts to a factor of about 2 to 3 at low metal loading (supercages) and roughly two orders of magnitude at high metal loading (small cages). The apparent discrepancies between selectivity sequences at low loading and those obtained from ΔG° values, as alluded to in the results section, is due to the participation of a fraction of the crystal charge for which transition metal ions show very poor selectivity which, in addition, may vary markedly with the nature of the ingoing and outgoing ions.

The additional fact that such equilibria at high loading are further complicated by kinetic effects and that there is a general lack of data at high metal loading add to the lack of insight into the factors determining ion equilibria. However, since the kinetic restrictions may be removed at high enough temperature (as indicated by the fact that 100% exchange levels lay be achieved at 45°C for Co, Cd, Zn) it should be possible to elucidate the purely thermodynamic aspects at very high metal loading.

This has been accomplished in the case of Co-Na at 45°C by studying the "trace exchange" of Na in a "homoionic" CoY prepared by exhaustive saturation (see more below). The CoY zeolite is equilibrated with Co-Na mixtures (0.01 total normality) in which the equivalent fraction in Na is varied up to 0.05. The solid/liquid ratios are chosen as to cover a range of Na-contents in the systems up to about 25% of the crystal charge; changes in Na-concentration in the equilibrium solution are monitored with a ^{22}Na-label. The results are shown in figure 4a.

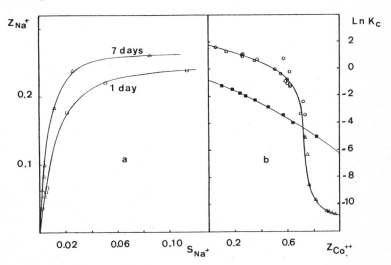

Figure 4. a) low loading portion of Co-Na exchange isotherm at 45°C on
CoY zeolite after 1 day($\square$) and after 1 week(Δ) and b) the
corresponding selectivity data(Δ) referring to 7 days exchange.
The K-Co selectivity data on KY($\blacksquare$) and the previously published
Na-Co data on NaY($\bigcirc$)[3] are incorporated

It is seen that an equivalent fraction of 0.02 in Na is sufficient to reduce
the Co-loading to about 0.08 and that the major part of this shift takes place
within one day; in other words, the Co-loading jumps from 0.8 to 1 when
changing the equivalent fraction in Co from 0.98 to 1. The selectivity coefficients, calculated from the data obtained after 7 days are shown in the
Kielland plot in figure 4b in which earlier data[3] on the same system, but only
at loadings up to .75, have been included.

The most striking feature of this result is that K_c(Co-Na) covers a range of
more than five orders of magnitude, from about 5 at $Z_{Co} \to 0$ to 2.10^{-5} at
$Z_{Co} \to 1$. If we take it that the fraction of the crystal charge exhibiting
this extremely high selectivity for Na^+ w.r.t. Co^{2+} amounts to about 20%
then the intrinsic selectivity coefficient for this group of sites K_c(Na-Co)
would amount to about 2.5×10^5. Evidently, the sites concerned here are those
in the small cages, as demonstrated by comparing with the Co-K-Y data at
$45°C$ [5], included for the sake of comparison in figure 4b. It clearly shows
the preference of K^+ to Na^+ ions in the supercages and the reverse effect in
the small cages. A consistency test of these data is provided by applying
the triangle rule on the three binary systems $NaK^{5,11}$, Na-Co and K-Co at 45 °C,
a procedure which is justified since all three systems refer to complete
exchange levels. Figure 5 shows the results of these calculations, in which
the complete data on the Na-Co (shown in figure 4b) have been used.
Agreement between calculated and experimental K-Na data $(1.55 \text{ kJ Equiv.}^{-1})^3$
is good.

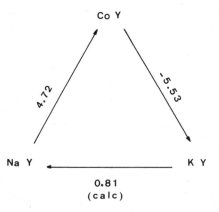

Figure 5.   Consistency test of the Na-K, Na-Co and K-Co
            equilibria in zeolite Y at 45°C: standard free
            energies are given in kJ Equiv.$^{-1}$.

These findings may well explain some of the diverging maximum exchange levels reported in the literature. It appears that, when saturating a NaY zeolite by exhaustive saturation, little is to be gained by using (small volumes of) concentrated solution since minor amounts of Na^+ in the equilibrium solution (being displaced from the solid phase) are sufficient to prevent further exchange (of transition metal ions). Moreover, the apparent advantage of using concentrated solutions is offset by the "electroselective effect" which retains the ions of lower valence in the crystal. Therefore maximum exchange levels may only be achieved by using relatively high dilutions combined with high liquid/solid ratios so as to minimize the effect of the amount of Na displaced on the composition of the solution, and displacing a sizeable fraction in each saturation step. The lower selectivity for K ions in the small cages (as clearly shown by the shape of the K-Cd isotherm at 45°C) would further explain the apparently higher maximum exchange levels obtained when using KY[1], which however disappear when saturating under conditions just described.

The lower maximum exchange levels, observed at lower temperatures may result from kinetic and thermodynamic effects. The kinetic effect, probably the most important one, relates to the fact that the migration of divalent ions into the small cages requires a partial stripping of hydration water. This is substantiated by the experimental fact that "apparent" maximum exchange levels are correlated with free energies of hydration[12]. In the extreme, the process may proceed at an immeasurably slow rate so that the system appears to be in a state of "quasi-equilibrium". This is illustrated by experimental findings showing that such ion exchange reactions appear to be "reversible"[3,10,12]. The thermodynamic effect relates to the fact that the reason for the low selectivity at high loading is predominantly energetic in origin, shifting the isotherms at high (transition) metal content in the liquid phase to lower levels.

In conclusion, it should be added that, besides kinetic and purely thermodynamic factors, hydrolysis of the zeolite, as induced by prolonged saturation with dilute, slightly acidic, electrolyte solutions, may be a reason for incomplete exchange. This effect, at least under the conditions described in this paper, is usually not higher than 2 ions per unit cell[2].

The financial support of the Belgian Government (Programmatie van het Wetenschapsbeleid) and the "Fonds voor Kollektief Fundamenteel Onderzoek" (F.K.F.O.) is gratefully acknowledged.

References

[1] M.Costenoble, W.Mortier and J.B.Uytterhoeven, J.C.S.Faraday Trans.I, 1976, 72, 1877.

[2] A.Maes and A.Cremers, Adv.Chem.Ser. 1973, 121, 230.

[3] A.Maes and A.Cremers, J.C.S.Faraday Trans I, 1975, 71, 265.

[4] V.M.Radak and I.J.Gal, Proc.3rd Int.Conf.Molecular Sieves, 1973, 192.

[5] A.Maes, J.Verlinden and A.Cremers, J.C.S. Faraday Trans I, 1979, 75, 440.

[6] A.Maes and A.Cremers, Proc.3rd.Int.Conf.Molecular Sieves, 1973, 192.

[7] R.M.Barrer, L.V.C.Rees and M.Shamsuzzoha, J.Inorg.Nuclear Chem., 1966,28,629. H.S.Sherry, J.Phys.Chem, 1966, 70, 1158.

[8] G.L.Gaines and H.C.Thomas, J.Chem.Phys., 1953, 21, 714.

[9] A.Cremers, Molecular Sieves II, A.C.S. Symposium Series, 1977, 179.

[10] R.M.Barrer and R.P.Townsend, J.C.S. Faraday Trans.I, 1976, 72, 661.

[11] R.M.Barrer, J.A.Davies and L.V.C. Rees, J.Inorg.Nuclear Chem. 1968, 30,3333.

[12] I.J.Gal and P.Radovanov, J.C.S.Faraday Trans.I, 1975, 71, 1671.

The Mobility of Cations in Synthetic Zeolite A

A.Dyer and H. Enamy
Department of Chemistry and Applied Chemistry,
University of Salford,
M5 4WT

Introduction

The ion-exchange properties of zeolites were the first attribute of these minerals to be subjected to scientific investigation but, despite this, relatively little use has been made of them as ion-exchangers.

Recently there has been a resurgence of interest in their ability to replace cations from solution. It has arisen mainly from their potential use as detergent 'builders', but also from their ability to remove radioisotopes from nuclear power station waste waters. In both instances the synthetic zeolite A has been considered for use.

Consultation of the literature shows that ion-exchange equilibria in A have been researched extensively[1] whilst the kinetics of the ion-exchange processes remain comparatively unexamined. This reflects the complex mathematical approach which is required to make progress towards an understanding of the intricate mechanisms whereby ions inside a solid matrix can exchange with ions from a solution in contact with it.

A primary requirement for the elucidation of ion-exchange kinetics is an accurate knowledge of the rates at which a particular cation will replace itself from solution into a solid exchanger over a range of temperature. The self-diffusion coefficients so obtained can be manipulated to provide an insight into the energy barriers encountered by an ion inside a porous exchanger together with some measure of the involvement of water during the exchange process.

So far as zeolite A is concerned work from these laboratories has furnished information on the self-diffusion of Ca^{2+}, Sr^{2+} and Ba^{2+}[2,3] but studies on the alkali metals pose some questions.

Ames[4] first measured self-diffusion parameters of Na^+ and Cs^+ in A at two temperatures using "bundles of cemented zeolite crystals". Hoinkis and Levi[5] examined Cs^+ self-diffusion in powdered A and their results were later reanalysed together with some information on Na^+ self-exchange[6].

The values quoted for the energy barrier to Cs^+ self diffusion were in good agreement but self-diffusion coefficients were at variance for both Na^+ and Cs^+. Brown and Sherry[7] analysed self-exchange data for Na^+ in A and calculated an energy barrier for the diffusion of mobile sodium ions. Their data differed from that of Ames and Hoinkis. None of these authors gave results for K^+ and Rb^+ migrations except that Hoinkis and Levi observed that Rb^+ was very like Cs^+.

A further point at discussion in the literature is the exact composition of the sodium form of A. Some authors assume that the proportions of sodium to aluminium are in the expected stoichiometric ratio of 1:1 whilst others report an excess of sodium which it is suggested is trapped in the framework as sodium aluminate[8]. No crystallographic evidence exists for the presence of occluded aluminate in A and it is relevant to record that excess washing of the zeolite causes the exchange of Na^+ for H_3O^+ species. This contribution uses radio-chemical methods to measure self-diffusion parameters for the ions Na^+, K^+, Rb^+ and Cs^+ in A powder as well as some simple exchange experiments intended to contribute further information on the presence of sodium aluminate as an occluded species in A.

Thermal analysis gave water contents and other information.

Experimental

Self-diffusion experiments. Pure NaA powder (Laporte Industries) was sedimented and a suitable size fraction taken for exchange experiments. The fraction chosen was labelled with 22-Na and repeatedly exchanged with 0.2M solutions, of the appropriate alkali metal salt, to obtain maximum possible exchange levels. Care was taken to keep washing with water at a minimum. Exchange levels were measured by monitoring 22-Na activity in the solution phase. The results from these experiments were used to prepare isotopically labelled potassium, rubidium and caesium A zeolites. The isotopes 22-Na, 86-Rb and 137-Cs were as supplied by The Radiochemical Centre, Amersham, but 42-K was prepared from potassium nitrate (Analar) by thermal neutron irradiation at the Universities Research Reactor, Risley. Exchanged samples were characterized by X-ray powder and thermal analyses.

Self-exchange experiments were performed, in duplicate as previously described[3], in the temperature range 243–273°. Half-life corrections were made for 42-K and 86-Rb when appropriate.

Exchange experiments (i) Zeolite A was placed in contact with 0.1M 22-NaCl overnight, washed with a minimum of water, air dried, and stored over saturated NaCl solution. About 50 mg of the 22-Na A of known specific activity so prepared was equilibrated with separate portions of 1M NaCl, KCl and NH$_4$Cl in polythene vials for at least one week. The solution phase was assessed for radioactivity daily.

ii)   24-Na A was prepared by thermal neutron irradiation from NaA powder. The maximum time of irradiation was 1 hour. Short-lived species were allowed to decay away and exchanges with NaCl, KCl and NH$_4$Cl carried out again.

iii) 22-NaA was synthesized by a standard procedure[9] from a reaction mixture labelled with 22-Na. The product was characterized by X-ray powder diffraction and by wet chemical analysis for SiO$_2$ and Al$_2$O$_3$. Exchange for Na, K and NH$_4$ was again performed. The radiochemical purity of the synthesized 22-Na A was checked by an γ-ray spectrometer with a Ge-Li detector.

Results

Analyses. The results of chemical, radiochemical and thermal analysis gave the composition of the zeolites used as shown in Table 1. X-Ray powder diffraction showed no loss of crystallinity with exchange.

Table 1   Zeolite pseudo unit cell composition

NaA	Na$_{12}$ (SiO$_2$)$_{12}$ (AlO$_2$)$_{12}$ 26.7H$_2$O
K A	K$_{12}$ (SiO$_2$)$_{12}$ (AlO$_2$)$_{12}$ 22.3H$_2$O
RbA	Rb$_{9.6}$ Na$_{2.4}$ (SiO$_2$)$_{12}$ (AlO$_2$)$_{12}$20 H$_2$O
CsA	Cs$_{7.4}$ Na$_{4.6}$ (SiO$_2$)$_{12}$ (AlO$_2$)$_{12}$ 20 H$_2$O

Self-diffusion experiments. Rate plots were analysed using computer solutions to the Carman-Haul equation as previously described.[10] For sodium some long term experiments were carried out but even when the duration of the exchange was fifteen days good fits up to nearly 100% exchange were found. No evidence of deviation from one exchange process was noted in this or the shorter time

experiment with other isotopic forms. Arrhenius plots of $\log_{10} D$, the self-

diffusion coefficient, against the inverse of absolute temperature are in Fig.1.

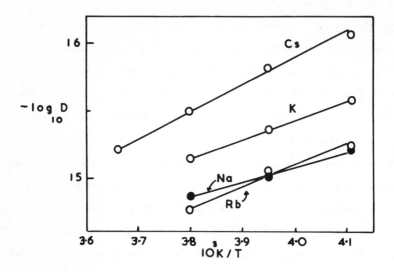

Fig.1. Arrhenius plots of cation self-exchange in A.

Elucidation of the data in Fig. 1 produced the further values listed in

Table 2.

Table 2.   Self-exchange plots of cation self-exchange in A.

Cation	D_0 $m^2 s^{-1}$	E_a kJ $mole^{-1}$	$\Delta S^{\neq}$ JK^{-1} $mole^{-1}$	$\Delta G^{\neq}$ kJ $mole^{-1}$
Na	4×10^{-11}	22.4	-99	49
K	1×10^{-10}	26.5	-89	53
Rb	6×10^{-10}	27.9	-77	48
Cs	5×10^{-9}	36.3	-58	51

Exchange experiments.   The results of these experiments are in Table 3.

Experiments (i) and (ii) were on samples of NaA from the same preparative

batch as supplied by Laporte Industries. A check on its Si/Al ratio by X-ray

fluorescence analysis gave Si/Al = 0.999. The Si/Al ratio of the NaA synthesised

for experiment (iii) was Si/Al = 0.98 by conventional wet analyses.

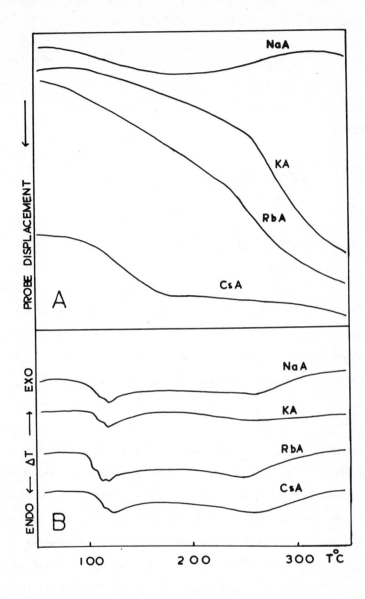

Fig. 2(a) Thermomechanical analysis curves

(b) Differential thermal analysis curves

Table 3   Recovery of radioactive sodium (22-Na or 24-Na) from zeolite A

labelled by different methods, after ion-exchange.

Exchanging Cation	Activity recovered (counts min^{-1} ml^{-1})	% Exchange at equilibrium	Number of Na* ions replaced per pseudo unit cell
(i) Labelling by self-exchange			
Na	2800	100	12
K	2855	100	12
NH$_4$	2671	95	11.4
(ii) Labelling by neutron irradiation			
Na	1567	100	12
K	1580	100	12
NH$_4$	1504	96	11.5
(iii) Labelling by synthesis from 22-NaOH			
Na	10259	100	12.24
K	10280	100	12.24
NH$_4$	9951	97	11.8

Thermal analyses.   The results of differential thermal analyses (DTA) and

thermomechanical analysis (TMA) are in Fig. 2. The thermogravimetric analyses

(TG) curves were smooth and showed that at temperatures up to about 523° water

loss virtually was complete. The total water losses observed at 1073° are given

in Table 1.

Discussion

Self-exchange experiments.   The variation of Ea with ion-size for the cations

studied is shown in Figure 3 with the results of other workers. The values

observed for Na$^+$, K$^+$ and Rb$^+$ are indicative of the dependence of the barrier to

ion movement in A zeolite upon bare ion size unlike those of Ca^{2+}, Sr^{2+} where

the migrating species is hydrated.[2] The behaviour of Cs$^+$ is anomalous and the

high resistance to ion movement may indicate that the progress of Cs$^+$ into the

zeolite cavity is hindered by its relative lack of ability to induce the

cooperative water movements which are necessary for self-exchange. Some

confirmation of this suggestion comes from the observation that the $\Delta S^{\neq}$ values

observed also do not have a simple relationship with ion size. An alternative

would be that the processes of self-exchange for Rb$^+$ and Cs$^+$ involve particip-

ation with the unexchanged Na$^+$ present in these forms (compare the full and

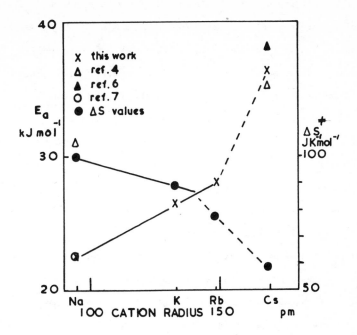

Fig.3. Variation of E_a and $\Delta S^{\neq}$ with cation radius for
ionic self-diffusion in zeolite A.

pecked lines in Fig.3). The value of E_a measured agree well with those of

other workers except that the value for Na^+ from Ames seems in doubt.

Exchange experiments. These demonstrate that Na^+ ions can be completely

exchanged by other Na^+ ions, or K^+ ions, even when there is an apparent excess

of Na^+ present. This would seem to rule out the presence of an occluded

$NaAlO_2$ species in the zeolites studied in experiment (iii). In our view reports

of occluded $NaAlO_2$ should be viewed with circumspection. Two other points arise

from the results in Table 3. Firstly it can be seen that the activity "recovered"

when K^+ is the exchanging ion is always higher than that for Na^+ counter-ion.

This arises from ^{40}K naturally occurring in potassium salts. The second point

concerns the inability of NH_4^+ entirely to replace Na^+. The results in Table 3

are only an example of one set of results, repetition of the exchange

experiments at least three times confirms the results in Table 3. It seems

that there is a true limit to the $Na^+ \rightleftharpoons NH_4^+$ exchange which does not stem from the presence of aluminate species. Of course limits to ion exchange are not new in zeolite species: indeed the CsA and RbA samples previously discussed are other examples. In other zeolite (e.g. X) the reasons for limited ion exchange can be discussed in terms of the inaccessibility of certain ion exchange sites at a particular temperature (See ref.4 pg.542). In A it might be that the limit to univalent ion exchange arises from a size restriction in that, for a particular ingoing ion the number allowed into the large cavity is a function of relative volumes. A corollary to this is that a bulky ingoing ion might cause Na^+ to reside in less accessible sites thus limiting ion exchange. This is the alternate suggestion of Barrer and Meier[11] to occluded aluminate.

Thermal analyses. Thermomechanical analyses in Fig.2(a) demonstrate shrinkages in a bed of powder with increasing temperature. Previous work[12] takes these changes in dimensions as representative of ion resitings as water is removed. Thus, after an initial shrinkage, NaA expands due to ion repulsions inside the large cavity. It might be that when water is removed from KA and RbA these repulsions are compensated by ions moving into sodalite cages causing shrinkage. Seff[13] has examined the ion sitings in $Rb_{11}Na_1A$ and shows, on the basis of X-ray crystallographic evidence, that Rb^+ ions are in sodalite sites in both the hydrated and dehydrated crystals. In a Cs_7Na_5A the cations remain in the large cavity after dehydration.

Fig.2(a) shows DTA curves. Those of the Na,K and Cs forms are very similar but that of RbA shows some indications of a more complicated water loss.

Acknowledgements

We thank Dr.R.Catterall for his invaluable help with computing procedures. One of us (HE) is grateful for a support grant from the Ministry of Higher Education, Iran. Laporte Industries Ltd. kindly supplied zeolite A powder.

Bibliography

1.   L.V.C.Rees, Ann.Rep., 1970, p.191.

2.   A.Dyer and J.M.Fawcett, J.Inorg.Nucl.Chem., 1966, 28, 615.

3.   A.Dyer, R.B.Gettins and A.Molyneux, J.Inorg.Nucl.Chem., 1968, 30, 2823.

4.   L.L.Ames, Jnr., Amer.Mineralogist, 1965, 50, 465.

5.   E.Hoinkis and H.W.Levi, Z.Naturforschung, 1968, 23a, 813.

6.   H.Gauss and E.Hoinkis, Z.Naturforschung, 1969, 24a, 1511.

7.   L.M.Brown and H.S.Sherry, J.Phys.Chem., 1971, 75, 3855.

8.   D.W.Breck, "Zeolite Molecular Sieves", John Wiley and Sons Inc., New York, 1974, p.87.

9.   P.D.Howes personal communication.

10. A.Dyer, R.B.Gettins and R.P.Townsend, J.Inorg.Nucl.Chem., 1970, 32, 2395.

11. R.M.Barrer and W.M.Meier, Trans.Faraday Soc., 1958, 54, 1074.

12. A.Dyer, T.R.Nowell and M.J.Wilson, Proc. 1st European Symp. Therm.Anal., D.Dollimore (Ed.), Heyden and Son, 1976, p.313.

13. K.Seff, Acc.Chem.Res., 1976, 9, 121.

Chairman: Dr. G.T. Kokotailo (Mobil, U.S.A.)

(Throughout this section, reference numbers correspond to those in the particular paper under discussion).

Paper 15 (L.V.C. Rees)

Dr. Breck: The Mg^{2+} exchange isotherms are shown to terminate at $Mg_z = 1$. Did you confirm that complete replacement of Na^+ by Mg^{2+} did indeed take place by analysis of the zeolite solid phase?

Dr. Rees: The final experimental points on the Na/Mg isotherms are at $Mg_z \sim 0.9$. In the determination of these isotherms the solution phase was analysed for Mg^{2+} to obtain the low Mg_z values; Na^+ in the zeolite phase was determined to obtain intermediate Mg_z values, and Na^+ in the solution phase was assessed to obtain the high Mg_z values. Where overlap in these measurements occurred, the isotherm was found to be continuous, indicating little or no hydrolysis.

In separate experiments, however, where we attempted to produce pure Mg–A by replacing the solution phase with numerous, fresh $MgCl_2$ solutions, some hydrolysis occurred for $Mg_z > 0.9$. In very recent experiments, where the solution phase pH was controlled at ~ 7.5 it seems as though we have produced some pure Mg–A.

Dr. Klinowski: I have two comments to make:

(i) I refer to your ion-exchange isotherms measured at different normalities of the aqueous solution. The "valency-concentration effect", whereby isotherms move to the right as the solution becomes more dilute, is a simple consequence of the fact that isotherms are normally expressed in terms of equivalent fractions and not in terms of molalities. Barrer and Klinowski (J.C.S. Faraday I 1974, **70**, 2080) gave a rigorous quantitative treatment of this effect. For instance, in a uni-divalent exchange the equilibrium condition is shown to be

$$K_a = \frac{2A_c \, B_s^2 \, f_A \, \gamma_B^2 \, T_N}{B_c^2 \, A_s \, f_B^2 \, \gamma_A}$$

where K_a is the equilibrium constant (dependent _only_ on the temperature and the pressure). A_c, B_c and A_s, B_s are the equivalent fractions of A and B in the crystal and the solution, and f_A, f_B, γ_A, γ_B are the respective activity coefficients and T_N is the total normality of the solution. The above relationship can be solved to give A_c in terms of A_s and T_N and thus, for any given K_a,

quantifies precisely the course of any isotherm. This conclusion is general
and does not require experimental verification. Likewise, isotherms measured
at different solution concentrations <u>must</u> result in the same Keilland plot.
The area under that plot leads to K_a, and that is constant for a given exchange
whatever the solution normality.

(ii) While it is true that the Glueckauf expression for the activity coefficients in mixed (A,B) solutions is only an approximation, the error attributable
to this source is usually small if one considers the accuracy with which the
isotherm itself is determined. In some simple systems it is possible to measure
these activity coefficients using ion-selective electrodes, but in most cases
(involving for instance complex cations) no such possibility exists. In other
words, in most systems one is forced to resort to Glueckauf's treatment.

<u>Dr. Rees</u>: The "concentration-valency" effect has been known for many years.
For examples, see the chapter on "Ion selectivity" by D. Reichenberg in "Ion
Exchange" Volume 1 (1966, ed. Marinsky) and Helfferich's book "Ion Exchange"
(1962). On p. 157 of the latter book numerous references are given where this
effect has been tested for resin exchange. However, I believe that the study
reported here is the first in which zeolites are used as the exchanger. Considering the factors which are generally considered to be insignificant (or are
ignored), *eg.* changes in water content on ion exchange, lack of experimental
data on the activity coefficients of even such a simple system as a low concentration, pure $MgCl_2$ solution, and zeolite hydrolysis, it is surely a worthwhile
exercise to check the effect for these extremely important systems. Regarding
the question on Glueckauf's expression, the point is that the work in this paper
has shown experimentally that the Glueckauf expression gives a poor approximation of the activity coefficient ratio in mixed electrolyte solutions. Hopefully, the chemical analysis of the exchange system will improve to the point
that it will become necessary to use accurate values of activity coefficient
ratios. This is especially the case if attempts are made to obtain enthalpies
of exchange from changes in free energy with temperature.

<u>Dr. Sherry</u>: I would like to ask the following questions:
(i) How do your newly determined Ca-Na-A isotherms compare with those of Sherry
and Walton?
(ii) Have you considered that excess sodium aluminate in the zeolites used by
Wolf and Furtig[6], Barrer, Ward and Rees[1], and Sherry and Walton[2] may have varied,
and that this difference in zeolite composition could account for part of the
difference in their results?
(iii) Have you considered that the Na/Al atom ratio of the zeolites used by
Barrer, Rees and Ward and Sherry and Walton may have been different and can you
speculate on how this difference would affect the selectivity plots and the

isotherms?

Dr. Rees:

1. A comparison of Figures 4 and 8 shows that good agreement between the data of Sherry and Walton and these new results exists over the range of $Ca_z=0 \rightarrow 0.6$, but significant differences are seen over the final 40% of the exchange.

2. Undoubtedly differences in the compositions of the starting materials are an important factor in explaining the differences reported in these three studies.

3. It is difficult to speculate on this point without having a better knowledge of the exact nature of the zeolites used in these earlier studies and also some knowledge of the H^+/Ca^{2+} exchange system. Intuitively one would think that the selectivity for Ca^{2+} would be degraded if the Na/Al ratio was very much less than unity.

Paper 16   (M.J. Schwuger)

Mr. Dulat: This paper shows that a trace of magnesium salt in the wash liquor can be beneficial in the absence of a significant percentage of sodium salts. However, the washing efficiency versus magnesium concentration curve has a maximum, and one would therefore expect that there is a limit to the proportion of magnesium ions that can be tolerated. I should therefore like to ask for a comment on optimum magnesium contents and on how these are related to the quantities usually found in mains water.

Dr. Schwuger: The location of the maximum depends on the detergent formulation, and especially on the surfactants used. All anionic surfactants presently used in production by the detergent industry reach their maximum at Mg-hardness above $1 \rightarrow 5°d$ which covers the range of hardness usually encountered in nature.

Mr. Laurent: What is the concentration equivalent to the unit that is used in this paper, the German degree of hardness?

Dr. Schwuger: $1°$ of hardness is equivalent to 10 mg dm^{-3} of CaO.

Dr. Vaughan: You have described the toxicological studies on A as being "very positive". What were these studies, and where can we find the appropriate reports?

Dr. Schwuger: The toxicological results, where published, can be found in the following papers:

1. P. Berth, Tenside Detergents 1978, **15**, 176.

2. Anhang 5 der Unfallvorschriften der Chemischen Industrie MAK-Werte, 1977

3. P.Gloxhuber *et al.*, in preparation.

Mr. Beard: Dr. Schwuger suggests the presence of magnesium ions has an overall beneficial effect on the washing efficiency of detergent mixtures. Such information as is available in the published literature points to the opposite conclusion in most cases. T.C. Campbell *et al.* (Household and Personal Products Industry, March 1978, p.32) show that the washing efficiency of polyester/cotton at 35°C, with an LAS based washing powder was considerably better in the presence of zeolite Na-A, when the water hardness was entirely due to magnesium (150 ppm magnesium hardness). In addition J.D. Sherman *et al.* (Soap/Cosmetics/Chemical Specialities, December 1978, pp 40, 64 and 66) give numerous washing results, which show that calcium is usually worse than magnesium at 35 or 50°C, but that using a wash mixture containing Zeolite A and Zeolite X (which they showed removed Mg ions in addition to Ca ions) was usually no worse, and was often even better, than using a wash mixture containing the same amount of zeolite, but all in the form of Zeolite A.

Dr. Schwuger: The statement that in most cases Mg^{2+} ions are considered to be detrimental in the laundering process, is mostly based on a scientifically unproved analogy to the behaviour of Ca^{2+} ions. Careful literature searches[21-23] reveal that favourable products can be realized even with magnesium salt containing formulations. Liquid laundering and cleansing agents containing magnesium salts are commercially available in the USA and in Brazil. The paper of Campbell *et al.* is not really pertinent in this context, because it is dealing with a special version of a high water-glass-containing detergent formulation (10% to 20% water-glass).

Modern detergents should contain only minor amounts of water-glass in order to achieve good secondary laundering properties. Therefore, the two systems described above are not comparable. Moreover, the chosen experimental conditions of 150 ppm magnesium hardness are unrealistic, because considerably lower concentrations are encountered in nature. The differences in laundering action between Zeolite A and mixtures of Zeolite A with Zeolite X (J.D. Sherman *et al.*) are very small and are generally within the limits of error of the test method.

Paper 18   (A. Maes, J. Verlinden and A. Cremers).

Prof. Barrer: There appears to be, in some isotherms, a considerable extrapolation in obtaining the limits for incomplete exchange. How confident are you about the values of these limits? Also, were you able to establish the reversibility of the exchanges, by following both the forward and reverse processes, and showing that they followed the same isotherm curve?

Dr. Maes: It is true that extrapolations were made from equivalent fractions in solution of 0.8 to pure 10^{-2} g equiv dm^{-3} solutions of M^{2+}. The values of the limits of incomplete exchange were obtained under the same conditions of temperature and exchange time as the isotherms and were therefore the correct end-members. Of course, the very small M^{2+}-K^{+} selectivities (see Keilland plots) at high bivalent ion loading indicate that very small fractions of K^{+} ions in solution are sufficient to maintain a sizeable fraction of the zeolite charge in the potassium form. Unfortunately potassium was not determined in solution. No extensive reversibility tests were made. It was however, shown in our work on Na-Y (Adv. Chem. Ser. 1973, **121**, 230, J.C.S. Faraday I. 1975, **71**, 265) and by Barrer and Townsend on NH_4-mordenite (J.C.S. Faraday I, 1976, **72**, 661) that reversibility exists if no temperature changes occur.

Dr. Sherry: Please describe the experimental procedure you used to determine the maximum degree of ion exchange for a particular isotherm.

Dr. Maes: Maximum exchange limits were obtained by contacting about 50 mg of K-Y zeolite suspended in 5 cm^3 10^{-2} g equiv dm^{-3} sodium nitrate in a dialysis bag with about 80 cm^3 portions of labelled 10^{-2} g equiv dm^{-3} transition metal ion solutions. The solution was renewed daily for 15 days. The contents of the dialysis bag were then dissolved in 0.1 g equiv dm^{-3} nitric acid and analysed radiochemically.

Paper 19   (A. Dyer and H. Enamy)

Dr. Rees: Did you try to measure the self-diffusion rate of the Na^{+} ions that had not been exchanged with Cs^{+} ions in your mixed Cs/Na-A?

Dr. Dyer: No, we did not do this experiment.

Dr. Rees: If NH_4^{+} ions do not remove Na^{+} ions sited in the sodalite cages one would expect (from our Szilard-Chalmers studies) to find larger quantities of ^{24}Na retained by the zeolite than observed after irradiation in a neutron flux, because of recoil into the sodalite cages.

Dr. Dyer: I agree, but our results show that it is the ingoing ion (NH_4^{+}) which causes sodalite cage sites to be occupied, due to size considerations. It can be seen from our experiments that if recoil into the sodalite cages has occurred on irradiation (experiment (ii)) this does not prevent complete exchange of ^{24}Na by ^{23}Na or by K. The two effects of

(i) occupation of the sodalite cage sites due to Szilard Chalmers recoils, and

(ii) the obstruction of ions, preventing complete exchange,

are not mutually exclusive.

Dr. Maiwald: I should like to make a comment on Dr. Dyer's paper, in support of the tentative proposal of Professor Barrer that there are occluded aluminium compounds in the zeolite A channel system. Using proton NMR, we have studied a large number of type A zeolites from different preparations and we have found occluded alumina compounds (mostly in the β cages but varying in quantity depending on synthesis conditions) in all of our samples. In addition we should like to make the following proposals, which are also our *resumé* of the proceedings concerning type A zeolites, at this symposium:

(a) If type A zeolite samples are subjected to investigation, it is necessary to try to reproduce the results with a sample of another batch, using another synthesis method if possible.

(b) Experimental results obtained from separate samples of one synthesis run often differ to a degree which is greater than that seen by comparing different samples from separate syntheses carried out under different experimental conditions. This conclusion is also supported by Dr. Ruthven's results which he reported in Session 1.

Dr. Dyer: There certainly are variations from one batch of A to another. You will see that in our studies the variations do not seem to be due to occluded aluminate species. May I draw your attention to the comments made by Mr. Roberts in Session 2 regarding occluded sodium hydroxide, and note that it is common practice not to overwash, so as to avoid "proton" exchange.

Industrial Uses of Zeolite Catalysts

By

D. E. W. Vaughan

Davison Chemical Division, W. R. Grace & Co.
Washington Research Center
7379 Route 32
Columbia, Md. 21044
U. S. A.

Introduction

The growth of interest in zeolite catalysts in recent years has triggered an outflowing of reviews on the subject, including several since the author agreed to present this talk. This paper will therefore be confined to discussing actual industrial catalytic processing using zeolites. To try and add a different dimension, and to satisfy the very diverse interests of this audience, I have included where possible the money factor - both in terms of catalyst values and products processed. It is a perspective that drives home the increasing indispensability of zeolite catalysts in a modern industrial economy.

Much of the earlier work in zeolite catalysis was described by Turkevich[1] and Venuto and Landis[2]. The composite volume edited by Rabo[3], and the triennial reviews in the proceedings of the International Zeolite Conferences[4-10] describe more recent developments. Faujasite is by far the most important zeolite promoter, and has been treated in detail by Rudham and Stockwell[11], and Jacobs[12]. The catalytic properties of mordenite have been discussed by Burbidge et al.[13], and of the natural zeolites in general by Vaughan[14]. This talk will tend to focus on industrial developments, giving particular attention to the patent literature.

Scope

Zeolite catalysts have been applied to the following four general areas of industrial processes:

- Petroleum Processing
- Petrochemicals

• Chemicals Synthesis

• Pollution Control

Except for the first, all are presently in their infancy but
seem poised for major expansion. As shown in Table 1, petroleum
cracking catalysts account for more than 95% of the market both
in terms of catalyst and value of products. For this reason
petroleum processing will be given major attention. However,
with the expiration of some of the basic zeolite catalyst patents
in the immediate future, the chemical industry seems to be
strongly evaluating zeolite catalysts in a range of processes.
Requests for zeolite samples from chemical companies, as distinct
from petroleum companies, are now commonplace, whereas even two
or three years ago such requests were rare. This upsurge of
interest is also reflected in the number of zeolite catalyst
patents issued for the preparation of specialised chemicals.
Because of the limited catalyst usage, and closely guarded
proprietary processing, it is difficult to assess the use of
zeolites in most of the specialised chemical businesses. At the
present time one assumes that most patented applications are
still in developmental stages.

Table 1
Free World Zeolite Catalyst Usage

Process	Annual Usage (Tons)	Catalyst Cost (Approx. $/Ton)	Zeolite Promoter
Catalytic Cracking:			RE-Y
Fluid CC.	200,000	900-2000	H-Y
Fixed Bed CC.	20,000	1200	Mg, RE-Y
Hydrocracking	1,200	10,000	Co, Mo, W or
		20-25,000	Pt-Pd on
			HY, Z14-US, Mordenite Erionite
Petrochemicals	200	4,000 / 80,000	RE-Y / ZSM-5
NO$_x$ Reduction	<100	5,000	H-Mordenite

Zeolite Properties

From a catalytic viewpoint, zeolites are molecular boxes having variable dimensions that are admirably suited to manipulating molecular rearrangements. Their main properties are:

- High density of "active" sites.
- Ingress - egress control - stereo specificity.
- Sites for occluded species.
- Controlable potential energy fields.

A manipulable active site, superimposed on a more or less fixed framework energy field, which can be modified by occluded or deposited salts or metals, is exposed to stereo specific reactive molecules; the products of the subsequent reactions are also sterically selected. With many different "molecular boxes" available, such materials are a catalyst chemist's dream.

Most zeolites are very acidic materials, and most investigations have focused on acid catalyzed reaction, and the reactivity of high acid site strength and site density, together with the methods of manipulating both factors. With the development of high silica/alumina ratio materials using organic base "template" synthesis methods, several zeolites have been prepared that have very low or no acidity - ZSM-5 is premier among these. Such products may develop hydrophobic properties[15], particularly as they approach the pure silica end member of the structure, as in silicalite[1e]. The emergence of these low acidity zeolite systems opens the door to catalytic processes that can benefit from the advantages of zeolites, but to which even moderate acidities are detrimental.

In reality few zeolites have been thoroughly evaluated catalytically, probably because such work has, with few exceptions, been the domain of the oil companies. Faujasite, with its large, three dimensional, pore system, is an excellent hydrocarbon conversion catalyst and has received more attention than all the other zeolites combined. Work on the relatively new zeolite, ZSM-5, in the last two or three years has overshadowed faujasite. The enormous research and development effort by the Mobil Company on ZSM-5 undoubtedly establishes this material as one of the most interesting and important catalytic materials to be found since the catalytic properties of faujasite were established in the early 1960's. Processes based on ZSM-5 are just beginning to come into operation. Mordenite is the third zeolite to have a niche in the catalyst marketplace. It is a very acidic material and shows excellent stability and catalytic activity in many reactions, but its one dimensional pore system is readily blocked by carbonaceous residues in hydrocarbon conversion systems, and use has been restricted to selected applications. Figure 1 illustrates the pore systems of these three catalytically important zeolites - coincidentally they are excellent examples of one, two and three dimensional pore systems.

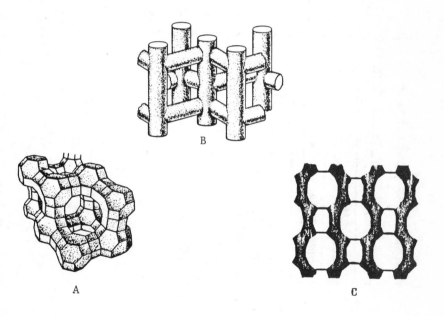

Figure 1:   Structures of (A) Faujasite (X, Y),
            (B) ZSM-5[86] and (C) Mordenite

Many other zeolites have been demonstrated to have good catalytic activities, and can be readily synthesized. These include offretite, erionite, ferrierite, chabazite, gmelenite, L, mazzite (Ω), and several ZSM and ZK types. In the future it is probable that some of these will find utility in specific

chemical and petrochemical applications. At the present time
they are completely overshadowed by the three 'real' catalysts -
faujasite, mordenite and ZSM-5. Important physical characteristics of these catalytically interesting zeolites have been
described by Breck[17].

Petroleum Processing Applications

Since the first application of zeolites as catalysts in
catalytic cracking in 1962[18],[19], they rapidly came to dominate
the petroleum refining industry. Since 1976 all catalytic
cracking in the United States and Canada has been done with
zeolites. In recent years, as other countries convert to gasoline economies, a similar zeolite catalyst demand pattern for
world usage is emerging. Figure 2 shows the present dominance
of zeolite cracking catalysts on a worldwide basis.

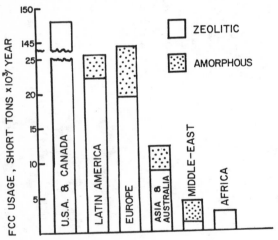

Figure 2: FCC Catalyst Usage

Hydrocracking was the second major process to convert to
zeolite based catalysts, and at the present time 50% of this
market uses zeolites. Hydroisomerization and hydrodewaxing are
specialised areas where specific companies have developed processes
using zeolite catalysts. Figure 3, which shows the general process schemes for a "typical" fuels refinery[20], highlights the
central position of zeolite processes. These are now examined
in order of catalyst consumption.

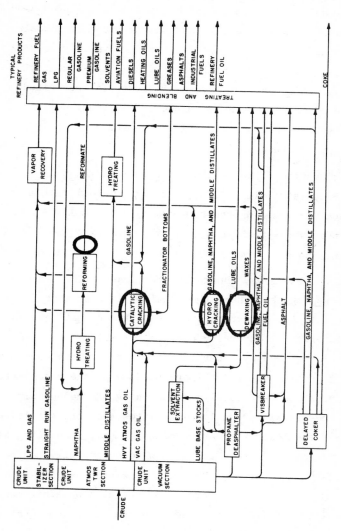

Figure 3: Outline of a fuels refinery flow scheme[20]

Catalytic Cracking

The fluid catalytic cracking (FCC) process in oil refining
is one of the major processing units and has great versatility.
The data in Figure 2 indicate the approximate FCC capacity in the
free world (zeolite usage is ~0.15 lb./barrel of oil processed),
and it represents one of the fastest growing processing tools
available to the refiner. Wherever there is a rapidly growing
demand for gasoline, FCC is the cheapest and fastest route to
obtain this (usually) premium priced product. In the United
States one third of all processed crude oil is reacted over
zeolite catalysts using FCC, amounting to about five million
barrels per day of crude oil.

A modern design FCC unit[21] is shown schematically in
Figure 4a. Pretreated oil (370°C) is contacted with the hot regenerated catalyst (raising the temperature to over 480°C) and
the slurry then passes through a reaction zone at between 480°C
and 520°C (this zone is now literally a pipe in the new "Riser"
FCC designs, allowing very short contact times). The reactor
joins a disengaging zone where the catalyst is separated from
cracked products, the latter passing on to fractionation, and
the former entering a "stripping section", in which the catalyst
is stripped of absorbed hydrocarbons with steam before continuing
on into the regenerator. The primary purpose of the regenerator
is to reactivate the used catalyst by burning off carbon deposited during the cracking reaction. Zeolites are coke sensitive
catalysts, and the lower the carbon on the catalyst reacting with
the fresh crude oil, the more active and selective will be the
subsequent cracking reaction. Recent regenerator designs and
catalyst formulations have therefore concentrated on achieving
reaction conditions suitable for reducing the carbon on regenerated catalysts to less than 0.1 wt.% C (conventional levels
were ~0.5 wt.% C).

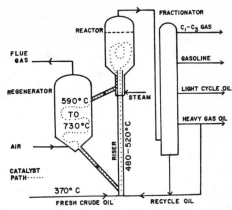

Figure 4a: Conventional Fluid Riser Cracking Design[21]

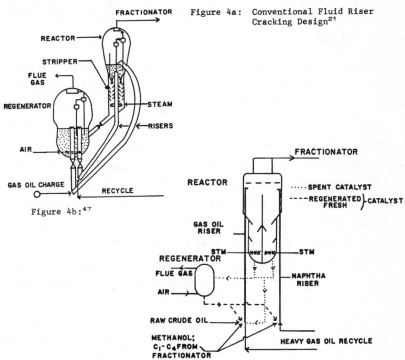

Figure 4b:[47]

Figure 4c:[50]

The FCC promoters are all structurally faujasite, but may have a variety of exchange cations. The commercial types available include:

- Rare-earth X (REX)
- Rare-earth Y (REY)
- Rare-earth hydrogen Y (REHY)
- Rare-earth magnesium Y
- Hydrogen Y (HY)
- Ultrastable Y (Z14-US)

Different manufacturers produce products having different silica/ alumina ratios and residual sodium levels, in a variety of matrices at different zeolite loadings. All these factors affect catalyst performance in terms of activity, selectivity and rates of deactivation[22].

The high process temperatures and steam partial pressures place a premium on stability, and sodium level is probably the most important single factor influencing catalyst stability; its effects on the steam and thermal stability of calcined rare earth exchanged Y faujasite have been discussed elsewhere[23]. Most exhaustively exchanged rare-earth exchanged promoters (<0.5 wt.% Na_2O) will survive thermal treatments at 815°C and high steam partial pressures up to 760°C without major degradation: above these temperatures structural collapse is rapid. At higher sodium levels degradation occurs more rapidly at lower temperatures. The nature of degradation may also change from straightforward lattice collapse to a more subtle reconstructive transformation resulting in ultrastabilization. This process, illustrated in Figure 5, is characterized by an initial loss in structural integrity (as shown by X-ray diffraction or sorption studies) that is regenerated as a function of increasing time in high temperature steam. The final stabilized, annealed product has a significantly lower unit cell than the starting material, and lower catalytic activity. This phenomenon has been extensively evaluated for $H(NH_4) \cdot Y$ faujasites[24] but not for the rare-earth exchanged materials. Examination of equilibrium FCC catalysts clearly shows that this is an important mechanism of FCC catalyst

deactivation, and its extent is very dependent upon the specific
exchange properties of the zeolite promoter, together with regeneration conditions.

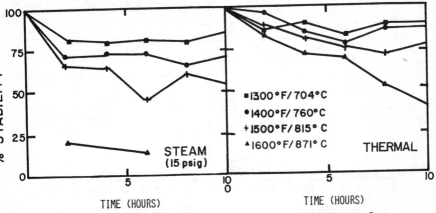

Figure :5: Hydrothermal and Thermal Stability of
REY as Function of Temperature and Time

Zeolite loading - simply the percentage of zeolite in a
catalyst - is a major factor in effecting selectivity changes.
Figure 6 shows how increasing the level of zeolite at otherwise
constant conditions effects the product distribution. As refiners
learned how to use these very active catalysts, zeolite loadings
have steadily increased and may vary between 5% and 40%. It is
a widespread industry practice to change from a high gasoline
summer operation to a high fuel oil winter operation merely by
changing from a high zeolite catalyst to a low zeolite catalyst.
The nature and amount of cation exchange also affects product
distribu tions by changing the acidity distribution on the catalyst, and the steam and thermal stability of the promoter. Particularly important is the search for catalysts that yield larger
amounts of gasoline range products, especially those having
high octane ratings - olefins and aromatics. Many proposals have
been made involving selective cation exchange of copper[25]
and rare-earths[26] ; use of multi-zeolite catalysts[27 -29] ; and
addition of reactive "fragments" to the reacting crude oil[30].
This is a major current field of research and development, with
new catalysts in various stages of field evaluation.

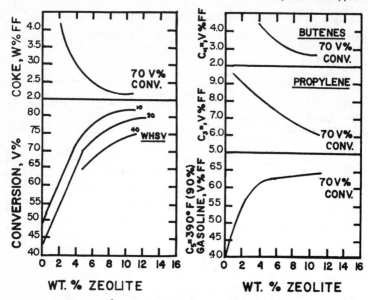

Figure 6:   Effects of Promoter Level (REY) on
FCC Product Distributions at 493 °C

There are at present no non-faujasite catalysts in use in
FCC, but there are many published research examples of multi-
zeolite catalysts that have interesting selectivity effects. The
addition of ferrierite[27], mordenite[28], ZSM-5[29] and erionite[28,29]
to a Re-Y catalyst improves coke and olefin selectivity.
Offretite[31] and gmelenite[32] promoted FCC swing the product distributions to an LPG or petrochemical feedstock range by the
addition of major shape selectivity effects. These selectivity
changes are shown in Table 2. Specific shape selectivity effects
for pure hydrocarbons have been extensively documented by several
authors[33-36].

One of the few certainties in the refining industry today
is the need to improve the efficiency and selectivity of crude
oil processing. Multi-zeolite FCC catalysts will inevitably be an
important future trend in selectivity control to achieve this end.

Table 2

Product Selectivities of Rare-earth Exchanged Zeolites and Zeolite Mixtures

Zeolite		RE-X	RE-Gmelenite	RE-Offretite	RE,HY/RE-Mordenite	Re,H-Y/H-Ferrierite	Re,HY/H-Erionite	REY/H-ZSM-5	REHY
Conversion, V%		55	54.5	56.4	68	71	68	72.8	69
$C_3^=$	V%	3.1	7.4	7.1	7.3	8.3	6.2	6.6	6.5
Total C_3	V%	4.2	18.8	20.8	9.1	9.8	7.2	8.7	7.7
$C_4^=$	V%	3.0	4.0	4.9	4.3	5.1	2.7	8.4	7.7
$i\text{-}C_4$	V%	6.0	8.0	5.5	8.5	8.4	4.7	9.2	4.4
Total C_4	V%	10.8	16.0	15.5	14.2	14.6	8.1	19.2	12.7
C_5^+ gasoline, V%		49.0	26.0	31.6	54.7	56.7	60.2	55.3	56.8
Conv./Coke		19.0	7.0	10.3	28.9	38.7	30.0	36.4	28.7
% Promoter		10	25	25	8Y/2M	8Y/2F	8Y/2E	10Y/5Z	10Y
Reference		31	31	32	28	27	28	29	27

Multi-function FCC

A developing trend in FCC is to "hang on" secondary pro‑ cesses that would otherwise require additional, more expensive, facilities. Most of these are related either to new environ‑ mental regulations or the need to process lower quality crude oils. Whether the 'add on' is matrix or promoter related is immaterial, as at the high temperatures of the process both will be influenced by any additional components in the catalyst. The future need will be to develop zeolite promoters that are either not influenced by the additional process components, or positive‑ ly add to the secondary process.

CO Oxidation. Concurrent with design changes to lower carbon on regenerated catalysts[37] were catalyst modifications that would catalytically enhance the achievement of low regenerated catalyst carbon levels. This was done by the addition of very low levels of noble metals[38,39] (<10ppm) which catalyzed the conversion of CO to CO_2:

$$C + \tfrac{1}{2}O_2 \rightarrow CO$$
$$CO + \tfrac{1}{2}O_2 \rightarrow CO_2$$

Not only is the residual coke minimized, but the carbon monoxide effluent is largely eliminated and a major improvement is achieved in the regenerator heat recovery system.

The result of these changes has been to increase the dense catalyst bed temperature in the regenerator from about 595°C to almost 730°C in some cases. As the regeneration takes place in the presence of steam, this represents a dramatic increase in the severity of the regeneration. As shown in Figure 5, this temperature is approaching the stability limit for most conven‑ tional promoter systems. Although the regenerator temperatures are more demanding of the catalyst, the controlled burning of all the CO to CO_2 in the regenerator catalyst bed minimizes the un‑ controlled "afterburning" of CO to CO_2 in the cyclone areas, that often resulted in temperature excursions well over 760°C, causing major catalyst structure collapse. The regenerator is the most severe section of the FCC unit from the standpoint of catalyst properties. Carryover of steam from the stripper, and addition of water sprays in the cyclones to control temperatures are major sources of steam partial pressures that have a high destabilization effect on zeolite catalysts at temperatures over 705°C - the higher the partial pressure the greater the deactivation. Time, temperature and steam pressure are all cumulative[22].

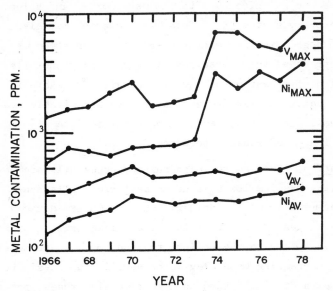

Figure 7: Contaminant Ni and V Level
Variations on Equilibrium
FCC (Averaged Over 110 Units)

S Oxidation - Reduction. Increasing sulphur levels in processed crude oils results in high SO_x emissions from FCC units. Desulphurization of FCC feed results in significant yield enhancements[40], but it is expensive. Recent developments are attempting to promote sulphur retention on the catalyst in the regenerator, and release as H_2S in the reactor where it can be recovered during product purification[41]. The general reaction scheme is:

(1) $S + O_2 \rightarrow SO_2$

(2) $2SO_2 + O_2 \rightarrow 2SO_3$

(3) $MO + SO_3 \rightarrow MSO_4$

(4) $MSO_4 + 4H_2 \rightarrow MS + 4H_2O$

(5) $MS + H_2O \rightarrow MO + H_2S$

Sulphur associated with coke is oxidized to SO_3 in the regenerator (1 and 2); the SO_3 then reacts with a component on the catalyst to form a stable sulphate (3). This passes with the catalyst into the reactor where the metal sulphate is reduced to sulphide (4). In the stripper the high partial pressure of steam oxidizes the sulphide back to the metal oxide and releases H_2S to the recovery system (5). This last reaction may also occur to some degree in the reactor when dispersion steam is used. It is presumed that these reactions are best controlled by impregnation of a base metal onto the catalyst[42], where it inevitably contacts the zeolite and may create undesirable stability problems. In such systems, matching of zeolite promoter with base cation is of great importance.

Metal Passivation. Just as the high sulphur crude oils are lower in cost than the premium low sulphur crudes, crude oils containing higher metals levels (primarily Ni, V, Cu, Fe derived from porphyrins) tend to be lower in price. This pricing differential is an incentive to the refiner to find improved processing methods. The increasing metal levels on equilibrium catalysts (Figure 7) show a distinct trend to higher metals crudes, particularly since the escalation in oil prices in 1973. Conventional methods of demetalization are either crude pre-treating with hydrogen, or chemical purification of the catalyst in use. Both are expensive, and therefore the next question is how to passivate metals using modified cracking catalysts. The metals themselves act as undesirable components by plugging or blocking the catalyst pore structure, and acting as dehydrogenation catalysts in the reactor, the latter resulting in higher levels of hydrogen and coke, with attendant losses in liquid products. Nickel is about four to five times more active than vanadium. Methods to passivate these undesirable metal components are:

- Compound formation with a catalyst component
- Ion exchange by the zeolite
- Separate addition of a "complexing" component

The focus of attention at present is on the third method, with Sb[43], Bi[44] or Sn[45] compounds being the preferred "agents".

These can be added by impregnation on the catalyst or by addition of solutions of the compounds to the regenerator. Presumably metal antimonates or stannates are formed on the catalyst and form inert residues. An alternative proposal, having much greater potential harmful effects on the zeolite, is to inject dry chlorine into the regenerator to react with the metal to form volatile metal chlorides that are recovered from the flue gas[46]. The first two approaches are the ones that offer significant opportunities for manipulation of the zeolite.

The ideal cracking catalyst is one that will give high gasoline yields with superior octane values (RON 90+), is insensitive to high metals loadings, and also minimizes the emissions of CO, SO_x and NO_x in the flue gases.

Process Developments. In viewing catalytic processes from the viewpoint of the chemist, it is easy to become overwhelmed by the importance of the chemistry of the system. However, the chemistry only operates within the engineering framework of the whole system, and it is engineering changes in the fluid catalytic cracking system that have, and will in the future, make major demands on the art and science of zeolite catalysis. Environmental demands, recent changes to riser cracking, and high temperature regeneration have led to increased demands for harder, more attrition resistant catalysts that contain more thermally stable zeolite promoters. If the patent literature is an indicator of future trends, the demand for new and improved catalysts will accelerate. The increasing value of petroleum products makes more selective crude oil processing financially attractive. To this end several interesting new engineering designs use variations on the basic concept of multiple riser reactors. Examples are:

- Different feed fractions fed to each riser, which uses the same catalyst and enter the reactor at the same point. Fresh feed passed through one riser and recycle gas oil to another is a specific case[47] (Fig. 4b).

- Fresh feed is split into different fractions and fed to separate risers using the same catalyst but enter the reactor at different points[48].

- As in the above example, except that a recycle stream comprising the C_4 and minus fraction is fed to one of the risers or into the reactor. A modification of this is to use a virgin low molecular weight hydrocarbon source in place of, or in combination with, the C_4 minus recycle stream, as shown in Figure 4c[49].

In all these cases it is possible independently to control the temperatures and pressures in the risers, together with the catalyst type. The simplest catalyst modification is to us a regenerated catalyst in one riser and a non-regenerated catalyst in another riser[51]. Further modifications can be made by using a multi-zeolite catalyst or a mixture of catalysts containing different individual zeolites that can be physically segregated prior to recycling. The installation of only one or two units using multi-zeolite catalysts will greatly stimulate research in the wider field of non-faujasite catalysts, and the synergism to be achieved with multi-zeolite catalysts. One anticipated a much wider interest in these systems in the future, with more attention being given to the large pore zeolites like gmelenite, ferrierite, offretite, mazzite, L and several ZSM materials.

Hydrogen - Processing

Hydrocracking

Hydrocracking is one of the most versatile processes available to the petroleum refiner, and is capable of converting inferior crude fractions into such desirable products as premium home heating oils, jet fuel, diesel fuel and petrochemical feedstocks[52-54]. As this is a hydrogen additive process, there is a net gain in volume of products which may be as high as 20% - a major economic advantage. This process uses a fixed catalyst bed through which are passed the oil and hydrogen at temperatures between 250°-430°C and hydrogen pressures between 200-2000 psi. The ability of hydrocracking to upgrade heavy and residual gas oils containing high sulphur, nitrogen and metals is of great importance. Without prior desulphurization such low quality oils are undesirable as catalytic cracking feedstocks. Ward has reviewed the flexibility of one of the many hydrocracking processes in converting a variety of feedstocks into a range of desirable fuel products, with emphasis on processing less desirable California and Middle East crude oils[55]. Recent developments, particularly the increasing demands for non-leaded gasoline, are promoting hydrocracking as an increasingly important process for making petrochemical feedstocks, with emphasis on high benzene, toluene and xylene (BTX) yields, and formation of excellent reformer feedstocks for even higher aromatics production[56]. Large price increases for BTX are already occurring because of competition between gasoline octane needs and petrochemical feedstock requirements (intermediates in the production of styrene, nylon, polyesters, platicisers, etc.). Environmental and manufacturing demands will continue to pressure production of these materials for many years. About 1,500,000 barrels/day of oil are processed using hydrocracking.

Hydrocracking catalysts are dual function catalysts comprising matched combinations of an acidity function with a hydrogenation - dehydrogenation function. The acidity function in
newer formulations is preferably provided by a zeolite, but other
hydrocracking catalysts use alumina or amorphous silica-alumina
as the acidic base. The hydrogenation-dehydrogenation function
is provided by combinations of Ni, Co, Mo, W, Pt or Pd (sulfided
or reduced) and is usually Co-Mo, Ni-W or Pt-Pd. Part of the
versatility of the hydrocracking process is the ability to ma-
nipulate the relative strengths of these functions to yield
catalysts having a spectrum of properties. The acidity function
increases in the following order:

  alumina   → silica-alumina → ultrastable Y → Ca,MgY → HY
  (inert base)   silica zirconia

and the hydrogenation-dehydrogenation function in the order:

  exchanged   sulfided   reduced
    metal   →   transition metal   →   noble metal

The reaction mechanism of blends of these components has
been reviewed by Langlois and Sullivan[57]; Turkevich[1], Venuto and
Landis[2] and Bolton[58] have evaluated systems where zeolites comprise the acid function, and Choudhary and Saraf[59] have reviewed
the recent hydrocracking literature in general.

Many of the patents describing these catalysts deal with the
subtle preparative differences that alter the relative function-
alities. The importance of this matching is illustrated in
Figure 8, which shows the importance of relative acidity (i.e.
promotion of cracking reactions) and hydrogenation activity in
converting polyaromatic molecules to a variety of products[60].
By suitable control of reaction conditions (H_2 pressure, temperature, contact time) and catalyst, crude oils having a wide range
of properties can be readily converted to useful products. Feeds
to hydrocracking catalysts are often used after desulphurization of the feedstock because of the susceptibility of these
catalysts to poisoning by sulphur and·nitrogen.

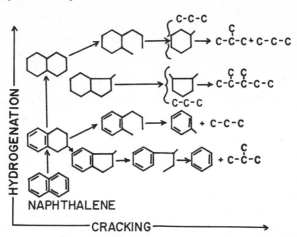

Figure 8:   Product Manipulation by Changing
            Hydrogenation and Cracking
            Functions on HCC[60]

The zeolite acid function is provided by faujasites of
various silica-alumina ratios, with an emphasis on low to moder‎
ate acidity levels. Although Ca and Mg exchanged Y (SiO_2/Al_2O_3 =
4 to 5) was preferred in earlier processes[61], high ratio "stabilized" faujasite is preferred in most catalysts used today. The
development of post-synthesis methods of manipulating framework
silica-alumina ratios in faujasite have become very important in
optimizing hydrocracking promoters[62,63]. The composition limit
for zeolite Y is a silica-alumina ratio of six, although in
practice 5.5 is a more realistic upper limit (i.e. a unit cell
of about 24.56Å). About ten years ago methods were described to
increase the effective structural silica-alumina ratio to yield
materials having unit cells down to 24.20Å[64]. The first method
comprised a series of ammonium exchanges followed by steam
annealing at high temperatures. Many subsequent methods are
refinements of this principle, with emphasis on the importance of
relative Na^+ - NH_4^+ exchange levels, calcination temperatures,

steam and ammonia partial pressures[65],[66]. Other techniques rely
on aluminum removal with complexing reagents (e.g. ethylene-
diaminetriacetic acid[67]), acid leaching, or exchange methods.
Essentially pure silica polymorphs of faujasite have been made
by the latter technique[68]. The mechanism is basically one proposed by Barrer and Makki[69] for clinoptilolite, in which framework aluminum ions are hydrolyzed, and replaced by "nests of
$(OH)_4$". The effective annealing of the structure by replacement
of the $(-OH)_4$ "nest" with mobile Si is the most important part
of the process, requiring control of temperature, residence time
and composition (H_2O, NH_3) of the calcining atmosphere. Detailed
reviews of some of these methods have been published by Kerr[70]
and McDaniel and Maher[24]. This field of research continues to
be a very active one. Mordenite is the only non-faujasite
zeolite extensively evaluated as a hydrocracking catalyst promoter, but it is not known to be in use for hydrocracking full range
crude oils.

The trend in hydrocracking is to the replacement of the
remaining alumina and silica-alumina base materials with zeolites.
Unfortunately (from the viewpoint of a catalyst supplier), hydrocracking catalysts have lifetimes of up to six years. The fixed
bed process allows periodic 'in situ' regeneration of deacti-
vated catalysts, and in the case of severe deactivation off-site
regeneration is a developing trend[71]. Old non-zeolite catalysts
will be replaced with zeolite catalysts slowly, and although the
total market will grow significantly, the zeolite usage will
remain relatively small. The main dollar value of hydrocracking
catalysts is in the metal component.

Shape Selective Hydrocracking

The broad field of shape selective catalysts has recently
been reviewed by Csicsery[72]. Such catalysts have been evaluated
for many years, but "Selectoforming"[73] is one of the few processes that has achieved a moderate level of utilization, with
at least six units in operation. The process is used primarily
to selectively sorb and hydrocrack C_5 to C_9 normal paraffins from
reformer product streams. The nickel exchanged or impregnated

erionite[74], or erionite-clinoptilolite[75], catalyst rejects the branched chain or aromatic molecules, but cracks the normal paraffin fraction to C_1 to C_3 (LPG) components which can be readily separated from the product effluent. As the low octane rating normal paraffins are eliminated, and the high octane components (isoparaffins, aromatics) unchanged, the octane rating of the liquid product stream is increased by 3 to 6 numbers, making Selectoforming a relatively low cost way to increase reformate octane rating. Most of the installed Selectoforming capacity treats reformate, but the process could be used with any product stream where a source of hydrogen is available. This is the only zeolite catalyst process known to use a natural zeolite, and is used in about six refineries.

Normal paraffins above C_{18} crystallize below about 100°C and present pour point problems in lubricating, diesel and fuel oils. Conventionally such components have been removed by crystallization, solvent extraction or adduct formation with urea, but in recent years selective hydrocracking has become a viable alternative[76,77]. Several proposed catalytic hydrocracking methods use the same selectivity principle as Selectoforming in separating the large naphthene and polyaromatic molecules from paraffins, then cracking the latter primarily to C_2 to C_4 fuel gas (LPG). Hydrogen and an active hydrogenation catalyst prevent the accumulation of coke deposits, by maintaining high levels of hydrogenation of the breakdown products.

Mordenite has been extensively investigated for hydrodewaxing, with reduced platinum on hydrogen mordenite being the favored catalyst[78]. Higher silica-alumina ratio (20 to 60) mordenites are preferred[79], and these are usually obtained by acid treatment of the sodium mordenite. ZSM-5 is also a good catalyst in this application[80], particularly when very small crystallite zeolite ($\sim 0.02\mu$) is used[81]. Presumably the small crystallites facilitate access to the pores in the presence of increasing coke levels during the catalyst aging. A major problem that is usually encountered with restricted pore systems of the mordenite type is that even small levels of coke "shut off"

the activity. Although mordenite has excellent cracking activity
this latter factor has restricted its use to hydrogen processes.

Synthesis methods that yield very small crystals should
give improved mordenite catalysts, in the same way that ZSM-5 is
improved. Other zeolites having good cracking activity, and
restricted channel systems that would be selective for n-paraffins in dewaxing applications are ferrierite, offretite,
zeolites L and mazzite.

Hydro-isomerization. Several years ago Shell Oil Company developed a mordenite catalyzed hydro-isomerization process that
converts low octane pentane and hexane feeds (RON ~70) to higher
octane products containing high yields of iso-pentane and di-
methylbutanes (RON ~80+)[82]. Known as the Hysomer Process[83] when
coupled with an i-/n-paraffin separation, it is now used to
enhance gasoline octane in about ten refineries worldwide. The
feed to the process may either be straight run or reformate
nC_5/C_6 gas fraction. The catalyst is a reduced platinum exchanged mordenite, having high stability, and high selectivity
for i-pentane formation. The chemistry of the process has been
reviewed by Kouwenhoven[84].

Non Petroleum Fuels Processing

The synthesis of gasoline from natural gas and coal has an
indeterminate future in a free economy. However, given present
rapid escalation of oil prices, the economics of such conversion
become increasingly positive, and the processing viable. At the
present time synthetic fuels can only be produced under government sponsorship, requiring major political or national security
considerations in assessing the economics (tax incentives, low
cost loans, etc.). In most cases such major projects, generally
considered to cost a minimum of ~10^9, can only be implemented
by governments. The recent South African announcement[85] that
they will build a new coal to liquid fuels complex at a cost of
more than 5.7×10^9 is a reflection of both present day project
costs and the enormous risks involved. These cannot be assumed
by private companies without government guarantees.

The development of the zeolite ZSM-5 has led to a number of important new options in considering synthetic fuels processing. The novel features of ZSM-5 have resulted in a series of high silica/alumina ratio (>30) catalyst promoters, comprising H-ZSM-5 impregnated or exchanged with half the elements of the periodic table, often mixed with other catalyst components. ZSM-5 combines the advantages of low acidity, high pore volume and moderate sized pores with a degree of hydrophobicity. Such products are ideal for petrochemical processing, and hitherto have been largely missing from the zeolite catalyst repertoire.

Methanol to "Gasoline"

One of the best publicized recent "events" of zeolite catalysis is the use of the zeolite ZSM-5 for production of synthetic gasoline. The discovery by Mobil Corporation research workers that ZSM-5 efficiently converts oxygenated hydrocarbons to valuable high octane gasoline components has led to the rapid development of fixed and fluid bed "synthetic gasoline" processing by joint Mobil-U.S. Government research projects[86-89].

The properties of ZSM-5 combine to make it an excellent dehydration and polymerization catalyst. The reactions of converting methanol to gasoline range products is shown in Figure 9, which essentially illustrates the sequential reactions taking place as a function of time in a catalyst bed promoted with ZSM-5. The range of possible reactions that may take place after the double dehydration of methanol to ethylene are shown in more detail in Figure 10, and also demonstrate the range of selective petrochemical conversions possible with ZSM-5. The enormous range of products possible starting with ethylene alone have been reviewed elsewhere[90].

The price of gasoline using this process is approximately given by:

$$\text{Gasoline } \$ = 2.37 \text{ (methanol price)} + 5c./\text{US gallon}$$

The main potential for the process depends on decisions to ship methanol rather than LNG to the U.S.A. from the gas fields of the Canadian Arctic, Africa and the Middle East. It also offers an

interesting alternative to flaring or re-injection of natural
gas associated with large oil fields, in cases where alternative
utilization is not economic[91],[92].

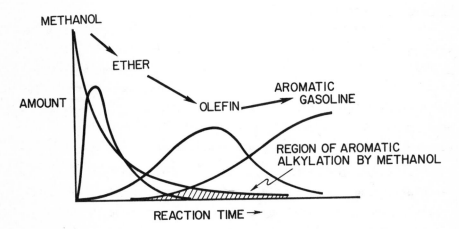

Figure 9:   Reaction Sequence for the Conversion
            of Methanol to Aromatic Gasoline[88]

Coal to Liquid Fuels

Coal to Gasoline

The Sasol processes[93] for the conversion of coal to gasoline range hydrocarbons are similar to those developed in the 1930's in Germany. South Africa is the only country presently using this processing in major facilities. The addition of ZSM-5 catalytic conversion steps after the Fischer-Tropsh reaction improves gasoline yield and quality by conversion of olefins and oxygenated products (alcohols, ketones, acids, aldehydes, creosoles, etc.) to gasoline range components[94-96]. The waxy oils are converted to improved fuel oils by reacting them with ZSM-5 under hydrodewaxing conditions. Process modification schemes, in addition to enhancing product yields, also simplify downstream purification of the products.

These changes significantly improve the process economics for the Sasol liquid fuels process, although not sufficient to make them competitive with crude oil processing in a "normal" environment. With the continued major expansion of South African synthetic fuels facilities, zeolite based process improvements should continue to develop.

Chemicals Manufacture

For over a decade the area of zeolite catalysis has been the province of petroleum companies, and it is only relatively recently that chemical companies have begun to give these materials the attention they deserve. This is reflected in the increasing number of zeolite patents issued to chemical companies, and requests for information and samples. Recent developments have pushed zeolite catalysts to the fore in benzene alkylation and xylene isomerization processes. Many process schemes have been proposed for petrochemical synthesis, and some of these are illustrated in Figure 10. At this time only those discussed below are commercial processes.

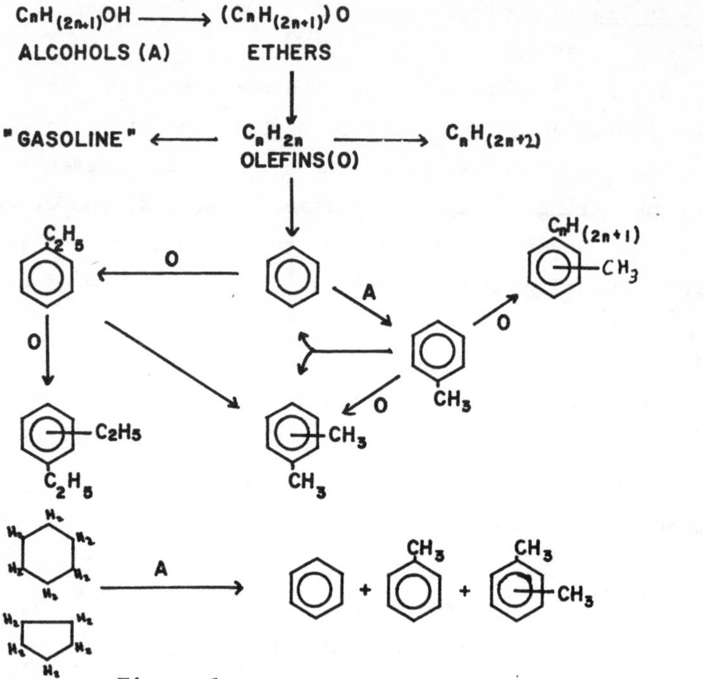

Figure 10: Methanol to Chemicals - Reaction
Routes

Benzene Alkylation

The alkylation of benzene with ethylene to yield ethylbenzene is a major intermediate process step in the production of
styrene.

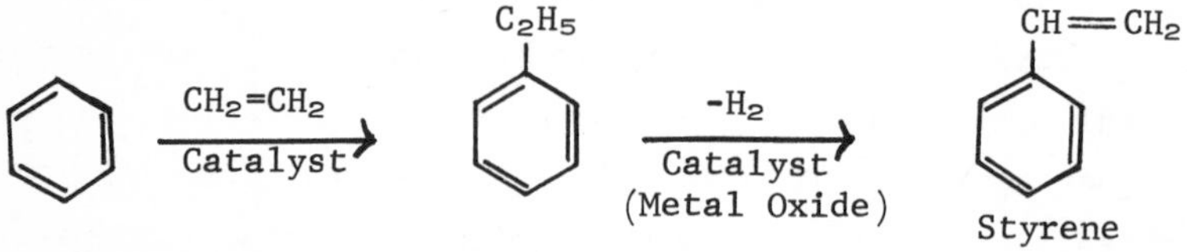

Aluminum chloride has been the traditional catalyst of choice for
this process, but the recent introduction of the Mobil-Badger
process using a zeolite catalyst based on zeolite ZSM-5 seems to
have major advantages[97,98]. U.S. capacity for ethylbenzene is
about 10^{10} pounds per year, but in 1978 three plants having combined capacities of 2.10^9 pounds have been announced using the
new ZSM-5 catalyst process. The reaction is a typical Friedel-
Crafts reaction, but avoids the aluminum chloride problems of
corrosion and product contamination.

A similar process is alkylation of benzene with propylene to yield cumene (intermediate in the production of phenol and acetone) over aluminum chloride or supported phosphoric acid catalysts. This reaction is also promoted by zeolites[99] and may be added to the zeolite catalyst business in the future. Product capacity for cumene in the U.S.A. is 3.5 x 10^9 lbs./yr.

Xylene Isomerization

Mixed xylene feed streams from petroleum processing are usually separated, by crystallization or zeolite selective sorption, into o-, m- and p-xylene components. p-Xylene has the largest demand as a precursor for terephthalic acid (polyester intermediate) and o-xylene is used to a lesser extent for phthalic acid manufacture (plasticizer intermediate). Isomerization of o-xylene, and to a lesser extent m-xylene to p-xylene is therefore desirable, and has been extensively investigated. Traditional catalysts were either $AlCl_3$ or BF_3-HF. The high shape selective character of zeolite catalysts to yield greater than equilibrium levels of p-xylene makes them catalysts of choice for this reaction. Early work developed rare-earth-Y as the best catalyst[100], but recently ZSM-5 has emerged as a major improvement[101]. More recent developments improve the shape selectivity by modifying pore blockage by the precipitation of low levels of Sb_2O_3 in the channels of the zeolite[102]. Similarly p-xylene, having favored diffusion rates in a restricted ZSM-5 pore system, is produced in higher yields on a catalyst comprising large ZSM-5 crystals[103]. These allow progressive escape of the p-isomer and continued shifting of the equilibrium to p-xylene formation.

Pollution Control Application

NO$_x$ Reduction

NO$_x$ reduction with ammonia over acid treated (H$^+$ form) mordenite is the only pollution control catalyst that is used industrially. Primary applications are for effluent control in nitric acid plants and nuclear reprocessing facilities. Although there are many recent patents covering zeolite promoted NO$_x$ reduction, the Gulf Oil process[104] seems to be the only one that

has achieved significant success. At 250°C (±50°C) ammonia
stoichiometrically reduces NO and NO_2 to nitrogen and water.
Process design and control details have been described by Thomas
and Munger[104]. A major increase in this processing may be
anticipated if NO_x emission limits are further tightened, to
require power plants to remove NO_x from their stack gases.

Summary and Outlook

In the near term major zeolite developments will continue
to be related to FCC of crude oils, with multi-zeolite catalysts
constituting a possible major development. Petrochemical zeolite
catalysts are just beginning to come into significant usage, and
will continue to expand their process coverage. This will be
particularly noticeable in the area of Friedel-Crafts reactions,
where conventional catalysts have major handling problems.
In the 1990's, as coal becomes a developing source of hydrocarbon
fuels and chemicals, zeolites should dominate the downstream
secondary catalytic processes used to upgrade products of the
Fischer-Tropsch converter.

Bibliography

1. J. Turkevich, Catal. Rev., 1967, 1, 1.

2. P. B. Venuto and P. S. Landis, "Advances in Catalysis", Academic Press, New York, 1968, vol. 18, p 259.

3. J. A. Rabo, ACS Monograph 171, 1976.

4. P. B. Venuto, E. L. Wu and J. Cattanach, "Molecular Sieves", Soc. Chem. Ind., London, 1968, p 117.

5. P. B. Venuto, Amer. Chem. Soc. Adv. Chem. No. 102, 1971, p 260.

6. J. A. Rabo and M. L. Poutsma, Amer. Chem. Soc. Adv. Chem. No. 102, 1973, p 284.

7. M. Minachev and Y. I. Isakov, Amer. Chem. Soc. Adv. Chem. No. 121, 1973, p 451.

8. D. Barthomeuf, Amer. Chem. Soc. Symp. Ser. No. 40, 1977, p 453.

9. J. H. Lunsford, Amer. Chem. Soc. Symp. Ser. No. 40, p. 473.

10. J. S. Magee, Amer. Chem. Soc. Symp. Ser. No. 40, p 650.

11. R. Rudham and A. Stockwell, Catalysis (Chem. Soc.), 1977, 1, 87.

12. P. A. Jacobs, "Carboniogenic Activity of Zeolites", Elsevier Sc. Publ. Co., Amsterdam, 1977.

13. B. W. Burbidge, I. M. Keen and M. K. Eyles, Amer. Chem. Soc. Adv. Chem. No. 102, 1971, p 400.

14. D. E. W. Vaughan, "Natural Zeolites", 1978, Pergamon Press, Oxford, p 353.

15. N. Y Chen, 1973, U. S. Patent 3,732,326.

16. E. M. Flanigen, J. M. Bennett, R. W. Grose, J. P. Cohen, R. L. Patton, R. M. Kirchner and J. V. Smith, Nature, 1978, 271, 512.

17. D. W. Breck, "Zeolite Molecular Sieves", Wiley, New York, 1974, Ch. 2, p 29.

18. C. N. Kimberlin and E. M. Gladrow, 1961, U. S. Patent 2,971,903.

19. C. J. Plank and E. J. Rosinski, 1964, U. S. Patent 3,140,249.

20. J. H. Gary and G. E. Handwerk, 1975, "Petroleum Refining", Marcel Dekker Inc., New York, p 3.

21. M. C. Bryson, G. P. Huling and W. E. Glausser, Hydrocarbon Process., 1972, 51 (50), 85.

22. J. S. Magee and J. J. Blazek, Amer. Chem. Soc. Monogr. No. 171, 1976, p 615.

23. W. S. Letzsch, R. E. Ritter and D. E. W. Vaughan, Oil Gas J., 1971, 69 (47), 130.

24. C. V. McDaniel and P. K. Maher, Amer. Chem. Soc. Monograph 171, 1976, p 285.

25. R. J. Lussier, J. S. Magee and E. W. Albers, 1975, U. S. Patent 3,929,621.

26. J. Scherzer and R. E. Ritter, Ind. Engr. Chem. Prod. R.&D., 1978, 17, 219.

27. J. Scherzer, D. E. W. Vaughan and E. W. Albers, 1975, U. S. Patent 3,894,940.

28. J. Scherzer and E. W. Albers, 1975, U. S. Patent 3,925,195.

29. E. J. Rosinski, C. J. Plank and A. B. Schwartz, 1973, U. S. Patent 3,758,403.

31. D. E. W. Vaughan, 1975, German Patent 2,420,850.

32. D. E. W. Vaughan, 1977, Brit. Patent 1,480,104.

33. C. L. Kibby, A. J. Perrotta and F. E. Massoth, J. Catalysis, 1974, 35, 256.

34. N. Y. Chen and W. E. Garwood, J. Catalysis, 1978, 52, 453.

35. L. D. Rollmann, J. Catalysis, 1977, 47, 113.

36. D. E. Walsh and L. D. Rollmann, J. Catalysis, 1977, 49, 369.

37. R. J. Shields and R. J. Fahrig, Oil Gas J., 1972, 70 (22), 45.

38. L. Rheaume, R. E. Ritter, J. J. Blazek and J. A. Montgomery, Oil Gas J., 1976, 74 (20), 103; 74 (21), 66.

39. W. A. Chester, A. B. Schwartz, W. A. Stover and J. P. McWilliams, Amer. Chem. Soc. Petrol. Div. Preprints, 1979, 24 (2), 624.

40. R. E. Ritter, J. J. Blazek and D. N. Wallace, Oil Gas J., 1974, 72 (41), 99.

41. I. A. Vasolos, E. R. Strong, C. K. R. Hsieh and G. J.
     D'Souza, Oil Gas J., 1977, 75 (26), 141.

42. R. J. Bertolicini, G. M. Lehmann and E. G. Wollaston,
     1974, U. S. Patent 3,835,031.

43. D. J. McKay and B. J. Bertus, Amer. Chem. Soc. Petrol.
     Div. Preprints, 1979, 24 (2), 645.

44. T. C. Readal, J. D. McKinney and R. A. Titmus, 1976,
     U. S. Patent 3,977,963.

45. B. R. Mitchell and H. E. Swift, 1978, U. S. Patent
     4,101,417.

46. R. M. Suggitt and P. L. Paull, 1977, U. S. Patent
     4,013,546.

47. D. P. Bunn, Jr., G. F. Gruenke, H. B. Jones, D. C.
     Luessenhop and D. J. Youngblood, Chem. Engr. Progress,
     1969, 65 (6), 88.

48. H. Owen and E. J. Demmel, 1975, U. S. Patent 3,894,934.

49. H. Owen, E. J. Rosinski, P. B. Venuto, 1976, U. S. Patent
     3,974,062.

50. H. Owen and E. J. Demmel, 1975, U. S. Patent 3,926,778.

51. A. B. Schwartz and H. A. McVeigh, 1974, U. S. Patent
     3,847,793.

52. G. E. Langlois and R. F. Sullivan, Amer. Chem. Soc. Adv.
     Chem. No. 97, 1970, 38.

53. J. W. Scott and A. G. Bridge, Amer. Chem. Soc. Adv. Chem.,
     No. 103, 1971, 113.

54. M. F. Stewart and J. T. Jensen, Amer. Chem. Soc. Adv. Chem.
     No. 97, 1970, 123.

55. J. W. Ward, Hydrocarbon Process., 1975, 54 (9), 101.

56. V. T. Mavity, Jr., J. W. Ward and K. E. Whitehead,
     Hydrocarbon Process., 1978, 57 (11), 157.

57. G. E. Langlois and R. F. Sullivan, Amer. Chem. Soc. Adv.
     Chem., 1970, No. 97, 38.

58. A. P. Bolton, Amer. Chem. Soc. Monogr. No. 171, 1976, 714.

59. N. Choudhary and D. N. Saraf, Ind. Engr. Chem. Prod. R.&D.,
     1975, 14, 75.

60. H. Clough, Ind. Engr. Chem., 1957, 49, 673.

61. R. C. Hansford, 1967, U. S. Patent 3,364,135.

62. R. C. Hansford, 1967, U. S. Patent 3,354,077.

63. J. W. Ward, 1977, U. S. Patent 4,036,739.

64. C. V. McDaniel and P. K. Maher, "Molecular Sieves", Soc. Chem. Ind., London, 1968, p. 186.

65. A. L. Hensley, 1975, U. S. Patent 3,894,930.

66. D. F. Best, A. P. Bolton and H. C. Shaw, 1975, British Patent 1,506,429.

67. G. T. Kerr, J. Phys. Chem., 1968, 72, 2594.

68. J. Scherzer, J. Catalysis, 1978, 54, 285.

69. R. M. Barrer and M. B. Makki, Canad. J. Chem., 1964, 42, 1481.

70. G. T. Kerr, Amer. Chem. Soc. Adv. Chem. No. 121, 1973, 210.

71. Chem. Week, 21/2/79, 53.

72. Csicsery, Amer. Chem. Soc. Monogr. 171, 1976, 680.

73. S. D. Burd and J. Mazuik, Hydrocarbon Process., 1972, 51 (5), 97.

74. R. C. Wilson, Jr. and A. B. Schwartz, 1978, U. S. Patent 4,077,910.

75. R. C. Wilson, 1972, U. S. Patent 3,640,905.

76. N. Y. Chen, R. L. Gorring, H. R. Ireland and T. R. Stein, Oil Gas J., 1977, 75 (23), 165.

77. J. D. Hargrove, G. J. Elkes and A. H. Richardson, Oil Gas J., 1979, 77 (3), 103.

78. B. W. Burbidge, I. M. Keen and M. K. Eyles, 1972, U. S. Patent 3,668,113.

79. H. C. Morris, 1972, U. S. Patent 3,663,430.

80. N. Y. Chen, S. J. Lucki and W. E. Garwood, 1972, U. S. Patent 3,700,585.

81. R. L. Gorring and G. F. Shipman, 1976, U. S. Patent 3,968,024.

82. H. W. Kouwenhoven and W. C. Van Zijll Langhout, Chem. Eng. Progress, 1971, 67 (4), 65.

83. <u>Hydrocarbon Process</u>., 1974, <u>53 (9)</u>, 212.

84. H. W. Kouwenhoven, Amer. Chem. Soc. Adv. Chem. No. 121, 1973, 529.

85. <u>Chem. Week</u>, 22/1/75, 36; 7/3/79, 23.

86. S. E. Voltz and J. J. Wise (Ed.), "Development Studies on Conversion of Methanol and Related Oxygenates to Gasoline", ORNL/FE-1 (Nov., 1977), p 51; FE 1773-25; NTIS, Springfield, Va., pp 340.

87. C. D. Chang, J. C. W. Kuo, W. H. Lang, S. M. Jacob, J. J. Wise and A. J. Silvestri, <u>Ind. Engr. Chem. Process R.&D.</u>, 1978, <u>17</u>, 255.

88. D. Liederman, S. M. Jacob, S. E. Voltz and J. J. Wise, <u>Ind. Engr. Chem. Process R.&D.</u>, 1978, <u>17</u>, 340.

89. C. D. Chang and A. J. Silvestri, <u>J. Catalysis</u>, 1977, <u>47</u>, 249.

90. L. F. Hatch and S. Matar, <u>Hydrocarbon Process</u>., 1978, <u>57 (4)</u>, 155.

91. D. G. Reynolds, <u>Oil Gas J.</u>, 1977, <u>75 (51)</u>, 68.

92. M. Ostby and S. S. Marsden, Jr., <u>Oil Gas J.</u>, 1977, <u>75 (46)</u>, 83.

93. J. C. Hoogendoorn and J. M. Salomon, <u>Brit. Chem. Engineering</u>, 1957, <u>2</u>, 238; 308; 368; 418.

94. H. R. Ireland and T. R. Stein, 1977, U. S. Patent 4,046,829.

95. J. C. Kuo, C. D. Prater and J. J. Wise, 1977, U. S. Patent 4,049,741.

96. H. R. Ireland and T. R. Stein, 1977, U. S. Patent 4,041,097.

97. P. J. Lewis and F. G. Dwyer, <u>Oil Gas J.</u>, 1977, <u>75 (40)</u>, 55.

98. G. T. Burress, 1975, U. S. Patent 3,751,506.

99. E. F. Harper, D. Y. Ko, H. K. Lese, E. T. Sabourin and R. C. Williamson, Amer. Chem. Soc. Symp. Ser. No. 55, 1977, p 371.

100. J. J. Wise, 1968, U. S. Patent 3,377,400.

101. W. O. Haag and D. H. Olson, 1974, U. S. Patent 3,856,871.

102. S. A. Butter, 1977, U. S. Patent 4,007,231.

103. P. Chutoransky, Jr. and F. G. Dwyer, Amer. Chem. Soc.
      Adv. Chem. No. 121, 1973, 540.

104. J. L. Carter, M. T. Chapman and B. G. Yoakam. 1975,
      U. S. Patent 3,895,094.

105. T. R. Thomas and D. H. Munger, "Evaluation of NO_x Abatement
      by NH_3 over H-Mordenite for Nuclear Processing Plants",
      1978, ICP-1133, NTIS, Springfield, Va., p 58.

      * U. S. Patents are obtainable from: Box 9, Patent
         Office, Washington, D. C. 20231 (Cost 50 cents).

Catalysis on X and Y Zeolites Containing Cupric Ions

F. Mahoney, R. Rudham and A. Stockwell

Department of Chemistry, University of Nottingham, Nottingham, NG7 2RD

Abstract

The oxidation of low concentrations of H_2, CH_4 or C_3H_6 in air was investigated in a flow system over a series of Cu,HX catalysts. All three reactions were first order, and the activity increased with Cu^{2+} content between 0.9 and 21.5 ions per unit cell. Increase of acidity at constant Cu^{2+} content had no effect on H_2 oxidation, but gave increased C_3H_6 oxidation and decreased CH_4 oxidation. Possible locations and mechanisms for the reactions are discussed.

To investigate the catalytic involvement of Brønsted acidity generated by cation hydrolysis alone, propan-2-ol dehydration to C_3H_6 and di-isopropyl ether was studied on a series of CuY catalysts. Zero order kinetics were observed using a flow of ~ 1% propan-2-ol in N_2, although the level of activity depended on the activation procedure. The steady-state activity initially increased with Cu^{2+} content, but was almost constant above 8 Cu^{2+} per unit cell. The activity was consistently below that for complete hydrolysis of Cu^{2+}, but was greatly increased by H_2S treatment. The reaction mechanism is believed to be the same as that for HX and HY catalysts.

Introduction

The exchange of the indigenous sodium ions in X and Y zeolites for other charge-balancing cations greatly enhances their catalytic activity for numerous reactions.[1-5] Many of the reactions proceed by carbonium ion intermediates, and are catalyzed by the strong Brønsted acidity of structural hydroxyl groups. Such groups are generated by deammoniation following NH_4^+ exchange, by cation hydrolysis, or by cation reduction with hydrogen or hydrogen containing compounds. For sustained catalytic activity in oxidation reactions X and Y zeolites normally contain transition metal ions or other cations which are readily involved in electron transfer processes. A number of investigations have been made of oxidation reactions on X and Y zeolites containing a range of transition metal ions, and activity sequences for equivalent or near-equivalent exchange ion concentrations have been reported.[5] Sequences found for the oxidation of H_2,[6,7] CO,[6] CH_4,[8] C_2H_4,[6] C_3H_6,[9,10] and cyclo-C_6H_{10},[11] show that the incorporation of Cu^{2+} invariably leads to high relative activity. However, few studies have been made of the variation in activity with the extent of Cu^{2+} exchange. In the present paper we compare the catalytic behaviour of a range of Cu,HX zeolites in H_2, CH_4 and C_3H_6 oxidation. Copper exchanged zeolites also exhibit activity in reactions involving carbonium ion intermediates. For the isomerization of butene[12] and dimethylbutenes[13] on CuX catalysts, it is believed that the necessary acidic centres are generated by reductive inter-action of Cu^{2+} and hydrocarbon. Whereas for alcohol dehydration on CuX[14] and CuY,[15,16] Cu^{2+} hydrolysis is held to be the source of the Brønsted acidity. Propan-2-ol dehydration occurs easily on weak acid sites,[4,17] and the rate can be taken as a measure of the Brønsted acidity available within the main pores of X and Y zeolites.[4,18]

Experimental

Oxidation reactions were studied in a flow system operating at a total pressure of one atmosphere.[7,10] For H_2 and CH_4 oxidation, the majority of experiments were made with a 25 cm^3 min^{-1} flow of 1% H_2 or 1% CH_4 in air, while for C_3H_6 oxidation a 50 cm^3 min^{-1} flow of 2.33% C_3H_6 in air was generally used. To determine kinetic orders, provisions existed for the variation of the partial pressures of H_2, CH_4, C_3H_6 and O_2 without altering the overall flow rate. Samples for analysis by gas chromatography were regularly taken from reactant and product streams by an automatic sampling valve. Precalibration of the chromatograph permitted the degree of oxidation α to be calculated from the peak heights for H_2, CH_4 or C_3H_6 in analyses of reactant and product samples. Catalyst beds were formed by supporting 1 g samples of zeolite on a No. 2 porosity glass sinter fused across the body of a glass reactor, and temperatures at the centre of the bed were measured with a chromel-alumel thermocouple.

Propan-2-ol dehydration was studied in a closely similar flow system, in which the total pressure of alcohol and nitrogen diluent was one atmosphere.[18] However, reaction was followed in terms of the rates of product formation from an 0.1g catalyst bed in a smaller reaction vessel.

The catalysts are listed in Table 1 and are designated $X/Cu(x)H(y)$ or $Y/Cu(x)$, where x and y are the percentages of Na^+ exchanged for Cu^{2+} or protons. Oxidation catalysts, 1-9, were prepared from NaX (Linde 13X, lot 0003940) by ion exchange with aqueous cupric sulphate following pH reduction of the zeolite slurry to pH = 6.[19] The additional acidity of catalysts 10 and 11 was obtained by further exchange of catalyst 5 with NH_4^+. To obviate acidity other than that arising from cation hydrolysis, catalysts 12-24 were prepared from NaY (Linde SK40, lot 3606-150) omitting pH reduction. Stoichiometric exchange occurred in all cases, since the amount of Cu^{2+} removed from solution was equivalent to the amount of Na^+ released from the zeolite.

The gases used were of the purest cylinder grades available, and no impurities could be detected by gas chromatography. Propan-2-ol was B.D.H. analytical reagent grade.

Results

Hydrogen, Methane and Propene Oxidation. The reactions were investigated on Cu,HX zeolites under conditions where H_2O and CO_2 were the only detectable products. Prior to catalytic measurements zeolite samples were dehydrated in a flow of N_2 for 16h; the dehydration temperatures were 723, 823 and 623 K for H_2, CH_4 and C_3H_6 oxidation respectively. Kinetic orders with respect to H_2 or CH_4 pressure over the range 0.2 - 1.2 kPa were determined for eight of the Cu,HX catalysts. The temperatures at which the measurements were made were selected to be as close to those at which the reactor operated in the differential mode compatible with calculating rates from differences in reactant pressure. Kinetic orders are the slopes of logarithmic plots of rate against reactant pressure, and the mean values were 0.94 for H_2 oxidation and 0.83 for CH_4 oxidation. For C_3H_6 oxidation the kinetic order was previously shown[10] to be 1.1. Kinetic orders with respect to O_2 over the range 10-25 kPa were also determined; the mean of three values was 0.17 for H_2 oxidation and 0.09 for CH_4 oxidation. Taking the kinetic orders in H_2, CH_4 or C_3H_6 to be unity and that in O_2 to be zero, first order rate constants, in molecules $g^{-1} s^{-1} Pa^{-1}$, are given by the integrated first order rate equation for a fixed bed flow reactor:

$$k_1 = \frac{f}{mp} \left[(1+y\delta)\ln(1/1-\alpha) - y\delta\alpha \right] .$$

In this expression, f is the reactant feed rate in molecules s^{-1} at pressure p in Pa, m the mass of catalyst in g, y the mole fraction of reactant in the reactant mixture stream, δ the increase in the number of molecules in the reaction per reactant molecule, and α the degree of oxidation. The complete oxidation presently observed leads to values for δ of -0.5, 0.0 and 0.5 for H_2, CH_4 and C_3H_6 respectively.

Table 1. Arrhenius parameters for hydrogen, and methane oxidation

No.	Composition	$E/$ kJ mol^{-1} (H_2)	$\log_{10}(k_1^o/$ molecules $g^{-1}s^{-1}Pa^{-1}$) (H_2)	$E/$ kJ mol^{-1} (CH_4)	$\log_{10}(k_1^o/$ molecules $g^{-1}s^{-1}Pa^{-1}$) (CH_4)
1	X/Cu(2)H(20)	65	18.99	132	21.75
2	X/Cu(5)H(20)	66	19.14	136	22.04
3	X/Cu(10)H(20)	64	19.48	138	22.64
4	X/Cu(15)H(20)	63	19.60	137	22.70
5	X/Cu(20)H(20)	64	19.88	135	22.63
6	X/Cu(25)H(20)	63	19.85	134	22.63
7	X/Cu(30)H(20)	63	19.99	129	22.33
8	X/Cu(41)H(20)	64	20.31	119	21.73
9	X/Cu(50)H(20)	67	20.92	115	21.51
10	X/Cu(20)H(30)	61	19.58	130	22.01
11	X/Cu(20)H(45)	62	19.66	138	22.31

The effect of temperature on the activity of each catalyst for H_2 and CH_4 oxidation was determined within the range $\alpha = 0.02 - 0.75$ with a constant initial reactant pressure of 1.01 kPa. Measurements were made at an average of 10 temperatures in random sequence, and after the initial measurement reproducibly stable activities were rapidly attained on changing the temperature in either direction. For each catalyst measurements were made over a range of ~100 degrees between 436 and 648 K for H_2 oxidation and between 691 and 823 K for CH_4 oxidation. All catalysts gave good linear plots of $\log_{10} k_1$ against T^{-1}, showing that the Arrhenius equation $k_1 = k_1^o \exp(-E/RT)$ adequately describes the effect of temperature on activity. Values of the apparent activation energy E and the pre-exponential factor k_1^o are given in table 1. Values for C_3H_6 oxidation on the same range of catalysts has been given in an earlier publication.[10]

It has been shown that the introduction of a low pressure of pyridine vapour into the reactant stream caused complete poisoning of C_3H_6 oxidation on HX and Cu,HX catalysts.[10] A similar pressure of pyridine was therefore introduced into the reactant stream used in studies of H_2 and CH_4 oxidation over catalyst 5. No poisoning effect was observed for either reaction over a period of 3h.

Propan-2-ol Dehydration. The reaction was investigated on CuY zeolites, where the only detectable products were C_3H_6 and di-isopropyl ether. The steady-state activities for the formation of both products obeyed zero order kinetics with respect to propan-2-ol pressures between 0.50 - 4.78 kPa in a total reactant-diluent flow of 30 cm^3 min^{-1}. It follows that the rates of ether and C_3H_6 formation expressed in molecules $g^{-1}s^{-1}$, r_e and r_p respectively, are the zero order rate constants k_e and k_p. Since the development of the full steady-state activities of freshly dehydrated catalysts took a considerable time, the effects of different dehydration pretreatments were investigated. Samples of Y/Cu(20) were dehydrated in a 40 cm^3 min^{-1} flow of N_2 for 2h at various temperatures, and the subsequent catalytic activity at 403 K monitored until it became constant. In these and all subsequent experiments a propan-2-ol pressure of 1.11 kPa in a total flow of 30 cm^3 min^{-1} was used. The steady-state values of k_e and k_p, together with the times taken to achieve these, are plotted against dehydration temperature in **Figure 1**. It is evident that the optimum activity is most rapidly achieved following dehydration for 2h at 523 K, and this procedure was followed in subsequent experiments. The influence of the nature of the dehydrating gas flow at 523 K on the subsequent activity at 403 K was investigated with Y/Cu(7.5). The activities (expressed in molecules $g^{-1}s^{-1}$) increased with increasing water content of the gas: He ($k_e = 9.2\times10^{14}$, $k_p = 5.8\times10^{15}$) < N_2 dried over 3A zeolite ($k_e = 1.5\times10^{15}$, $k_p = 8.2\times10^{15}$) < N_2-O_2 80/20 ($k_e = 3.2\times10^{15}$, $k_p = 1.9\times10^{16}$) < N_2 ($k_e = 4.0\times10^{15}$, $k_p = 2.2\times10^{16}$).

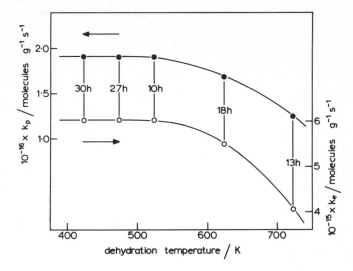

Figure 1. Activity of Y/Cu(20) at 403 K as a function of
dehydration temperature: o-k_e, $\bullet$-k_p. Times are those taken
to achieve steady-state activity.

It appears that too high a temperature or too dry a gas results in the
migration of Cu^{2+} ions to sites from which they cannot be extracted and
hydrolysed to give accessible Brønsted acidity. The reaction vessel was
replaced by an infra-red cell in which sample discs could be given the same
dehydration treatment as the catalysts. The spectrum of Y/Cu(30) dehydrated
in N_2 for 2h at 523 K showed a broad band between 3700-3200 cm^{-1} associated
with OH stretch, and a narrow band at 1641 cm^{-1} which is diagnostic of OH bend
in molecular water.

The effect of temperature on k_e and k_p was investigated on nine catalysts
of varying Cu^{2+} content. Measurements covering a range of ~30 degrees
between 360 and 420 K gave excellent Arrhenius plots, and values of the
activation energies E_e,E_p and pre-exponential factors k_e^o,k_p^o are given in
Table 2.

The catalytic activity of zeolites is frequently affected by reactive
gases, so that the activity of Y/Cu(15) at 377 K was determined following a
15 min treatment with ~20 cm^3 of CO_2, SO_2 or H_2S. The activities (expressed
in molecules $g^{-1}s^{-1}$) before treatment were k_e = 3.6×10^{15}, k_p = 1.3×10^{16}; after
CO_2 treatment the activities fell to k_e = 1.3×10^{15}, k_p = 5.3×10^{15} but partially
recovered over 24h. Treatment with SO_2 increased the activities to
k_e = 5.4×10^{15}, k_p = 1.9×10^{16}, while H_2S greatly increased the activities to

Table 2. Arrhenius parameters for product formation in propan-2-ol dehydration

catalyst		di-isopropyl ether		propene	
No.	composition	E_e/ kJ mol^{-1}	$\log_{10}(k_e^o/$ molecules g^{-1} s^{-1})	E_p/ kJ mol^{-1}	$\log_{10}(k_p^o/$ molecules g^{-1} s^{-1})
12	Y/Cu(5)	82	25.66	98	28.35
13	Y/Cu(7.5)	86	26.75	101	29.47
14	Y/Cu(7.5)-H$_2$S	101	29.37	111	31.38
15	Y/Cu(10)	98	29.00	105	30.54
16	Y/Cu(15)	100	29.41	109	31.22
17	Y/Cu(15)-H$_2$S	99	29.53	110	32.14
18	Y/Cu(20)	103	30.05	115	32.21
19	Y/Cu(20)-H$_2$S	104	31.06	113	32.67
20	Y/Cu(30)	121	32.77	128	34.17
21	Y/Cu(30)-H$_2$S	99	30.71	111	32.92
22	Y/Cu(40)	122	32.78	134	34.97
23	Y/Cu(48.2)	122	32.96	132	34.77
24	Y/Cu(57.5)	120	32.81	132	34.77

$k_e = 2.2 \times 10^{16}$, $k_p = 7.9 \times 10^{16}$. After SO_2 treatment trace amounts of acetone were produced, but activity remained exclusively dehydrating after CO_2 and H_2S treatment. The increase in activity brought about by H_2S possessed a similar stability to that in untreated CuY catalysts, since there was no diminution in activity at 377 K following heating at 523 K for 2 h in flowing nitrogen. Both the activity of untreated Y/Cu(15) and of gas treated samples can be associated with Brønsted acidity, since complete poisoning with pyridine readily occurred at 377 K. To further investigate the effects of H_2S, the activities of H_2S treated Y/Cu(7.5), Y/Cu(15), Y/Cu(20) and Y/Cu(30) were determined over a ~30 degree temperature range. As with untreated samples, the data yielded excellent Arrhenius plots, and values of E_e, E_p, k_e^o and k_p^o are given in **Table 2.**

Discussion

Activity sequences for H_2, CH_4 and C_3H_6 oxidation on X zeolites containing both transition metal ions and protons, M,HX, have been reported for samples in which both M and H concentrations correspond to ~20% exchange.[7,8,10] For H_2 oxidation[7] the sequence, based on the temperature necessary to maintain a fixed reaction rate, was: Pd,HX >> Cu,HX > Fe,HX > Ni,HX > Cr,HX > Zn,HX > Mn,HX > Co,HX > HX. The sequence for CH_4 oxidation[8] was similarly based and was: Pd,HX >> Cu,HX > Cr,HX > Ni,HX > Fe,HX ≈ Mn,HX > Co,HX > HX > Zn,HX. For C_3H_6 oxidation[10] the sequence based on the extent of reaction at a fixed temperature was Pd,HX >> Cu,HX > Co,HX > Zn,HX > Ni,HX > Mn,HX > Cr,HX > Fe,HX > HX. Although the sequences differ, they emphasize the high relative activity of Cu,HX in oxidation reactions, while the absence of poisoning effects with hexamethyldisiloxane in CH_4 oxidation[8] confirmed that reaction proceeded within the zeolite structure.

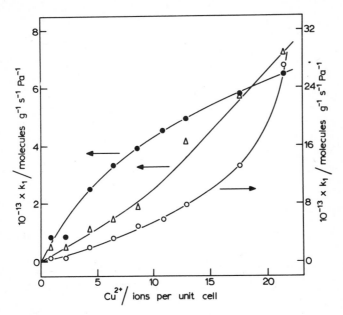

Figure 2. Activity of Cu,HX zeolites in oxidation reactions as a function of Cu^{2+} content: $o-H_2$ at 540 K, $\bullet-CH_4$ at 780 K, $\triangle-C_3H_6$ at 480 K.

The present results show that the three oxidation reactions are first order in H_2, CH_4 or C_3H_6, and effectively zero order in O_2 under conditions where this gas is in considerable excess of that required for stoichiometric oxidation to H_2O and CO_2. In **Figure 2**, first order rate constants at one temperature for each reaction are plotted against the Cu^{2+} content in ions per unit cell. The temperatures were selected to be within the experimental range over which all the catalysts were studied and to give comparable activities. It is clear that the activity for all three reactions increases with the extent of ion exchange, although the form of the plot depends upon the particular reaction. For H_2 oxidation, which shows the greatest increase in activity with Cu^{2+} content, analysis of the plot is justified since the apparent activation energy was constant. It was found that a logarithmic plot of k_1 at 540 K against Cu^{2+} content was linear, but the slope of 1.5 does not permit a decision concerning the number of Cu^{2+} ions constituting an active centre. The non-integer dependence probably arises from the distribution of Cu^{2+} between active and inactive sites being a function of the extent of exchange. The apparent activation energy for CH_4 oxidation is initially constant, but falls above a Cu^{2+} content corresponding to 25% exchange. It follows that the shape of the plot in **Figure 2** does not arise from energetic factors, but from a progressive decrease in the ability of Cu^{2+} ions to generate active sites. With C_3H_6 oxidation the opposite occurs, and the increasing effectiveness of Cu^{2+} ions in generating active sites more than compensates for a ~50% increase in the apparent activation energy.

Table 3. The effect of Brønsted acidity on the activity for catalytic oxidation

Catalyst	$10^{-13}xk_1$/molecules $g^{-1}s^{-1}Pa^{-1}$		
	H_2 at 540 K	CH_4 at 780 K	C_3H_6 at 480 K
X/Cu(20)H(20)	4.89	3.90	1.86
X/Cu(20)H(30)	4.78	2.02	2.96
X/Cu(20)H(45)	4.61	1.15	4.14

The effect of increased Brønsted acidity on catalytic activity at constant Cu^{2+} content can be assessed from the first order rate constants given for catalysts 5, 10 and 11 in **Table 3**. The temperatures for the three reactions are identical with those for the data plotted in **Figure 2**. For C_3H_6 oxidation the increase in activity with acidity, and sensitivity to poisoning by pyridine, lead to the proposal[10] that the initial step is the adsorption of C_3H_6 as a secondary carbonium ion by proton transfer at a Brønsted acid site in the α-cage network. The acid sites arise either from proton exchange, hydrolysis of Cu^{2+}, or possibly from reduction of Cu^{2+} with C_3H_6 or its partial oxidation products. Since Cu,HX was consistently more active than HX, it was considered that attack of the carbonium ions by oxygen activated by neighbouring Cu^{2+}/Cu^{+} was more likely than direct attack by molecular oxygen. The manner in which the activity for CH_4 oxidation decreases with increasing acidity, together with the absence of pyridine poisoning, suggests that the additional protons occupy α-cage sites that would otherwise be occupied by Cu^{2+} ions contributing to catalysis. However, at the high temperature necessary for measurable activity, and thus used in catalyst pretreatment, loss of Brønsted acidity by dehydroxylation is most likely. It follows that CH_4 oxidation does not, as might be anticipated, involve carbonium ion intermediates. This difference between CH_4 and C_3H_6 oxidation may account for the considerable differences in position of the less active catalysts within the activity sequences for the two reactions. In view of the high temperatures involved, the initial step in CH_4 oxidation might involve the reduction of a Cu^{2+} ion in an α-cage site:

$$Cu^{2+} + O_z^- + CH_4 \longrightarrow Cu^+-CH_3 + O_zH$$

Subsequent reaction of the methyl entity with oxygen can be expected to be rapid if it is retained or if it is released as a radical into the "oxygen containing atmosphere" of the α-cage.

With H_2 oxidation the absence of effects from increased acidity and pyridine vapour suggest that the reaction proceeds at a different location from CH_4 and C_3H_6 oxidation. Molecular size considerations indicate that reaction within the β-cages is a possibility. The present results compare favourably with those of Davydova et al[20] for H_2 oxidation on Cu,HY zeolites in a recirculatory flow system. First order kinetics in H_2 were observed in both studies, atomic catalytic activities at 573 K were consistently within a factor of seven, and the present mean activation energy of 64 kJ mol^{-1} is in reasonable agreement with their mean value of 71 kJ mol^{-1}. Since the activation energy was similar to that for H_2 oxidation on bulk CuO, partially covalent Cu-O-Cu bridges common to both catalysts were considered to be the catalytically active sites.[20] The agreement between our value for Cu,HX and that of 68 kJ mol^{-1} for H_2 oxidation on CuO supported on porous and non-porous silica-alumina[7] supports this suggestion. The stoichiometry of ion exchange in catalysts 1-9 indicates that activity arises from Cu-O-Cu bridges within the β-cages,[21] rather than from "CuO-like phases" within the α-cages.[22] The first-series transition metal ions which significantly enhance the activity of X zeolite for H_2 oxidation, Cr^{3+}, Fe^{2+}, Ni^{2+} and Cu^{2+}, are known to form M-O-M bridges within the β-cages of the faujasite structure common to X and Y zeolites.[21,23,24,25] Results from studies of temperature programmed reduction of X zeolites containing these ions[7] suggest that it is the reactivity of such bridges to H_2, rather than their concentration, which controls the pattern of catalytic activity.

The formation of the structural O_zH groups yielding the Brønsted acidity of CuY zeolites does not occur until the $Cu^{2+}(H_2O)_n$ ions are dehydrated to the extent n = 1, when the polarizing field of the partially unscreened Cu^{2+} is sufficient to dissociate the remaining water molecule:

$$Cu^{2+}(H_2O) + O_z^- \rightarrow Cu^+OH + O_zH$$

Further dehydration may occur under more stringent conditions, but no further acidity is generated:

$$2Cu^+OH + 2O_zH \rightarrow Cu^+-O-Cu^+ + 2O_zH + H_2O$$

The exact location of the various copper species from $Cu^{2+}(H_2O)_6$ to Cu^{2+} has been the focus of considerable attention, and the distribution between sites I, I', II', II and III depends upon the extent of dehydration and the presence of reactive gases.[5] In the present measurements of propan-2-ol dehydration the observed zero order kinetics show that the reaction rate may be taken as a direct measure of the Brønsted acidity within the α-cages, while the absence of acetone as a product precludes the intrazeolitic "CuO-like phases" responsible for dehydrogenating activity.[22] Consideration of the experiments made to determine the optimum pretreatment conditions shows that dehydration at 623 and

723 K, or in gases of lower water content than N_2, gives a lower steady state
activity at 377 K than pretreatment in N_2 at 523 K. This can be ascribed to a
larger proportion of the copper being located as Cu^{2+} in I sites from which it
is not easily extracted. A similar dependence of activity on pretreatment
temperature has been observed for ethanol dehydration on CuY.[15] Dehydration
in N_2 at 423, 473 and 523 K results in the same final activity, but the times
taken to reach this suggest that 523 K produces an acidity nearest to that
established under reaction conditions at 377 K. Infra-red spectroscopy showed
the retention of appreciable amounts of molecular water by Y/Cu(30) dehydrated
at 523 K, so that the presence of both $Cu^{2+}(H_2O)_n$ and H_3O^+ within the α-cages
is a possibility.

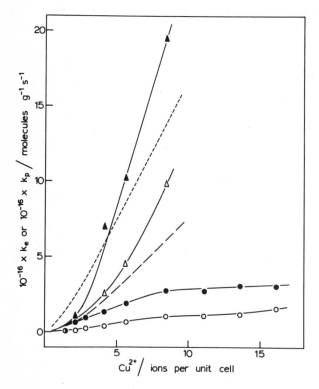

Figure 3. Activity of CuY zeolites for propan-2-ol
dehydration at 377 K as a function of Cu^{2+} content: CuY,
$o-k_e$, $\bullet-k_p$; H_2S treated CuY, $\Delta-k_e$, $\blacktriangle-k_p$. Activity of HY
zeolites with extents of exchange equivalent to the Cu^{2+}
contents: —k_e, ----k_p.[16]

Figure 3 shows that the activities for di-isopropyl ether and propene
formation at 377 K initially increased with Cu^{2+} content, but were almost

constant above 8 Cu^{2+} per unit cell. At all concentrations the activities per Cu^{2+} ion fell below those per O_zH in NaY.[16] It follows that the proportion of Cu^{2+} ions in the hydrated state, or located where they cannot contribute to the α-cage acidity, increases with the extent of exchange. Although zero order kinetics were observed, table 2 shows that E_e increased from 82 to 122 kJ mol^{-1} and E_p increased from 98 to 132 kJ mol^{-1} with increasing Cu^{2+} content. For a range of HY catalysts made from the same batch of NaY, E_e and E_p were all within 4 kJ mol^{-1} of the respective mean values of 99 and 111 kJ mol^{-1}.[26] Thus for CuY zeolites the concentration of accessible acidic centres is a function of temperature, and E_e and E_p differ from those for HY zeolites by a contribution from the ΔH values associated with changes in location and hydration of the Cu^{2+} ions.

The experiments with Y/Cu(15) at 377 K show that considerable changes in activity can be brought about by reactive gases. However, selectivities for ether formation, $S_e = 2k_e/(2k_e + k_p)$, remained effectively constant; 0.36 before treatment and 0.33, 0.36 and 0.36 after CO_2, SO_2 and H_2S treatment. This, together with the poisoning effects of pyridine, show that activity changes arose from changes in the concentration of acidic centres rather than the introduction of alternative reaction paths. The increase in activity following H_2S treatment is in agreement with the observation that H_2S considerably increased the activity of CuY for cumene cracking at 573 K.[27] Figure 3 shows that the activities of H_2S treated CuY zeolites compare favourably with those of HY zeolites of nominally equivalent proton content. This, together with the pronounced blackening generated by H_2S treatment, suggest that the increase in activity arises from 2 O_zH groups generated from every Cu^{2+} by the reaction:

$$Cu^{2+} + 2O_z^- + H_2S \rightarrow 2O_zH + CuS$$

In addition, existing acidity would be enhanced by:

$$Cu^+OH + O_zH + O_z^- + H_2S \rightarrow 2O_zH + CuS + H_2O$$

The stable catalytic activity indicates that the neutral CuS resists hydrolysis, and it probably exists as agglomerates within the α-cages. Infra-red spectroscopy proved to be unsatisfactory due to the black colour, but it showed that the OH bending band of molecular water at ~1640 cm^{-1} had been eliminated. Such an observation is in accord with all the Cu^{2+}, irrespective of the extent of hydration, being converted to CuS. Unfortunately, conventional X-ray powder diffraction failed to detect CuS, but it showed that the zeolite structure was unaffected by H_2S treatment. Table 2 shows that the values of E_e and E_p for H_2S treated catalysts are almost identical with the corresponding values of 99 and 111 kJ mol^{-1} for HY zeolites.[26] This may be taken as evidence for the same reaction mechanism as that proposed for HY zeolites,[16] and that the concentration of acid sites does not vary with temperature as in untreated CuY.

In the steady catalytic state water is present so the O_zH groups are ionised to form hydroxonium ions:

$$\overset{|}{\underset{O}{H}} + H_2O \quad \longrightarrow \quad \overset{+}{H_3}\underset{O}{\overset{O}{:}}{}_-$$

Proton transfer to propan-2-ol gives an oxonium ion, which yields a carbonium ion by the E_1 elimination of H_2O:

$$\overset{+}{H_3}\underset{O}{O}{:}{}_- + (Me)_2CHOH \quad \longrightarrow \quad (Me)_2\overset{+}{CH}\underset{O}{\underset{-}{O}H_2} + H_2O$$

$$(Me)_2\overset{+}{CH}\underset{O}{\underset{-}{O}H_2} \quad \longrightarrow \quad (Me)_2\overset{+}{CH}\underset{O}{:}{}_- + H_2O$$

Subsequent proton transfer to water yields propene and regenerates the hydroxonium ion:

$$H_2O + (Me)_2\overset{+}{CH}\underset{O}{:}{}_- \quad \longrightarrow \quad \overset{+}{H_3}\underset{O}{O}{:}{}_- + MeCHCH_2$$

Alternatively, interaction of the carbonium ion with propan-2-ol gives an oxonium ion, which on proton transfer to water yields di-isopropyl ether and regenerates the hydroxonium ion:

$$(Me)_2CHOH + (Me)_2\overset{+}{CH}\underset{O}{:}{}_- \quad \longrightarrow \quad (Me)_2\overset{+}{CH}OHCH(Me)_2 \underset{O}{:}{}_-$$

$$H_2O + (Me)_2\overset{+}{CH}OHCH(Me)_2 \underset{O}{:}{}_- \quad \longrightarrow \quad \overset{+}{H_3}\underset{O}{O}{:}{}_- + (Me)_2CHOCH(Me)_2$$

The E_1 elimination of water probably makes the largest contribution to the overall activation energy, and this accounts for the similarity in E_e and E_p. This mechanism, although basically similar to those previously suggested for propan-2-ol dehydration,[17,18] differs in that only a single site is involved in ether production.

References

1   P.B. Venuto and P.S. Landis, Adv. Catalysis, 1968, 18, 259.

2   M.L. Poutsma, "Zeolite Chemistry and Catalysis", A.C.S. Monograph 171, Washington, 1976, Chapter 8, p.437.

3   M.L. Poutsma, "Zeolite Chemistry and Catalysis", A.C.S. Monograph 171, Washington, 1976, Chapter 9, p.529.

4   P.A. Jacobs, "Carboniogenic Activity of Zeolites", Elsevier, Amsterdam, 1977.

5   R. Rudham and A. Stockwell, "Catalysis" (Specialist Periodical Reports), The Chemical Society, London, 1977, Vol. 1, p.87.

6   S.Z. Roginskii, O.V. Al'tshuler, O.M. Vinogradova, V.A. Seleznev and I.L. Tsitovskaya, Doklady Akad. Nauk. SSSR, 1971, 196, 872.

7   F. Mahoney, R. Rudham and J.V. Summers, J.C.S. Faraday I, 1979, 75, 314.

8   R. Rudham and M.K. Sanders, J. Catalysis, 1972, 27, 287.

9   I. Mochida, S. Hayata, A. Kato and T. Seiyama, J. Catalysis, 1971, 23, 31.

10   S.J. Gentry, R. Rudham and M.K. Sanders, J. Catalysis, 1974, 35, 376.

11   I. Mochida, T. Jitsumatsu, A. Kato and T. Seiyama, Bull. Chem. Soc. Japan, 1971, 44, 2595.

12   C. Kemball, H.F. Leach and I.R. Leith, J. Chem. Research (S), 1977, 4.

13   C. Kemball, H.F. Leach and B.W. Moller, J.C.S. Faraday I, 1973, 69, 624.

14   M. Baile, A. Cortes and J. Soria, Anales de Quim., 1974, 70, 305.

15   S. Yoshida, K. Akimoto, Y. Koshimidzu and K. Tarama, Bull. Inst. Chem. Res. Kyoto Univ., 1975, 53, 127.

16   R. Rudham and A. Stockwell, Acta. Physica et Chemica Szegediensis, 1978, 26, 281.

17   P.A. Jacobs, M. Tielen and J.B. Uytterhoeven, J. Catalysis, 1977, 50, 98.

18   S.J. Gentry and R. Rudham, J.C.S. Faraday I, 1974, 70, 1685.

19   R. Rudham and M.K. Sanders, "Chemisorption and Catalysis", Institute of Petroleum, London, 1971, p.58.

20   L.P. Davydova, G.K. Boreskov, K.G. Ione and V.V. Popovskii, Kinetics and Catalysis, 1975, 16, 91.

21   C.C. Chao and J.H. Lunsford, J. Chem. Phys., 1972, 57, 2890.

22   R.A. Schoonheydt, L.J. Vandamme, P.A. Jacobs and J.B. Uytterhoeven, J. Catalysis, 1976, 43, 292.

23   I.D. Mikheikin, O.I. Brotikovskii and V.B. Kazankii, Kinetics and Catalysis, 1972, 13, 481.

24   T. Kubo, H. Tominago and K. Kunugi, Bull. Chem. Soc. Japan, 1973, 46, 3549.

25   R.A. Dalla Betta, R.L. Garten and M. Boudart, J. Catalysis, 1976, 41, 40.

26   R. Rudham and A. Stockwell, unpublished observations.

27   M. Sugioka, T. Hosotsubo and K. Aomura, J.C.S. Chem. Comm. 1976, 54.

Catalysis by Highly Siliceous Zeolites

M S Spencer and T V Whittam

Imperial Chemical Industries Limited, Agricultural Division
Billingham

Introduction

In the last ten years many new zeolites have been discovered
with SiO_2/Al_2O_3 greater than 15. These materials, especially
ZSM5, have been shown to possess unique catalytic properties,
and we believe that they will open up a new catalyst era.

In this paper we compare our two novel zeolites Nul and Ful
with ZSM5 and mordenite in relation to their sorptive and
catalytic behaviour.

Experimental

Zeolite Nul.   Zeolite Nul is a novel material readily
identified by its unique X-ray of diffraction and infra red
data.   It can have SiO_2/Al_2O_3 ratios from 15 → 180, and is
preferably synthesised with tetramethylammonium (TMA) as
a major cation.   Detailed information is available in patents[1]
and in a recent paper[2].   Typical X-ray data are given in
Table 1.   At least five different varieties of Nul have been
identified by subtle variations in X-ray data.

Table 1 - Zeolite H-Nul

d(A)	100I/Io	d(A)	100I/Io	d(A)	100I/Io
8.87	18	4.30	51	3.81	22
8.28	69	4.08	37	3.687	16
6.53	43	4.03	100	3.508	29
6.19	75	3.965	73	3.256	27
4.45	52	3.845	74	2.858	15

The Nul used in this work was prepared as reported elsewhere[2],
and had a crystallite size of $\sim$ 300 nm.   Its composition
after drying at 120° was :- 0.73 Na_2O, 2.1 $(TMA)_2O$,
Al_2O_3, 50 SiO_2, 10.4 H_2O.

H-Nul Preparation.   The Nul was exchanged twice with 3 mls
1.5 N hydrochloric acid per gram (5 hours at $95^{\circ}C$).   After
filtering, washing and drying it was calcined in 50° steps
from 150 - $450^{\circ}C$ for 1 hour/step.   The temperature of $450^{\circ}C$
was maintained for 17 hours and then raised quickly to $550^{\circ}C$,
which was maintained for 3 hours. The product was:-
0.01 Na_2O, Al_2O_3, 50 SiO_2 and contained 0.3% w carbon and
< 0.05% w nitrogen.

Sorption Studies on H-Nul.   To obtain reproducible sorption
data it was necessary to thoroughly "clean up" the TMA zeolites
by calcining in ammonia at 450°C for up to 24 hours followed
by a coke burn off in air.   The products contained less than
0.1% w carbon.   Failure to "clean up" gave rise to non-
reproducible water sorption and variable hydrophobic
behaviour, along with reduced capacities for larger molecules.
Sorption data were obtained on McBain Bakr type balances and
typical results after 2 hours at 23°C and p/po = 0.5
(p/po = 0.25 for water) are given in Table 2, where σ
is the minimum kinetic diameter[3] of the sorbate.

Table 2 - Sorption on H-Nul

Sorbate	σ nm	% w/w sorbed	Apparent Voidage cc/100g
Cyclohexane	0.60	< 0.05	–
p-xylene	0.585	3.8	4.4
pyridine	0.58	4.4	4.5
isobutane	0.50	3.0	4.6
n-hexane	0.43	4.0	6.1
water	0.265	7.8	7.8

Sorption isotherms for all sorbates were rectangular as were
kinetic curves for n-hexane and isobutane.   However, plots
of sorption rates for p-xylene, pyridine and water were as
shown in Figure 1.

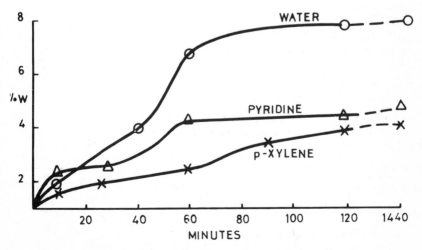

FIGURE 1. SORPTION   RATES ON HNul

Zeolite Ful.   A second novel zeolite designated Ful has
been synthesised in the TMA/Na field.   Data have so far only
been published in patent form[4].   Ful can be prepared with any
alkali cation including ammonium, but a methylated quaternary
compound or certain degradation products are an essential
feature of any synthesis.   The most typical products have
SiO_2/Al_2O_3 ratios between 15 and 35, and the morphology is
very unusual.   Typically, as seen on electron micrographs,
Ful consists of angularly interlocking platelets which are
from 5 - 10 nm thick, and are usually from 50 - 100 nm in
the other dimensions.   These platelets always agglomerate
in packs of from 300 $\rightarrow$ 10^4 nm.   X-ray analysis confirms very
small crystallites because very substantial line broadening
is evident on diffractometer traces.   However, these are
very reproducible and characteristic of the whole range of
products.   Attempts to grow larger crystals of Ful have
failed so far.   Typical X-ray data for H-Ful are given in
Table 3.

Table 3 - Zeolite Ful in Hydrogen Form

d(A)	100I/Io	d(A)	100I/Io	d(A)	100I/Io	d(A)	100I/Io
16	∿ 12	5.86	3	3.65	∿ 23	2.61	∿ 2
14.3	∿ 7	5.68	∿ 2	3.40	100	2.53	∿ 2
11.4	∿ 4	5.19	3	3.20	∿ 3	2.45	∿ 2
9.21	56	4.54	46	3.12	∿ 2	2.42	∿ 2
8.00	6	4.20	∿ 2	3.03	∿ 5	2.37	∿ 3
6.85	63	4.06	∿ 15	2.88	11	2.08	2
6.6	∿ 5	3.71	∿ 24	2.64	∿ 2	2.00	5

Relative intensities marked ∿ are approximate because of
interference from adjacent lines.   The X-ray and infra-red
data and its morphology place Ful clearly as a distinct
species.   It is more robust than Nul and air calcination
to remove occluded TMA presents no serious problems.   The
hydrogen form does not lose all vestige of crystallinity
until temperatures in excess of $900^{\circ}C$.

In as made Ful up to 2.5 moles of TMA per Al_2O_3 can be
incorporated as well as alkali cations.   Usually about 20%
of the TMA can be removed rapidly by heating to $400^{\circ}C$, and
this is evolved as trimethylamine and methanol.   The
remaining TMA can only be removed by oxidative burn out at
temperatures in excess of $400^{\circ}C$.   If carbonaceous residues
are not fully removed then hydrogen forms are extremely
hydrophobic, but eventually become more hydrophilic on
multiple regenerations and re-exposure to water.   The samples
used in this work were prepared in a stainless steel 44 litre
autoclave with turbo stirring at $180^{\circ}C$ for 17 hours, from
reaction mixtures of composition:- 3 Na_2O, 1.5 $(TMA)_2O, Al_2O_3$
25 SiO_2, 900 H_2O.   After filtration, washing and drying at
$120^{\circ}C$ for 17 hours, the product had a composition:-
0.3 Na_2O, 1.3$(TMA)_2O$, $Al_2O_3, 24.6 SiO_2$, 5.8 H_2O.

H-Ful Catalyst Preparation. The Ful was exchanged once with 3 mls N HCl per g at 90°C for 1 hour. After filtering, washing and drying, the pelleted product was fired at 450°C in air for 48 hours. Ignoring H the product was:- 0.02 Na_2O Al_2O_3 24 SiO_2 and contained 0.3% w carbon and ⩤ 0.05% w nitrogen.

Sorption Studies on H-Ful. Samples of Ful for sorption studies were "cleaned up" in the same way as Nul. All isotherms and rate curves were determined on Microforce electronic balances at 23°C. Data given in Table 4 were after 2 hours at p/po = 0.5 (water at p/po = 0.25)

Table 4 - Sorption on H-Ful

Sorbate	σ nm	% w/w sorbed	Apparent voidage filled cc/100 g	No molecules per g
Triethyl-benzene	0.8	0.2 .	0.28	7.4×10^{18}
cyclohexane	0.6	3.7(10mins) 5.7(17 hrs)	4.7 7.2	2.7×10^{20} 3.7×10^{20}
p-xylene	0.585	7.7	8.9	5.0×10^{20}
n-hexane	0.43	6.8	10.4	5.0×10^{20}
methanol	0.38	6.0	10.0	1.4×10^{21}
water	0.265	8.5	8.5	2.8×10^{21}
ammonia	0.26	7.7	10.0	2.7×10^{21}

Full isotherms are given for ammonia and water in Figure 2 and rate curves at p/po = 0.5 (water 0.25) are given in Figure 3.

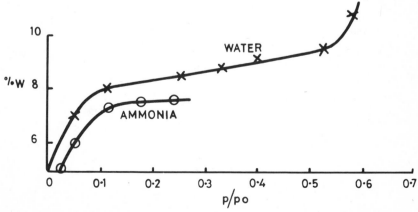

FIGURE 2. ISOTHERMS AT 23°C FOR H Ful

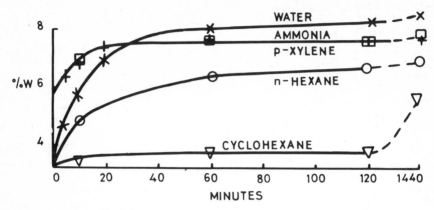

FIGURE 3. SORPTION RATES ON HFul

Catalytic Reactions

Conventional pulse microreactor systems were used to obtain
the initial activity of zeolites in the catalytic conversion
of methanol or other organic compounds to hydrocarbon
products.   1 or 2 $\mu\ell$ injections of reagent were vaporised
in a nitrogen stream, and passed over a bed containing
0.3 - 0.4 g of 3mm catalyst pellets.   The reaction products
were analysed by on-line gas chromatography : 3m column of 15%
OV101 silicone on Chromasorb, programmed for 60° - 300°C,
was used for C_5^+ hydrocarbons, especially aromatics; 2m
column of Carbopack C/0.19% picric acid + 1.5m and 6m columns
of phenyl isocyanate on Porosil C (with intermediate back-
flush) were used for $C_1 - C_4$ hydrocarbons.   The calculated
product distributions do not include involatile deposits
(eg coke) on the catalyst or CO, CO_2 etc.   Fast decay of
initial activity was studied by the injection of much larger
titres of reagents, followed by further 1 or 2 $\mu\ell$ injections.
Slow decay and steady state performance were investigated with
flow reactors (10 ml catalyst bed of 3mm pellets) with
appropriate off-line analysis.   On-line activation/regeneration
of catalysts was possible in all units, typically 18 hr at
450°C under fast air flow.

H-mordenite was obtained from Norton Co (Zeolon 100).   HY and
HZSM-5 were synthesised by published procedures.   The
amorphous silica-alumina was a commercial cracking catalyst
containing 12% Al_2O_3.

Methanol Conversion

Several different samples of HNu-1 were tested in the pulse microreactor at 375° and 450°C.   Wide variations in activity for the conversion of methanol to hydrocarbons were found, but these could not be correlated with differences in either preparative methods or observable properties (eg crystallite size, X-ray diffraction pattern, residual sodium content) of the catalyst.   Nevertheless the yields of aromatics were consistent over a range of conversions and a similar product distribution was obtained from dimethyl ether feed.   Aromatics and light hydrocarbon distribution from two samples of HNu-1 are compared with those from other zeolites in Tables 5 and 6.

Table 5 - Initial Aromatics Product Distribution (%) 450°C

Zeolite Catalyst	HNu-1 A*	HNu-1 B*	HZSM-5	H-mordenite	HY†
Benzene	1	8	14	41	9
Toluene	17	12	33	38	45
Ethyl benzene, m-, p- xylene	33	20	24	7	22
o-xylene	10	5	8	2	7
C_9	21	23	9	-	-
C_{10}	14	26	6	-	-
C_{11}^+	4	6	6	13	18

* Activity of sample A about 10 x activity of sample B

† Less than 5% total products

Table 6 - Initial C_1 - C_4 hydrocarbon product distribution (%) 450°C

Zeolite Catalyst	HNu-1 A*	HNu-1 B*	HZSM-5	H-mordenite
CH_4	91	25	4	7
C_2H_6		1	3	5
C_2H_4	2	16	12	8
C_3H_8		3	36	74
C_3H_6	5	10	10	1
$i-C_4H_{10}$	2	1	22	3
$n-C_4H_{10}$		1	6	3
$1-C_4H_8$		7	1	
$i-C_4H_8$		11	1	
$t-2-C_4H_8$		10		
$c-2-C_4H_8$		14	1	
C_4H_6		1	4	

* Samples A and B of HNu-1 as Table 5

Very fast decay of hydrocarbon-forming activity by HNu-1 was found. The injection of 20 µℓ CH_3OH (10 x initial pulse) decreased total aromatics yield (from a subsequent 2 µℓ CH_3OH) by about 60% and after 60 µℓ CH_3OH all hydrocarbon formation was negligible. Calcination in air restored the original activity. A sample of the same batch B of HNu-1 was used with methanol feed in the flow reactor at 375° and 450°C, liquid space velocity of 1 - 2 hr^{-1}. Dimethyl ether was the only gaseous product and no hydrocarbons were found.

Similar tests with HZSM-5 and H-mordenite gave contrasting results. In agreement with other work[7,8], negligible change in performance was observed with HZSM-5. The activity of H-mordenite for hydrocarbon formation decayed much more slowly than that of HNu-1 in the pulse reactor experiments, but the rate of decay was still sufficiently fast for only traces of hydrocarbons to be found in flow experiments.

Hydrocarbon Cracking

The initial catalytic activity of HNu-1 and HFu-1 in cracking various hydrocarbon feeds was investigated with a pulse microreactor at 450°C. The decalin feed contained 52% cis and 48% trans-decalin. The light gas oil was from the Ekofisk field. Conversions are given in Table 7. Differences in product distribution with feed and catalyst were also found but these will be discussed elsewhere.

Table 7 - Initial Conversion (%) in hydrocarbon cracking 450°C

Feed	Hexadecane	Decalin	Tetralin	Light Gas Oil
Catalyst:				
HNu-1, A*	95	20	33	46
HNu-1, B*	60	20	35	22
HFu-1	90	85	75	78
HZSM-5	> 99	60	86	66
H-mordenite	> 99	> 99	> 99	99
Silica-alumina (amorphous)	80	78	78	68

* Samples A and B of HNu-1 as in Table 5

Discussion

Sorption on HNu-1. The sorption results in Table 1 suggest that HNu-1 has ports close to 0.60 nm. Assuming sorbed molecules have typical liquid densities then p-xylene, pyridine and isobutane appear to fill the same voidage, about 56% of that available to water, whereas n-hexane fills about 76% of the voidage available to water. The weight ratio $\frac{\text{n-hexane}}{\text{water}}$ is 0.5 typical of hydrophilic zeolites such as zeolite A and X, and quite unlike other siliceous zeolites which are distinctly hydrophobic eg ZSM5 and dealuminated mordenites[5,6]. ZSM5 and siliceous mordenites give $\frac{\text{n-hexane}}{\text{water}}$ ratios greater than 1; indeed Chen[5] found for SiO_2/Al_2O_3 ratios greater than 30 a rapid fall occurred in water sorption of mordenite, whereas hydrocarbons still filled the zeolite channels. The overall picture for Nul appears to be a dual channel system. The primary channels probably have distorted 10 ring ports allowing access of simple aromatic molecules and isoparaffins as well as smaller molecules. The secondary channels are probably distorted 8 rings allowing restricted access to n-hexane. Rate curves for sorption of water, pyridine and p-xylene all suggest an energy barrier is present requiring a definite energy of activation for full penetration to occur. For water this may mean that at least part of the channel system is somewhat hydrophobic, but that this property can be overcome by the presence of water.

Sorption of HFu-1. Ful is a very unusual material and the results obtained are difficult to interpret. Typically the small platelets tend to stack perpendicularly in a house of cards manner, but the bulk density is about 0.5 g/cc, which means most of the platelets must stack in a more compact manner within the 300 - 10^4 nm agglomerates. Assuming a typical zeolite density, one might expect the small platelets to have a surface area of about 150 - 200 m^2/g. Indeed the nitrogen area for dried Na TMA Ful is about 100 m^2/g. On calcining in NH_3 followed by air calcination the nitrogen area is then typically about 350 m^2/g. Of course, much of this additional access may be within macro pores within the agglomerates.

Large sorbate molecules eg triethylbenzene, quinoline or tripropylamine give coverage which corresponds to only 5 - 10 m^2/g, presumably they sorb only on the outer surfaces of the agglomerates and cannot penetrate. On the other hand molecules up to about 0.5 nm have substantial access with threshold levels corresponding to about 10 cm^3 hg^{-1} of Ful for n-hexane, ammonia and methanol. For water the access is to 8.5 cm^3 hg^{-1}, perhaps this latter reflects some degree of hydrophobicity. Indeed the $\frac{n\text{-hexane}}{water}$ ratio is 0.8 which places Ful closer to the hydrophilic zeolites than to the very hydrophobic ones.

Medium sized molecules eg p-xylene apparently have access to about 90% of the voidage available to n-hexane. The slow penetration of cyclohexane illustrated in Figure 3 suggests access for almost half of the cyclohexane via ports of size just greater than 0.60 nm. It is possible that initial sorption of cyclohexane occurs on the external surface of the platelets and that penetration of the platelets follows slowly.

Again, the evidence suggests at least a dual channel system with ports near 0.5 and 0.6 nm respectively. However, this may be an over simplification, because the medium to small molecules are probably also sorbed on the platelets' external surfaces. Water sorption at $p/po > 0.4$ shows definite capilliary condensation is setting in, presumably between platelets. There is no similar evidence for condensation of hydrocarbons below $p/po = 0.7$.

Catalysis of Methanol Conversion to Hydrocarbons. HNu-1 resembles other highly-siliceous zeolites, eg H-mordenite and HZSM-5, in catalysing the conversion of methanol to hydrocarbons including aromatics, so it probably has broadly similar strongly acidic sites. HY gave much lower yields of aromatics. The sorption results show that the largest ports of HNu-1 at about 0.6 nm are comparable in size to those of HZSM-5. Product selectivity in the aromatics, with the upper limit at about C_{10} for both zeolites, is in agreement with the sorption results, and also with earlier work[7] on HZSM-5. Both sorption and X-ray diffraction show, however, that the internal pore structures of these zeolites are different, so selectivity differences are to be expected within the range of possible products. Two effects contributed to this transition state selectivity[9]: (a) Physical constraints due to the zeolite structure at the acid site limit possible configurations and size of the transition state. (b) Zeolites can be regarded as rigid ionising solvents [10-12], in which the solvation effects of the oxide ions in the structure on the cationic reaction intermediates vary with the zeolites. The variations in benzene yield (table 5) and $C_1 - C_4$ yields (table 6) can be attributed to these effects.

The failure of H-mordenite and HY to give high yields of C_8 aromatics in this work is in contrast with results under other conditions[13,14] when higher aromatics were found. This may be due to transition-state selectivity, but the fast transport of aromatics out of these large port zeolites must also tend to decrease the extent of methylation. Karge[15] also found monalkylbenzenes to be formed selectively over mordenite catalysts.

The mechanism of conversion of methanol to hydrocarbons over highly siliceous zeolites has been investigated[2,8,16] but the initial reaction is still obscure. The formation of linear butenes over HNu-1 could be via process B of Derouane et al[16] or alternatively from a dehydration

$$C_2H_5OC_2H_5 \rightarrow C_4H_8 + H_2O$$

analogous to the initial formation of ethene from dimethyl ether.

The rate of coking of zeolites is a function of zeolite structure, not port size only[17]. HNu-1 deactivates much faster than any of the other zeolites tested: complete deactivation needed only about 5 molecules CH_3OH feed/Al atom. Even if no volatile products had been formed there would have been barely sufficient material to block all acid sites. Deactivation by pore-mouth blocking or the participation of only a few sites in catalysis are possible explanations. This fast deactivation may indeed be related to the variability in initial activity of different samples of Nu-1. The continued formation of dimethyl ether probably occurs on unpoisoned weak acid sites, including those on the crystal exterior.

Catalysis of Hydrocarbon Cracking. The sorption results on HNu-1 and HFu-1 indicate that hexadecane should have access to internal sites of these zeolites as well as of the other zeolites tested. Except for the inactive (B) sample of HNu-1, all the zeolites (table 7) showed, as expected, a higher activity than amorphous silica-alumina. The differences in activity between the zeolites can be attributed to geometric and solvation effects at the acid sites.

A different pattern of reaction is seen with reactants of larger molecular size (decalin, tetralin). Mordenite continues to be very effective because its large ports allow these molecules access to the active interior sites, but reaction on the other zeolites can occur only on the surface (or at pore mouths). The geometric and solvation effects at the exterior of zeolite crystals must be broadly comparable with those of an amorphous silica-alumina surface and there is experimental evidence[18,19] that catalytic performance is similar. HFu-1, which has a large external surface, gave conversions close to those by amorphous silica-alumina. Much lower conversions were found with both samples of HNu-1 because of the larger crystal size. These results also show that the differences between the more and less active forms of HNu-1 lie in the crystal interior. Although HZSM-5 is much less active in the cracking of decalin and tetralin than hexadecane, this activity is still high relative to silica-alumina. Light gas oil contains a range of components and consequently the overall cracking activity of the zeolite catalysts falls between the values for cracking the pure hydrocarbons.

REFERENCES :

1   T.V.Whittam and B.Youll, US Patent, 1977, 4,060,590.

2   M.S.Spencer and T.V.Whittam, Acta.Phys.et Chem, 1978,
    24, 307.

3   D.W.Breck, "Zeolite Molecular Sieves", Wiley International
    New York, 1974, Chapter 8 p 634.

4   T.V.Whittam, German Patent Appl. 1978, 2,748,276

5   N.Y.Chen, J.Phys.Chem, 1976, 80, 60.

6   W.O.Haag, D.H.Olson, US Patent, 1978, 4,097,543, p 5.

7   N.Y.Chen and W.E.Garwood, J.Catalysis, 1978, 52, 453.

8   C.D.Chang and A.J.Silvestri, J.Catalysis, 1977, 47, 249.

9   S.M.Csicsery, "Zeolite Chemistry and Catalysis",
    ed J.A.Rabo, A.C.S.Monograph 171, American Chemical Society,
    Washington, 1976, p 680.

10 M.L.Poutsma, ref 9, p 437.

11 J.A.Rabo and P.H.Kasai, Prog Solid State Chem, 1975, 9, 1.

12 J.A.Rabo, R.D.Bezman and M.L.Poutsma, Acta.Phys.et Chem,
    1978, 24, 39.

13 M.S.Spencer, unpublished.

14 Mobil Oil Corp, US Patent, 4052472.

15 H.G.Karge, "Molecular Sieves - II", ed J.R.Katzer,
    A.C.S.Symposium Series 40, American Chemical Society,
    Washington, 1977, p 584.

16 E.G.Derouane, J.B.Nagy, P.Dejaifre, J.H.C.van Hooff,
    B.P.Spekman, J.C.Vedrine and C.Naccache, J.Catalysis, 1978
    53, 40.

17 L.D.Rollman, J.Catalysis, 1977, 47, 113.

18 N.Y.Chen, "Proceedings of the 5th International Congress
    on Catalysis", North Holland, Amsterdam, 1973, p 1343.

19 E.J.Detrekoy, P.A.Jacobs, D.Kallo and J.B.Uytterhoeven,
    J.Catalysis, 1974, 32, 442.

The Catalytic Activity of Transition Metal Exchanged Mordenites Towards Methane Oxidation

P. Fletcher, P.R. Lower and R.P. Townsend*
Department of Chemistry, The City University,
Northampton Square, London EC1V 0HB

Introduction

The development of the microcalorimetric bead technique[1] as a means of catalytically detecting and quantitatively estimating the presence of flammable gases in an atmosphere, initially involved using alumina beads as support and noble metals as catalysts[2]. Since then, interest has developed in the use of transition metal exchanged zeolites as catalysts for the detection of methane as conventional beads are rapidly poisoned by chemicals that are ubiquitous in the industrial environment, especially silicone oils and lead compounds. Firth and Holland[3] carried out initial studies using zeolite X exchanged with rhodium, iridium, palladium and platinum, and found initial evidence of resistance to silicone oil poisoning. A further study by Rudham and Sanders[4] again involved zeolite X, but a range of transition metal ions (mainly first-row) were examined in this case.

This paper describes the initial results of a further study on methane oxidation using zeolite catalysts, with synthetic mordenite as the support. Mordenite was chosen for its high thermal and chemical stability, the latter allowing the use of solution conditions under which transition metal exchange can occur reversibly[5,6], thus facilitating reproducible preparations of any potential catalyst. Ideally, a catalytic detector should have the maximum sensitivity for the flammable gas that is consistent with reliability; the mordenite was therefore maximally exchanged for the transition metal ions where possible, which gave weight per cent compositions slightly higher than those used by Rudham and Sanders[4]. In addition, some catalysts with lower exchange levels were examined.

Experimental

Materials: Synthetic sodium mordenite (sodium Zeolon) was supplied by the Norton Company. Manganese(II), nickel(II), copper(II) chloride, vanadyl(IV) sulphate, uranyl(VI) acetate and silver(I) nitrate were of AnalaR grade. Tetrammine-platinum(II) and tetramminepalladium(II) chloride were prepared in the laboratory according to standard methods in the literature.

Oxygen, nitrogen, helium and methane, supplied by B.O.C., were of the purest grade available. In addition, cylinders of 1% methane in air were used. Samples of methane were analysed chromatographically; in the undiluted methane cylinder 0.9% ethane, but no other hydrocarbons or carbon dioxide, were detected. Trace hydrocarbons were not detected in the 1% methane/air samples.

Preparation of the Catalysts: Before the transition metal exchanged forms of mordenite were prepared, the sodium form was converted to the ammonium form

(NH₄-MOR) by exchange with ammonium chloride for the reasons, and by the method, described before[5].

Transition metal exchanged mordenites were then prepared by equilibrating 20 g samples of ammonium mordenite four or five times with solutions of the metal salts at room temperature, each equilibration lasting two days. The choice of which solution concentrations to use was determined by pH; it was important to avoid solutions too acid to cause dealumination of the zeolite[5,6]. In the cases of vanadyl(IV) sulphate and uranium(VI) acetate, even solution concentrations of 0.1 mol dm^{-3} exhibited low pH's (3.2 and 4.45 respectively) and precipitation occurred on raising the pH only slightly (4.15 and 5.9 respectively). Exchanges were of shorter duration in these cases, especially in the case of the vanadium solution (two exchanges of ~ 30 minutes). For other catalyst preparations, 0.5 mol dm^{-3} manganese(II) and nickel(II) chloride, and 0.1 mol dm^{-3} silver nitrate solutions were employed, but in the case of copper(II), tetrammineplatinum(II) and tetramminepalladium(II) salts, the solutions were prepared in dilute ammonia (pH ~ 10.5). After the exchanges were complete the samples were washed four times either with water or dilute ammonia (pH ~ 10.5), then dried at 80°C overnight. Finally, the samples were equilibrated over saturated ammonium chloride solution for at least two weeks in a desiccator before analysis.

Analyses of the Catalysts: Samples of NH₄-MOR, Mn-MOR, Ni-MOR and Cu-MOR were analysed by standard methods described previously[5,6]. Uranium content was determined by dissolving the residue remaining after hydrofluoric acid treatment, followed by polarographic analysis of the solution. Palladium and platinum contents were determined gravimetrically, weighing the residues as PdO and Pt respectively[7]. These results could be counter-checked for consistency using ammonium analyses of the samples by the Kjeldahl method. Dissolution of samples containing silver for analysis by atomic absorption spectroscopy proved to be difficult; gravimetric estimation of the residue as Ag seemed satisfactory. Analyses on each catalyst were carried out in quadruplicate on separate samples.

In addition, the catalysts were examined by thermogravimetry, differential thermal analysis, and by X-ray. On the basis of the thermal analyses, samples were heat-treated to 720K under vacuum before determination of adsorption capacities using a Carlo-Erba Sorptiometer.

Catalytic Apparatus: Oxidation of methane was accomplished in a copper and glass flow system operating at atmospheric pressure. Accurate control of gas flow rates was ensured by employing Watts pressure controllers in conjunction with Brooks model 8744A mass-flow controllers. The mass-flow controllers maintained a steady flow-rate through the reactor, even when the reactor back-pressure varied. A transducer to measure this back-pressure was included in the system. Gas samples from the product or reactant streams were taken by means of an automatic solenoid valve, and analysed by gas chromatography. Detection was by means of a Servomex microkatharometer, bridge network and chart recorder with analogue integrator.   Several types and combinations of separating columns were tested; finally a half metre long column of Poropak Q-S was employed using helium as the carrier gas.

Catalyst beds were prepared by accurately weighing out approximately 0.25 g of catalyst, then lightly pressurising the powder into a 4 mm inner diameter glass tubular reactor between pads of acid-washed silica wool.   The final catalyst bed length was about 30 mm in all cases.   Before examining any catalyst, dehydration *in situ* was accomplished by passing a stream of nitrogen for 16 h over the catalyst at 673K.   This had the additional effect of removing the ammonium ion as ammonia, yielding either mixed metal/H mordenite catalysts, or, in the case of pure NH_4-MOR, the hydrogen form of the zeolite.   Under any given set of reaction conditions, samples were taken for analysis by gas chromatography until results were reproducible (this usually required ~ 30 min). Life tests were also undertaken with the catalyst continuously operating under one set of reaction conditions for one week; no significant change in catalytic activity in these circumstances was taken to indicate stability.

Results

Analyses of Catalysts: The results of chemical analyses of the catalysts are shown in Tables 1 and 2.   As previously reported[5] the nickel and copper mordenites remained stable after ion exchange; a series of copper mordenites were prepared by successive exchanges and heat treatments at 350K for reasons described in the Discussion section.   Again, as noted before[5] the manganese mordenite was observed to darken in colour, finally to yield manganese dioxide occluded in the zeolite.   This occurred even at room temperature on equili-brating samples over saturated ammonium chloride solution.   Silver mordenite was initially white, but only remained so if stored in the dark.   On heat-treating Ag-MOR before testing for catalytic activity at 673K, the catalyst turned yellow, whilst at the higher temperature of 773K, the final colour was red.   The heat-treated and stabilised palladium and platinum catalysts were red-brown and black respectively; in the former case it is probable that the mordenite contained a small proportion of Pd(III) ions in addition to Pd(II)[8]. Uranyl(VI) ions could be reversibly exchanged into NH_4-MOR, and the zeolite remained yellow after heat-treatment.   Vanadyl(IV) mordenite, initially blue, turned green even during an exchange time of ~ 20 minutes and exchange was not reversible.   The short exchange times used in the preparation of VO-MOR explain   the low vanadium loading in the zeolite (Table 1).

X-ray powder diffraction patterns indicated that the catalysts had retained their crystallinity.   Adsorption capacities were measured at 77K using nitrogen as the sorbent; values are included in Tables 1 and 2.   The substantially lower value obtained for Ag-MOR (x/m = 0.086 compared to

Table 1 : Analyses of Less Active Mordenite Catalysts

Zeolite Component	Composition mol hg^{-1}					
	H-MOR	Mn-MOR	Ni-MOR	Ag-MOR	VO-MOR	UO$_2$-MOR
SiO$_2$	1.2247	1.1702	1.1687	1.0165	1.1932	1.1238
Al$_2$O$_3$	0.1078	0.1030	0.1029	0.0894	0.1050	0.0989
(NH$_4$)$_2$O	0.0972	0.0552	0.0558	0.0023	0.0845	0.0689
Na$_2$O	0.0069	0.0069	0.0069	0.0056		0.0064
M$_m$O$_n$*		0.0443	0.0436	0.0816	0.0195	0.0266
H$_2$O	0.5927	0.6238	0.6260	0.6737	0.6238	0.5050
NH$_3$		0.0786	0.0958			0.1226
% exch.(M)		44.5	43.9	91.3	18.7	27.8
Adsorption † capacity(^{x}/m)	0.165	0.156	0.151	0.088	0.166	0.153
NH$_3$:M ratio		1.77	2.20		4.61	

Table 2 : Analyses of Most Active Mordenite Catalysts

Zeolite Component	Composition mol hg^{-1}					
	Cu-MOR (1)	Cu-MOR (2)	Cu-MOR (3)	Cu-MOR (4)	Pd-MOR	Pt-MOR
SiO$_2$	1.2004	1.1513	1.1469	1.1182	1.1483	1.0905
Al$_2$O$_3$	0.1068	0.1027	0.1020	0.0995	0.1011	0.0960
(NH$_4$)$_2$O	0.0900	0.0612	0.0386		0.0443	0.0385
M$_m$O$_n$*	0.0168	0.0415	0.0634	0.1067	0.0568	0.0575
H$_2$O	0.5466	0.6287	0.5860	0.5411	0.4684	0.4212
NH$_3$	0.0600	0.1418	0.1647	0.2469	0.2110	0.2259
% exch.(M)	15.7	40.4	62.2	107.2	56.2	59.9
Adsorption † capacity(^{x}/m)	0.162		0.143	0.153	0.138	
NH$_3$:M ratio	3.57	3.42	2.60	2.31	3.71	3.93

* where M is the metal ion in question . m = n = 1, except with

　　Ag-MOR (m = 2, n = 1),   VO-MOR (m = 1, n = 2) and

　　UO$_2$-MOR (m = 1, n = 3).

†^{x}/m expressed as g_{N_2}(adsorbed) $g^{-1}_{catalyst}$.

x/m = 0,165 in H-MOR) suggests that in the silver catalyst nitrogen was excluded from the side-pockets which are situated each side of the main c-axis channels in mordenite[9].

Catalytic Activities: The fractional degree of oxidation of methane (α) at a particular temperature was determined by comparing the methane peak height in the product stream with that observed in the feed, as described previously[4]. Initially, peak areas were determined using the integrator, but this was not found to improve significantly the accuracy in measuring α.

The simple plug flow equation, which is applicable in the absence of boundary layer diffusion control (film diffusion) between the gas stream and the catalyst particle, is

$$^{m}/_{f} \quad = \quad \int_{o}^{\alpha} \frac{d\alpha}{r_m} \tag{1}$$

where m is the mass of catalyst in the bed, f the volumetric flow rate and r a volumetric rate per unit mass of catalyst. Equation (1) implies that provided $^{m}/_{f}$ is kept constant whilst f is varied, the integral should remain constant if film diffusion is absent. If film diffusion is significant, r will change with f and α will be a function of f for a given $^{m}/f$ value. To ensure absence of film diffusion control, detailed initial experiments were undertaken using different masses of Cu-MOR catalyst and a 1% methane feed for a range of gas flow rates from 2 to 20 cm^3 min^{-1} (flow rates as measured at 298K). Plots of α against $^{m}/f$ were coincident within the range 4 < f < 20 cm^3 min^{-1}; activity measurements were therefore undertaken in the lower part of this range, in order to maximise contact times.

Thus the activities, as a function of temperature, of all twelve catalysts were examined over a flow rate range of 5 < f < 8 cm^3 min^{-1} for different gas compositions. Stable and reproducible activities were obtained as evidenced by life tests (descussed in Results section) or by changing reaction temperatures in either direction. Products other than carbon dioxide and water were not detected, and in most cases agreement between methane conversion and carbon dioxide yield was good. A small quantity of formaldehyde production was suspected with two of the least active catalysts (UO$_2$-MOR and H-MOR) where a significant discrepancy between methane conversion and carbon dioxide production occurred; the analysing system could not detect this compound, but small quantities of formaldehyde have been reported before[10].

The flow reactor model, discussed by Hougen and Watson[11] can be transformed to yield

$$k_v(c) \quad = \quad \frac{\rho_c}{m} \frac{F}{c^n} \int_{o}^{\alpha} (\frac{1}{1-\alpha})^n d\alpha \tag{2}$$

where $k_v(c)$ is the observed rate constant for unit volume of catalyst in terms

of concentration units, ρ_c is the catalyst particle density, F the molar flow
rate (mol s^{-1}) and c the concentration of methane in the feed.   n is the
order of the reaction with respect to methane.   Equation (2) is only applicable
to the case where reactants are transformed to products with no net change in
the number of moles, as is the case in the complete oxidation of methane.   Also
an Arrhenius relationship should hold, *viz*

$$k_v(c) \;=\; B_c \, \exp \, (^{-E_a}/_{RT}) \qquad\qquad (3)$$

where E_a, in conformity with $k_v(c)$, is an observed energy of activation.   Then

$$\ln k_v(c) \;=\; \ln(\frac{\rho_c \, F}{m}) \;+\; \ln \psi_n(c) \;=\; \ln B_c \;-\; ^{E_a}/_{RT} \qquad (4)$$

where

$$\psi_n(c) \;=\; \frac{1}{c^n} \int_o^{\alpha} (\frac{1}{1-\alpha})^n \, d\alpha \qquad\qquad (5)$$

The determination of $k_v(c)$, and hence B_c and E_a, is however dependent on the
value which is assigned to the order with respect to methane n.   This is
a priori unknown. Rudham and Sanders[4] overcame this problem by choosing the
best straight line fit to the Arrhenius plots for the orders 0, 0.5, 1 and 1.5.
The procedure used here involved a computer program which iteratively solved
for n, E_a and B_c.   Using initially eleven values of n in steps of 0.2 from 0
to 2, for each n value the integral of equation (5) was evaluated; then a plot
of $\ln k_v(c)$ against $^1/_T$ was constructed from the supplied experimental data.   A
linear regression analysis followed, yielding a value for the correlation coefficient R, where

$$R \;=\; (m\sigma_x)/\sigma_y \qquad\qquad (6)$$

m is the gradient of the best fit, and σ_x, σ_y are standard deviations (N
weighting) in the abscissa and ordinate data respectively.   Since the Arrhenius
plots have negative gradients, $-1 < R < 0$, with $R = -1$ for a perfect correlation.
The value of n two steps below that yielding the best correlation was then taken,
and the procedure repeated over 21 steps of $\Delta n = 0.02$.   A third cycle, with
$\Delta n = 0.002$ followed, and then finally the value of n giving the best fit,
together with corresponding E_a, B_c, ψ_n(observed), ψ_n(fit) and $^1/_T$ values were
outputted.

In all cases analysed except two, values of $-R > 0.998$ for the best fitting
value of n were obtained, indicating good correlation.   The exceptions were
H-MOR and VO-MOR, and even in these cases $-R > 0.997$.   Again, with the
exception of H-MOR (which showed no maximum since zero order kinetics gave the
best correlation) all catalysts showed one clear maximum when $-R$ was plotted as
a function of n.   Examples are shown in Figures 1 and 2, where the dependencies
of E_a and B_c on n are also shown.   The importance of choosing a correct value

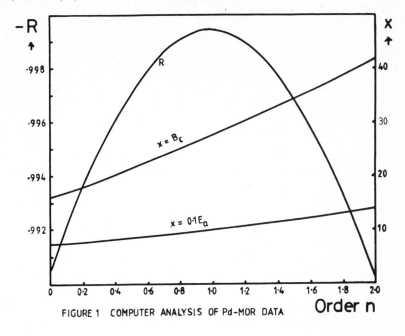

FIGURE 1   COMPUTER ANALYSIS OF Pd-MOR DATA

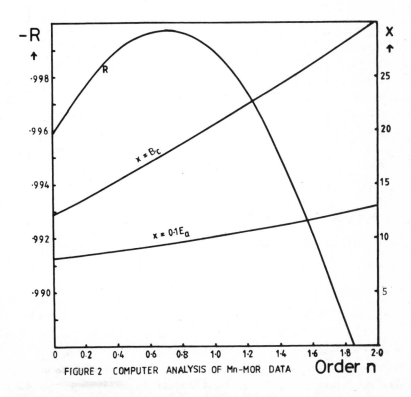

FIGURE 2   COMPUTER ANALYSIS OF Mn-MOR DATA

for n is evident.

In addition to calculating E_a and B_c, free energies of activation were calculated using transition state theory, since

$$k_v(c) = (kT\,K^{\ddagger})/h = (kT/h)\exp(^{-\Delta G^{\ddagger}}/_{RT}) \qquad (7)$$

where $K^{\ddagger}$ is the constant (in concentration units) for the equilibrium between reactants and activated complexes and $\Delta G^{\ddagger}$ is the free energy of activation (with the standard state defined by the choice of the concentration units of $k_v(c)$, *viz* 1 mol dm^{-3} methane in feed over 1 cm^3 of catalyst at 600K). Calculations of entropies of activation $\Delta S^{\ddagger}$ from B_c involve the following assumptions:

(i) the apparent energy of activation E_a must be equal to the true energy of activation;

(ii) the change $\Delta v^{\ddagger}$ in the number of moles when passing from reactants to activated complex must be known in order to relate $\Delta H^{\ddagger}$ to E_a.

Uncertainties regarding (ii) preclude calculations of $\Delta S^{\ddagger}$ without further work.

Table 3:   Catalytic Activities of Ion Exchanged Mordenites

Catalyst	$E_a/kJmol^{-1}$	$\Delta G^{\ddagger}/kJ\ mol^{-1}$	Methane Order n	Correlation Coefficient R	T/K (α=0.5)
H-MOR	99.7	255.6	0.000	0.99774	828
Mn-MOR	94.3	218.2	0.698	0.99975	757
Ni-MOR	103.4	202.5	1.120	0.99815	748
Ag-MOR	84.9	218.4	0.670	0.99892	743
VO-MOR	142.5	229.2	0.924	0.99722	829
UO₂-MOR	142.4	230.0	0.918	0.99935	820
Cu-MOR(1)	117.5	199.2	0.994	0.99986	676
Cu-MOR(2)	121.8	185.3	1.182	0.99893	635
Cu-MOR(3)	120.8	180.6	1.286	0.99918	633
Cu-MOR(4)	109.1	187.7	1.100	0.99982	632
Pd-MOR	99.2	173.9	0.978	0.99941	517
Pt-MOR(i)	96.9	183.5	1.116	0.99948	599
Pt-MOR(ii)	87.4	193.5	0.874	0.99986	597

The results obtained for all twelve catalysts are shown in Table 3, where the values of E_a, B_c and $\Delta G^{\ddagger}$ correspond to the maximal value in $-R$ (also given). The copper mordenite catalysts are numbered to correspond with the analytical data in Table 2.   With the exceptions of H-MOR (which was not expected to be active, and was examined in order to enable comparisons between the different exchanged forms and the unexchanged sieve) and also Mn-MOR and Ag-MOR, all catalysts showed orders with respect to methane that were close to unity. Catalyst activities were not sensitive to oxygen concentration in the feed (indicating an order with respect to oxygen of zero, as noted before[3,4]) with the exception of Pt-MOR, where higher oxygen concentrations were found to be deactivating. This effect has been noted before both on platinum supported on alumina and in Pt-X zeolite[2,3]. The data given in Table 3 for Pt-MOR correspond to feed compositions of (i) CH_4 0.84%; N_2 80.25%; O_2 18.91%, and (ii) CH_4 0.84%; N_2 70.25%; O_2 28.91%. The marked change in methane order results from equation (2) being inapplicable since the dependency of $k_v(c)$ is not in this case only a function of methane concentration. Further studies are in progress on Pt-MOR.

Arrhenius plots corresponding to the data in Table 3 are shown in Figure 3.

Discussion

Assessment of Validity of Data: Although experiments outlined earlier in the Results section had ensured that film diffusion processes were not influencing the observed kinetics, equation (2) is also rendered invalid if sufficiently high axial and radial concentration and temperature gradients exist within the bed. In particular, axial diffusion processes and inefficient removal of heat arising from the exothermicity[12] of the methane reaction must be considered. Obviously these effects would be most influential under conditions when methane conversion was highest. Model calculations were therefore undertaken with the most active catalyst, Pd-MOR, using temperatures at which $\alpha \rightarrow 1$. These indicated that at $\alpha \sim 0.9$, axial diffusion would still contribute only minimally to the total flux of methane through the bed, and that a radial temperature difference of $\Delta T < 2K$ would occur. Ignoring these effects would cause an error in E_a of $< 2\%$.

Direct measurements of ΔT in the microreactor, although difficult, were accomplished; under similar conditions ΔT was found to be between 2 and 2.5K, in satisfactory agreement with the calculations. The calculated values are also comparable with data of Rudham and Sanders[4], who, using a different reactor design, found a temperature rise of 1K for an oxidation rate of 3.9×10^{16} molecule g^{-1} s^{-1}. Insertion of this value into the data for Pd-MOR gave a predicted rise for $\alpha = 0.91$ of $\Delta T = 1.67K$.

In their work on Pd-X, Rudham and Sanders[4] found clear evidence of diffusion limitation (rather than diffusion control) affecting reaction rates at high temperatures. The apparent energy of activation at high temperatures was only half the value initially found, in accord with theoretical predictions for diffusion limitation[13] within catalyst particles. Such an effect (also observed by Paetow and Riekert[14]) was expected in this work, but not found, the Arrhenius plots being linear over the whole range of α values for even the most active catalysts. In order, therefore, to ensure that the observed kinetics for Pd-MOR were not diffusion limited over the whole temperature range, values of the effectiveness factor η for the catalyst were calculated. Using diffusion data available in the literature[15] values of the dimensionless modulus

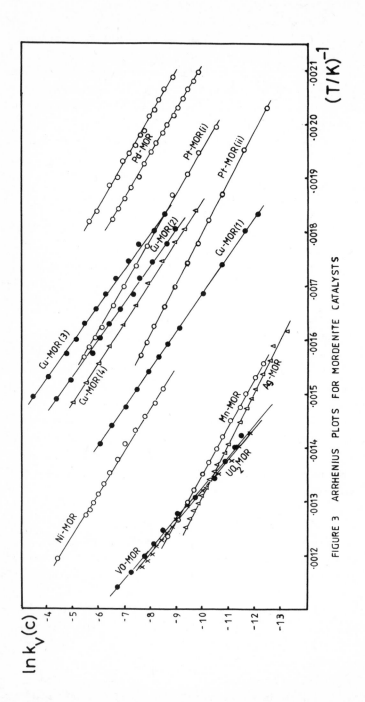

FIGURE 3   ARRHENIUS PLOTS FOR MORDENITE CATALYSTS

Φ were calculated, where[16]

$$\Phi = \eta\phi_1^2 = (R^2R_v)/(Dc) \qquad (8)$$

ϕ_1 is the first order Thiele modulus (spherical particle), R the average catalyst particle radius, R_v the rate per unit volume of catalyst and D the diffusivity. An electron microscopy study of the mordenite gave an average particle radius of 1.08×10^{-6}m ($\sigma_r(N) = 5.9 \times 10^{-7}$m). Allowance for non-isothermicity was made by calculating the further dimensionless parameters

$$\beta = (\Delta HDc)/(HT_o)$$
$$\gamma = E_a/RT_o \qquad (9)$$

where ΔH is the heat of combustion of methane, H the thermal conductivity of the catalyst particles and T_o the bath temperature. The criterion given by Weisz and Hicks[16] for conditions under which mass and heat transfer limitation effects are absent is

$$\Phi \exp\left[\gamma\beta/(1+\beta)\right] < 1 \qquad (10)$$

At the highest temperature used with Pd-MOR (550K), $\Phi = 0.31$, $\beta = 2.92 \times 10^{-8}$ and $\gamma = 21.57$. Thus

$$\Phi \exp\left[\gamma\beta/(1+\beta)\right] \rightarrow 0.31 \qquad (11)$$

which satisfies the above criterion. Use of a Weisz-Hicks diagram[16] for $\beta = 0$ and $\gamma = 20$ further confirmed that η could be taken as unity under all conditions examined and that the estimated value of Φ could be underestimated in the above calculations by an order of magnitude without η departing significantly from unity.

Thus it is concluded that the observed energy of activation was not affected by diffusion limitation or heat transfer effects within the catalyst particles. The above checks justified the calculation of $\Delta G^{\ddagger}$ from E_a values.

Catalytic Activities and Exchange Levels: From the values of $\Delta G^{\ddagger}$ given in Table 3 the activity series for the catalysts is

$$Pd > Pt \sim Cu > Ni > Mn \sim Ag > VO \sim UO_2 > H$$

This order is similar to that obtained from comparing the temperatures at which $\alpha = 0.5$ for each catalyst, except that this comparison indicates that Pt-MOR was distinctly more active than the copper mordenite series. Because of the marked dependency of the activity of Pt-MOR on oxygen composition in the feed, $\Delta G^{\ddagger}$ values for Pt-MOR may well be in error since equation (2) is not strictly applicable (see comments in the Results section). Where comparisons are possible the series is in agreement with those found by Firth and Holland[3] and Rudham and Sanders[4] for transition and noble metals in zeolite X, and differs from that found over bulk oxides[17] for reasons discussed before[4]. Orders of ~ 1 with respect to methane are also in agreement with those found before for zeolite X^4 with the exception of Mn-MOR, where the observed order of ~ 0.7 was also found for manganese oxide[17].

Table 4 : Comparisons of Catalytic Activities with Zeolite X

Catalyst	% w/w Metal Oxide	Temperature Range of Activity $/_K$	$E_a/_{kJ\ mol^{-1}}$	Reference
Cu-MOR(1)	1.34	546 → 713	117.5	This work
Cu-MOR(2)	3.30	545 → 672	121.8	(")
Cu-MOR(3)	5.04	545 → 670	120.8	(")
Cu-X	3.55	732 → 838	142.9	(4)
Pd-MOR	6.95	477 → 549	99.2	This work
Pd-X	4.81	526 → 778	104.1	(4)
Pd-X	4.53	571 → 658	90.0	(3)
Ni-MOR	3.26	390 → 565	103.4	This work
Ni-X	3.34	473 → 557	136.7	(4)

Comparisons of activities with those found for the most active metals in zeolite X show that the mordenite catalysts are more active for methane oxidation, especially Cu-MOR, where the temperature range of activity is substantially lower, even after allowing for the different weight percentages of copper in the two zeolites (Table 4). Comparisons of exchange levels are not valid as mordenite has a much lower exchange capacity than X. In the case of the palladium zeolites, the differences in activity may only be due to the difference in the quantity of palladium present; however the activity must also depend on the concentration of *accessible* ions rather than on the total metal ion content. The large number of different ion site types in X make it particularly difficult to decide what proportion of the total active ion content in the framework is contributing to the observed activity[4,18], especially as the ions may migrate and redistribute reversibly on raising and lowering temperatures.

Mordenite is a simpler case. In sodium mordenite, half the ions are found in relatively inaccessible side-pockets[19], and previous work[5,6] has shown that many transition metals showed a limiting exchange level of ~ 50% in NH₄-MOR provided that heat treatments were not carried out between the room temperature equilibrations. Successive heat treatments allow exchange to proceed to higher levels, especially with copper[5]; this method was used to produce catalysts shown in Table 2. The foregoing evidence leads to the reasonable conclusion that only ions in the main channels in mordenite will significantly contribute to the catalytic activity; if this is so, exchange beyond 50% should not significantly alter this activity. This effect is seen with the copper mordenites.

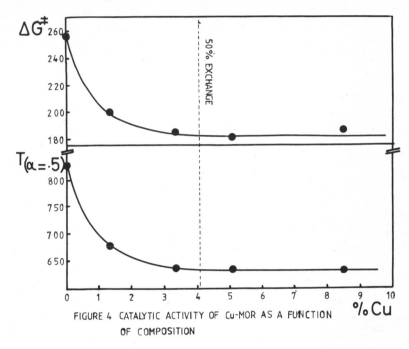

FIGURE 4 CATALYTIC ACTIVITY OF Cu-MOR AS A FUNCTION OF COMPOSITION

In Figure 4 are plotted $\Delta G^{\ddagger}$ values and the temperatures at which $\alpha = 0.5$ for each Cu-MOR catalyst, with H-MOR as comparison and standard. The slight decrease in the activity of Cu-MOR(4) may be due to some precipitation having occurred within or on the crystals, partially hindering access to active sites. This conclusion is supported by the chemical analysis (> 100% exchange) and the lower adsorption capacity (Table 2).

<u>Comparisons with Other Work:</u> Boreskov *et al*[18] studied the oxidation of carbon monoxide over a series of Cu-Y zeolites containing levels of copper from zero to ~ 16% by weight, and found a marked increase in activity when zeolites contained > 5% by weight of copper. This was attributed to the formation of associates between the initially separate Cu^{2+} ions within the framework. Ione *et al*[20] extended this work, showing that exchange from Cu^{2+} solutions of low pH or from solutions containing amminated copper complexes led to isolated ions within the zeolite, whereas exchange from solutions of intermediate pH resulted in cation association and partially hydrolysed species being found in the crystals. Even with the amminated copper solutions, higher levels of exchange led to associated cations which were thought to be the species

primarily responsible for oxidative catalytic activity, through oxygen atoms forming bridges between copper ions. Further work[14,21,22] has been in accord with these suggestions; however, these conclusions are partly disputed by other workers[23], who show that, in the case of carbon monoxide oxidation at least, cuprous ions are responsible for the activity. Paetow and Riekert[14] interpret their data for the oxidation of carbon monoxide and propylene over Cu-MOR in terms of lattice oxygen bound to copper as suggested by Maksimov and co-workers[18,20,21]; it is difficult to be definite in this matter in this present work, especially as the catalyst pretreatment must have partially reduced the catalyst to Cu(I)[24,25], a process which in mordenite may be reversible (as in Y[25]) or not (as in X[25]). Further work on the redox behaviour of copper ions in mordenite, and on the degree to which autoreduction of Cu(II) can occur in a dynamic inert atmosphere, would therefore be helpful.

Comparisons of this work on Pd-MOR and Pt-MOR with that of Firth and Holland[3] and Rudham and Sanders[4] are in agreement in demonstrating that palladium zeolites are the most active, and that platinum catalysts are adversely affected by an increasing oxygen current in the feed. Catalytic mechanisms are therefore likely to be similar in mordenite to those in X. Firth and Holland showed that the rate determining step with noble metals in zeolites did not involve the formation of a metal-carbon bond with the methane, and suggested that instead the step was "possibly the formation of adsorbed vapour by removal of a hydrogen atom from the adsorbed methane fragment to an adsorbed hydroxyl group"[3]. This suggestion accords with work by Naccache et al[8] and Schmidt et al[26], who have shown that in an oxidising atmosphere at elevated temperatures, the Pd^{2+} ions are partially oxidised to Pd^{3+}, probably by the reaction[8]

$$2Pd^{2+} + H_2O + \tfrac{1}{2}O_2 \rightarrow 2(Pd-OH)^+ \qquad (12)$$

The suggestions in the literature that, for some oxidations at least, activity in zeolites can be correlated with oxygen attached to copper[18,20,21,22], iron[27], or nickel[20,28] ions prompted the examination of uranyl and vanadyl ions, the former of which especially is known[29] readily to form partially hydrolysed species such as $[UO_2(OH)]^+$, $[(UO_2)_2(OH)_2]^{2+}$ and $[(UO_2)_2(OH)_5]^+$ in solution, even at low pH. Although a substantial degree of partial oxidation occurred with UO_2-MOR, initial results are disappointing; further investigations are at present underway on UO_2-MOR, as well as on the silver and platinum mordenite catalysts.

Acknowledgement

We acknowledge the generous support of the Safety in Mines Research Establishment which enabled P.R.L. to take part in this programme.

References

1 A. Jones, J.G.Firth and T.A. Jones, *J.Phys.E*, 1975, **8**, 37.

2 J.G. Firth and H.B. Holland, *Trans. Faraday Soc.*, 1969, **65**, 1121.

3 J.G. Firth and H.B. Holland, *Trans. Faraday Soc.*, 1969, **65**, 1891.

4 R. Rudham and M.K. Sanders, *J. Catalysis*, 1972, **27**, 287.

5 R.M. Barrer and R.P. Townsend, *J.C.S. Faraday I*, 1976, **72**, 661.

6 R.M. Barrer and R.P. Townsend, *J.C.S. Faraday I*, 1976, **72**, 2650.

7 F.A. Cotton and G. Wilkinson, "Advanced Inorganic Chemistry", Interscience, 1966, p. 984.

8 C. Naccache, J.F. Dutel and M.Che, *J. Catalysis*, 1973, **29**, 179.

9 R.M. Barrer and D.L. Peterson, *Proc. Roy. Soc.*, 1964, **280A**, 466.

10 A. Matsui and M. Yasuda, *Kogyo Kagaku Zasshi*, 1940, **43**, 116B; C.F. Cullis, D.E. Keene and D.L. Trimm, *Trans. Far. Soc.*, 1971, **67**, 864.

11 O.A. Hougen and K.M. Watson, "Chemical Process Principles", Wiley, New York, 1955, Vol. 3.

12 M. Prettre, Ch. Eichner and M. Perrin, *Trans. Faraday Soc.*, 1946, **42**, 335.

13 A. Wheeler, *Adv. Catalysis*, 1951, **3**, 249.

14 H. Paetow and L. Riekert, *Acta Univ. Szegediensis*, *Acta Phys. Chem.*, 1978, **24**, 245.

15 D.W. Breck, "Zeolite Molecular Sieves", Wiley, New York, 1974, p. 683; P. Eberly Jr., in "Zeolite Chemistry and Catalysis", 1976, *ACS Monograph 171*, 392.

16 P.B. Weisz and J.S. Hicks, *Chem. Eng. Sci.*, 1962, **17**, 265.

17 T.V. Andrushkevich, V.V. Popovskii and G.K. Boreskov, *Kinetika i Kataliz*, 1965, **6**, 860.

18 G.K. Boreskov, N.N. Bobrov, N.G. Maksimov, V.F. Anufrienko, K.G. Ione and N.A. Shestakova, *Dokl. Akad. Nauk SSSR*, 1971, **201**, 887.

19 W.M. Meier, *Z. Krist.*, 1961, **115**, 439.

20 K.G. Ione, N.N. Bobrov, G.K. Boreskov and L.A. Vostrikova, *Dokl. Akad. Nauk. SSSR*, 1973, **210**, 388.

21 N.G. Maksimov, K.G. Ione, V.F. Anufrienko, P.N. Kuznetsov, N.N. Bobrov and G.K. Boreskov, *Dokl. Akad. Nauk SSSR*, 1974, **217**, 135.

22 L.P. Davydova, G.K. Boreskov, K.G. Ione and V.V. Popovskii, *Kinetika i Kataliz*, 1975, **16**, 117.

23 H. Beyer, P.A. Jacobs, J.B. Uytterhoeven and L.J. Vandamme, *Proc. 6th Int. Cong. Catalysis, London*, 1976, **1**, 273; see also *Disc. 4th Int. Conf. Mol. Sieves, Chicago*, 1977, p. 159.

24 P.A. Jacobs, W.de Wilde, R.A. Schoonheydt and J.B. Uytterhoeven, *J.C.S. Faraday I*, 1976, **72**, 1221.

[25] S.J. Gentry, N.W. Hurst and A. Jones, Acta Univ. Szegediensis, Acta Phys. Chem., 1978, 24, 135.

[26] F. Schmidt, K. Naumann and W. Gunsser, Acta Univ. Szegediensis, Acta Phys. Chem., 1978, 24, 287.

[27] N.N. Bobrov, G.K. Boreskov, K.G. Ione, A. Terletskikh and N.A. Shestakova, Kinetika i Kataliz, 1974, 15, 413.

[28] N.N. Bobrov, A.A. Davydov and K.G. Ione, Kinetika i Kataliz, 1975, 16, 1272.

[29] R. Arnek and K. Schlyter, Acta Chem. Scand., 1954, 8, 1907.

Catalysis on Transition-Metal Ion-Exchanged Zeolites

F.O. Bravo, J. Dwyer[*] and D. Zamboulis
Department of Chemistry, UMIST, Manchester, United Kingdom.

Abstract

Properties of NaCuX zeolites are examined in relation to methods of preparation and pretreatment. Additionally, some results for oxidation of propene over zeolites containing Pd and Cu or Pd and Ni ions are presented.

Thermal analysis of NaCuX zeolites confirms reports that hydrolysed cation species, incorporated during ion exchange, can be detected and demonstrates that copper ions destabilise the structure. Reduction of the catalysts with carbon monoxide demonstrates differences in site reactivity resulting from variation in methods of preparation and pretreatment. Ion-exchange in conditions favourable to cation hydrolysis produces sites which react very rapidly with CO. Generally, reaction rates are second order to Cu(II) ions and initial rates are decreased by the presence of water vapour but not by carbon dioxide. Subsequent reduction of zeolitic Cu^{2+} ions is enhanced by water. Kinetic results are discussed in terms of chemisorption. The stoicheiometry of the reduction is examined by esr and some results of surface studies using SIMS and ESCA are presented.

The selectivity of reactive oxygen species is examined by reaction of propene with NaCuX zeolites in different stages of reduction. At the temperatures used, no selectivity is observed either with fresh catalysts or with catalysts modified by redox processes. Reaction of propene and oxygen over zeolites containing palladium ions results in good yields of acetone at lower temperatures where cations are hydrated.

Introduction

Differences in catalytic properties arising from variation in ion-exchange procedures have been emphasised by Ione et al [1] and by Schoonheydt et al [2]. Ion-exchange of copper from solutions containing hydrolysed cations produced agglomerations of copper species which provided very active sites for the catalytic oxidation of carbon monoxide. Differences in thermal behaviour (DTA, TGA) were observed between zeolites stoicheiometrically exchanged with Ni (II) or Cu (II) and zeolites containing an excess of Ni (II) or Cu (II) ions. Infrared intensities of OH bands and intensities of Cu (II) esr signals were also dependent upon ion-exchange conditions. Surface species produced during ion-exchange and which may be present when non-stoicheiometric exchange is not extreme were suggested by Schoonheydt and Velghe on the basis of conductivity measurements [3] and by Dyer and Barri on the basis of X-ray powder diffraction [4].

More extreme modifications of transition-metal ion-exchanged zeolite catalysts following evacuation, heat treatment and redox reactions have been observed by ESCA [5], by X-ray diffraction [6][7] and by esr [6].

There have been only a few studies of oxidation by zeolites in the absence of gas phase oxygen. Roginski et al [8] considered the oxidation of ammonia, carbon monoxide and ethylene by a range of transition-metal ion-exchanged zeolites. Reactive oxygen in a series of NaCuX zeolites was examined by Benn et al [7].

This paper examines some properties of NaCuX zeolites, including the activity (reaction with CO) and selectivity (reaction with propene) of reactive oxygen in the zeolites, in relation to methods of preparation and pretreatment. Additionally some results concerning reactions over zeolites containing Cu(II) and Pd(II) or Ni(II) and Pd(II) ions are presented.

Catalysts

NaCuX zeolites were prepared from Linde NaX (ex BDH). The parent zeolite
was sieved through a 200 mesh sieve, treated with excess NaCl solution (2M),
washed free of chloride ion using de-ionised water and dried at $100^{\circ}C$.
Portions of each sample were stored over saturated calcium nitrate solution.

Problems associated with ion exchange of NaX concern instability in
solutions of low pH and non-stoicheiometric exchange at higher pH. Two series
of catalysts were prepared, both in the absence of buffer solution. Preparation
of the first series (A) involved dropwise addition of aqueous copper acetate
($\approx$ 0.3M, pH 5.2) to a suspension of zeolite in de-ionised water (25g litre^{-1}).
Addition was slow and after completion the mixture was stirred for several
hours. At all times the concentration of aqueous Cu^{2+} was very low but pH was
allowed to vary. Initially pH was alkaline (suspension of the zeolite in
water), and decreased on addition of copper acetate solution. Consequently
these catalysts contained a mixture of copper species resulting from variation
in pH during ion exchange. A second series (B) was prepared using copper
nitrate ($\approx$ 0.005 M) and a constant solid/solution ratio (1 g litre^{-1}). In the
case of NaCuX(36) it was necessary to add nitric acid (0.5 M) to keep pH below
6.5. Additionally one sample NaCuX(N) was prepared from $Cu(NO_3)_2$ solution
(0.02 M) at pH 8.5, obtained by addition of NaOH (1 M). This sample contained
mainly hydrolysed copper species and chemical analysis confirmed a significant
excess of Cu^{2+} ions.

All the samples were washed with de-ionised water to remove occluded salts
and the filtrate and washings from series B were tested for aluminium. There
was no evidence of de-alumination and final washings were free of counter
anions. The solids were analysed for sodium and copper. X-ray powder
diffraction patterns showed all the hydrated samples to be crystalline.
Microporosity, as measured by nitrogen sorption, decreased with increase in the
extent of ion exchange. For NaCuX(N) most of the copper was not exchanged into
the zeolite and microporosity was retained (Table I).

TABLE I   NaCuX Catalysts

		Unit cell parameter a_o/$\overset{\circ}{A}$	Si/Al	SSA/ $m^2 g^{-1}$	%. Cu exchange[c]
SERIES A	NaCuX(3)	–	1.23[a]	–	33.1[c]
	NaCuX(4)	24.92	"	734	53.8
	NaCuX(5)	24.92	"	662	62.3
	NaCu(30)	24.91		768	27.1
	NaCuX(40)	24.90		700	38.4
SERIES B	NaX	24.92	1.23[b]	854	
	NaCuX(36)	24.90	1.23[d]	824	36.5
	NaCuX(76)	24.87	"	577	76.0
	" (80)	24.83	"	476	80.5
	" (N)	24.87	"	830	(62.8)

(a) Average of several samples by electron probe analysis
(b) Determined by wet chemical analysis
(c) Calculated from analysis of solids
(d) Assumed equal to that in parent NaX (Al absent in filtrate)

Thermal Analysis

A Dupont 900 thermal analyser was used for DTA. For the lower temperature
DTA a sample ($\approx$ 20 mg) was heated at 15 K min^{-1} under a stream of nitrogen
(0.2 litre min^{-1}) using glass beads as reference. For high temperature DTA the

reference was alumina and heating was at 20 K min^{-1}.

DTA and TGA profiles of NaCuX zeolites have been given by Dyer and Barri (4) and Schoonheydt et al (2) have described profiles for NaNiY and NaCuY.   In samples containing excess Ni(II) or Cu(II) the latter authors observed additional endotherms (613 ± 10 K for Cu(II)) with associated weight losses.

Detailed results of the present work (9) will be published separately. However the main conclusions are that Cu^{2+} ions destabilise NaX zeolites and that some of the catalysts prepared under conditions of ion-exchange involving alkalinity at some stage showed an extra endotherm around 600 K.   This endotherm which was observed with NaCaX(N), NaCuX(4) and NaCuX(5) agrees well with that assigned by Schoonheydt et al (2) to water loss from non-stoichiometric species.   An endotherm at 566 K was observed with a catalyst NaCu(I)X-(50) which was prepared (10) by ion-exchange of Cu(I) ions and was subsequently partially oxidised.   This sample contained non-zeolitic oxygen.

Results for reaction of several of these catalysts with CO are presented subsequently and they show that all the catalysts for which some of the ion exchange occurred under alkaline conditions (all series A and NaCuX(N)) reacted very rapidly initially.   The active sites associated with this rapid reaction are attributed to the presence, in the catalysts, of agglomerated copper species which can arise from non-stoicheiometric ion exchange.   Since for some of these active catalysts (NaCuX(30), NaCuX(40) NaCuX(3)) no endotherm around 600 K was detected it appears that DTA is not a very sensitive method for discriminating between the types of copper species in the zeolites but that extensive non-stoicheometric exchange can be detected.

Typical DTA profiles are shown in Fig.1.

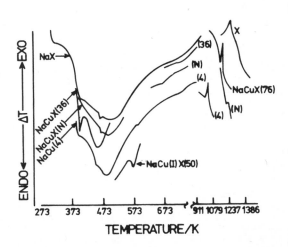

FIG. 1   Thermal Analysis of NaCuX Zeolites (DTA)

Reaction of NaCuX Zeolites with Carbon Monoxide

Previous work (7) using catalysts prepared as in series A established a
pattern for the dependence of reactive oxygen on the concentration of copper
and on temperature.   Following a standard activation procedure the catalysts
were oxygenated and then reduced with CO (4-12 Torr).   In all cases reaction
was stoicheiometric (1CO→1CO$_2$).   Rates were initially rapid but decreased to a
slow "tail" within a few minutes.   The reactive oxygen content was taken to be
the amount reacted in thirty minutes.   After the first reaction it was
reproducible through several redox cycles and was a kinetic rather than an
equilibrium effect (7).

Generally it has been assumed that CO reduces Cu(II) ions in zeolites to
Cu(I) rather than to the metal.   This was found to be the case for the mild
conditions used to measure reactive oxygen.   However, prolonged reduction with
relatively low partial pressure of CO led to an increase in the molar ratio of
active oxygen/copper to values >0.5 which implies some reduction to metallic
copper.   Whereas reduction (Cu^{2+}→Cu$^+$→Cu^0) of NaCuX(5) in a flow reactor
resulted, on oxidation, in formation of a separate copper oxide phase, no such
phase could be detected following prolonged reduction of similar catalysts with
CO in the static reactor.   Clearly the particular reduction procedure can
influence the degree of aggregation of metallic copper and of the oxide phase
formed from it (7).   Indeed it seems that careful reduction may lead to
isolated copper atoms.

Series 'B' catalysts showed somewhat different behaviour in mild reduction
with CO.   Catalysts prepared using dilute solutions of Cu(NO$_3$)$_2$ under mildly
acidic conditions did not show the very rapid initial reaction with carbon
monoxide associated with series 'A' catalysts (where initial ion exchange
involved pH > 7.0), and the catalyst prepared from Cu(NO$_3$)$_2$ at pH 8 showed a
much enhanced region of rapid reaction, Fig. 2.

Rates were measured in static systems described previously (7).   The ratio
of cool volume to heated volume was very small and gas-phase temperature
gradients were absent.   Initial reaction rates were in some cases very rapid
and not easily defined.   However, variations in catalyst bed geometry and in
particle size of pelleted catalysts did not reveal any diffusion resistance
either in the gas phase or within or between catalyst pellets.

Calculations using appropriate procedures (11)(12) also confirmed the
absence of diffusion resistances external to the crystallites.   Assuming a
first order process and taking, as a conservative estimate, the diffusivity of
CO at 400^{0}C as 4.6 x 10^{-7} cm^{2}s^{-1} (value for CO in NaA (13)) it seems that, for
the higher rates, pore resistances could not be completely neglected.

From Fig. 3 it seems that more than one reduction process may be involved.
Initially rates can be very rapid but subsequently reduction is very slow.
This is particularly noticeable for catalysts where exchange involved alkaline
pH (NaCuX(N), NaCuX(4), NaCuX(30), NaCuX(40)).   Moreover, with several
catalysts it was observed that the first reduction of the oxygenated catalyst
with CO was somewhat slower than subsequent reductions which were completely
reproducible.   This implies catalyst modification in the first redox cycle
which produces some sites more active than those present initially.   In the
case of NaCuX(80) reduction at 400^{0}C followed by oxidation produced a catalyst
which reacted very rapidly initially, more like the catalysts prepared using
alkaline ion-exchange conditions.   In this case it seems that more extensive
irreversible changes took place, perhaps with loss of zeolite structure, to
produce active sites similar to those found in series A catalysts.

Several kinetic models were applied to the data but no single model was
satisfactory over the whole reduction.   Experimental results correlated well
with the Elovich equation (14) and correlation was improved using two Elovich

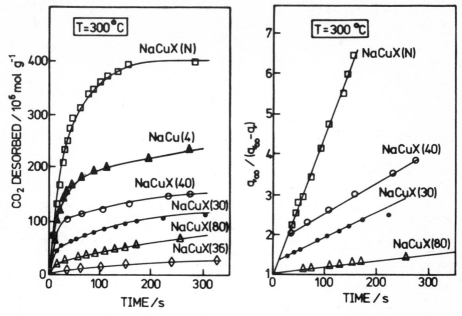

FIG. 2 Reaction of CO with oxygenated catalysts

FIG. 3 Application of Two Site Model

equations which can imply two types of site or process (15) although agreement is not general.

A simple kinetic model proposed recently for chemisorption (16) and desorption also correlated results well over a limited range and was used to compare the catalysts.

For chemisorption the model is:

$$q_{\infty}^{n-1}/(q_{\infty}-q)^{n-1} = (n-1)\,\alpha t + 1 \quad (n \neq 1) \qquad (1)$$

where q and q_{∞} are the amounts adsorbed at t and t_{∞} respectively,
    n is the number of sites occupied by each molecule of adsorbed gas
and α is the rate constant.
Extrapolation of the plot 1/q against 1/t is used to find q_{∞}.

In the present work a value of n=2 gave a good fit, over a limited range, as shown in Fig. 3. E.s.r. measurements, discussed subsequently, demonstrate that, under the reaction conditions used, the stoicheiometry of the reduction is

$$2Cu^{2+} + O^{2-} \xrightarrow[-CO_2]{+CO} 2\,Cu^{+} + CO_2 \qquad (2)$$

in agreement with a molar ratio, active oxygen/copper $\leqslant$ 0.5 (7). Based on this stoicheiometry it can be shown that equation (1) is equivalent to a rate (at constant pressure of CO) proportional to $\left(Cu_t^{2+} - Cu_{\infty}^{2+}\right)^2 / \left(Cu_0^{2+} - Cu_{\infty}^{2+}\right)^2$ where subscripts imply Cu^{2+} concentration at times 't', '∞' and 'o'. Since the term $\left(Cu_0^{2+} - Cu_{\infty}^{2+}\right)$ is constant within a given kinetic run, equation (1) implies a second order dependence on "available" copper. A reduction process second order in "available" Cu^{2+} was also observed in a separate study of NaCuX(4) (54% exchanged).   In this study [17] the concentration of Cu^{2+} in the catalyst was varied by pre-reduction to set levels of Cu^{2+}. Initial rates were proportional to $\left(Cu_i^{2+} - Cu_\infty^{2+}\right)^2$, where Cu_i^{2+} represents the pre-set level of Cu^{2+} concentration, in accord with second order dependence on available Cu^{2+}. This work also supported first order dependence in CO and is expected if reaction (2) is rate controlling.   A second order reduction of Cu^{2+} in zeolites was reported by Maxwell and Drent [18] for reduction with butadiene and by Naccache and Ben Taarit for reduction with hydrogen [19].

The amount of Cu^{2+} remaining after reduction (Cu_∞^{2+}) depended on extent of ion exchange and conditions of reduction.   These remaining Cu^{2+} ions are presumably located in the small-pore region.

Fig. 2 shows considerable differences in reactivity arising from differences in preparation.   Samples prepared from very dilute $Cu(NO_3)_2$ solution in acid conditions (NaCuX(36) and NaCuX(80)) react much more slowly than the sample prepared from $Cu(NO_3)_2$ in alkaline conditions, (NaCuX(N)), or the samples prepared from copper acetate solution which involved some alkaline ion exchange. This pattern parallels the reactivities observed in the CO/O_2 reaction (1) and is also, presumably, due to the formation of highly reactive copper species by agglomeration.   Furthermore Fig. 3 shows that plots of $q_\infty/(q_\infty-q)$ do not extrapolate to unity, as required by equation (1), for the catalysts made from copper acetate solution.   The increased values for these intercepts parallel the extent to which the desorption curves follow that for the non-stoicheio-metrically exchanged catalyst (NaCuX(N) prepared at pH 8.0) in Fig.2 and arise as a result of a relatively small proportion of highly reactive catalytic sites similar to those in NaCuX(N).   In this last sample, NaCuX(N), these active sites are present in sufficient quantity to produce a line, over the experimental region selected, with intercept close to unity.

Some reactions were made on catalysts NaCuX(30), NaCuX(N), NaCuX(40), NaCuX(80), NaCuX(36) and NaCuX(4) in the presence of water vapour ($\ast$ 2 Torr). In all cases initial reaction rates were reduced except for one reaction on catalyst NaCuX(36) at 375°C where rate was not affected by water, presumably because at this temperature the amount of water sorbed would be very small. Maxwell and Drent (**18**) reported an increase in the rate of reduction of Cu(II) ions, in NaCuY zeolites, by butene in the presence of traces of water but concluded that higher levels of water would cause blocking of Cu(II) sites and reduced rates.   The present work supports the conclusion concerning reduction in rate due to blocking of sorption sites, so far as the initial reaction rates are concerned.   Following the initial decrease, reaction rates were enhanced by the presence of water vapour,   and for all samples examined (except NaCuX(N)) there was increased reduction of Cu^{2+} ions.   Increased reduction is almost certainly due to the effect of water on the transport of Cu^{2+} ions from small to large pore sites.   Increased amounts of available Cu^{2+} increase both rates and extent of reaction.   For catalyst NaCuX(N) most of the reactive copper is in the form of agglomerates at least some of which is on the outside surface of the crystals.   Water appears to compete with CO for sorption on these sites (presumably by coordination to Cu^{2+}).

The nature of the active sites within the zeolite crystals depends on preparation and pretreatment.   Active sites Cu^{2+} O^{2-} Cu^{2+} (7), where the oxygen is non-zeolitic, can be formed in zeolites (Z) by reactions proposed by Jacobs et al (20)

$$2Cu^{2+} + ZO \rightleftharpoons Z^+ + Cu^{2+} \, O^{2-} \, Cu^{2+} \tag{3}$$

$$2Cu^{2+} + H_2O \rightleftharpoons 2H^+ + Cu^{2+} \, O^{2-} \, Cu^{2+} \tag{4}$$

producing either Lewis or Brønsted sites, as well as by dehydration of hydrolysed species

$$2Cu(OH)^+ \rightleftharpoons Cu^{2+} \, O^{2-} \, Cu^{2+} + H_2O \tag{5}$$

and perhaps by dissociation of coordinated water

$$Cu(H_2O) \rightleftharpoons Cu(OH)^+ + H^+ \tag{6}$$

Reaction of CO with active sites produced by (3) or (4) would be consistent with second order in Cu^{2+} and first in CO. The slow reduction of zeolitic Cu^{2+} involves less active sites which utilise lattice oxygen, presumably of the type removed during autoreduction (21) of NaCuY (i.e. O_1 atoms). The active sites are again probably of the type $Cu^{2+} \, O_Z^{n-} \, Cu^{2+}$ where O_Z^{n-} is lattice oxygen and the Cu^{2+} ions form (on the basis of esr work) a weakly interacting non-linear copper pair (g = 2.16).

The effect of temperature has been examined in a few cases but data are not yet sufficient to define activation energies precisely. Parameters based on the two-site chemisorption model gave low values for apparent activation energies (E_a) particularly in the case of zeolitic Cu^{2+}. Low values for E_a may arise as a result of diffusion interferences although calculations suggest this is unlikely. Because the two-site model used here for comparison has been applied over particular time intervals the Arrhenius parameters are not in all cases based on the same changes in degree of reduction and this tends to give apparent activation energies lower than the 'true' values.

Reductions of NaCuX(4), NaCuX(N) and NaCuX(36) were made in the presence of a small partial pressure of CO_2. Reaction rates were in no cases decreased and there was evidence for a slight increase in rate. Consequently, at the pressures used, CO_2 does not compete with CO for sorption sites and does not significantly restrict diffusion of CO to active sites. More detailed work will form the basis of a separate publication.

Electron Spin Resonance Measurements

Esr measurements were made using a Varian E9 (X-band) spectrometer with manganese(II) in strontium oxide as standard.

For one sample, NaCuX(3), the stoicheiometry of the mild reduction of series 'A' catalysts with CO was confirmed by measuring the esr spectrum and the cumulative amount of oxygen removed as CO_2 at several levels of catalyst reduction (7). A plot of relative signal intensities, obtained by double integration of the esr spectra, against oxygen removed as CO_2 was linear (Fig. 4). The least squares line indicated that 2.0 copper ions were reduced for each oxygen removed, in accordance with the stoicheiometry.

$$2Cu^{2+} + O^{2-} \xrightarrow{\;CO\;} 2Cu^+ + CO_2$$

and the involvement of a single oxygen rather than an O_2^{n-} species was supported by results using N_2O (7).

Electron spin resonance has been used to study cation coordination in zeolites. In the present work (22) the hydrated NaCuX signal appeared to be mainly due to a single line (parameters $g_{11} = 2.369$, $A_{11} = 127$ gauss; $g_1 = 2.08 \pm 0.01$; $A_1 \approx 20$ gauss) similar to that of Leith and Leach (23). On dehydration a signal (signal I) similar to that described as spectrum A by Leith and Leach was obtained in addition to a broad signal, g = 2.16 ($\Delta H \approx 170$ gauss), which may be assigned to non-linear copper pairs (24). Calculations showed this broad signal to be associated with about ten times the copper giving signal I. On reduction with CO, the broad signal, g = 2.16, disappeared preferentially; it could be seen after 7 but not after 31 pulses of CO. After 31 pulses of CO signal I remained along with a second signal (signal II) which had parameters similar to those of spectrum B of Leith and Leach. Signal II was about twice the intensity of signal I.

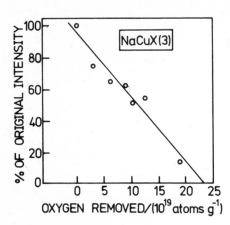

FIG. 4   Decrease in esr Signal on
Removal of Oxygen by CO→CO$_2$

The esr parameters for NaCuX(36), prepared from dilute Cu(NO$_3$)$_2$ solution, were similar to those of NaCuX(3). The signal g = 2.16 was evident in the spectrum of the hydrated as well as of the dehydrated sample. Approximate calculations showed that most of the copper ions in the dehydrated sample existed as weakly interacting non-linear pairs (g = 2.16). The esr spectrum of NaCuX(N) (prepared at pH = 8.0) differed substantially from that of NaCuX(3) and NaCuX(36). Although NaCuX(N) contained more than twice the amount of copper in NaCuX(36) the signal was of lower intensity, as expected when non-stoicheiometric exchange is extensive (1), and there was little or no evidence for non-linear copper pairs (g = 2.16).

It seems clear that for the levels of exchange used, the zeolitic copper is mainly present, on dehydration, as the species giving rise to the g = 2.16 signal assigned to weakly interacting non-linear copper pairs (more recent work by Conesa (25) assigns an isotropic signal g = 2.14 to copper pairs bridged by oxygen with at least one copper in the supercage   of NaCuY). In the present work, the pairs (g = 2.16) seem to be removed preferentially during reduction with CO and presumably are the precursors of the active CuOCu species. For extensive non-stoicheiometric exchange most of the copper is present as agglomerates which are not easily detected by esr. In the case of NaCuX(N) (exchanged at pH 8) a limited amount of zeolitic copper is also present largely as isolated ions.

Reaction of NaCuX Catalysts with Propene

Many metal oxides and mixed metal oxides which show selective oxidation of propene at higher temperatures react with propene to produce oxygenated organic intermediates. Product distribution frequently depends on the degree of reduction of the catalyst. The state of the transition-metal cations in zeolites can vary according to the method of preparation and pretreatment. Exchange may be stoicheiometric or otherwise and, following a sequence of redox reactions, new phases (e.g. metal oxides) may be formed. Consequently, in considering selectivity patterns it is necessary to examine catalysts in different states of reduction and after different pretreatment.

NaCuX(5) (62% exchanged) which contained mainly ion-exchanged Cu(II) ions, as shown by analysis and DTA, and some very reactive species, as indicated by reaction with CO, and samples of NaCuX(5) which had undergone several redox cycles with TME and O$_2$ and contained a copper oxide phase were reacted with propene.

A series of samples of each of these catalysts was pre-reduced to selected levels using pulses of CO, a procedure which removes oxygen without permanently altering the catalyst and which must have altered the relative amounts of different types of site. One pulse of propene was then reacted with the catalyst, at 400°C, and the products analysed.

In all cases products were ethylene, CO_2 and, for the fresh catalyst, CO. Fractional conversion changed with degree of reduction but relative proportions of products did not. No oxygenated organic intermediates were desorbed. Carbon balances, especially on the fresh catalysts, indicated considerable retention of carbon but retained products were not examined. It appeared that variation in the types of copper species present in NaCuX zeolites did not lead to significant variation in products of reaction with propene at 400°C (Figs. 5 and 6).

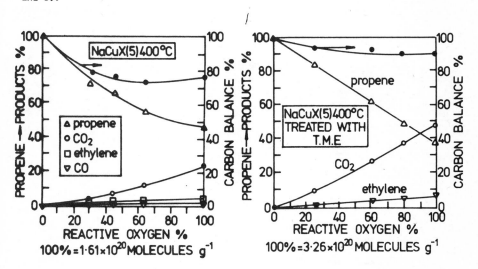

FIG. 5 Reaction of Propene with Fresh Catalyst

FIG. 6 Reaction of Propene with Treated Catalyst

Separate experiments with a fixed bed reactor (26) demonstrated that NaPdCuX zeolites were selective in oxidation of propene with oxygen to acetone and similar results were obtained with NaPdNiX. The feed gas stream was 63% He, 6.3% O_2, 4.3% propene and 26.4% H_2O.

These results (Fig. 7) are similar to previous reports (27) concerning NaCuY zeolites. Good selectivities to acetone were obtained around 120 to 150°C but as temperature increased to 200°C deep oxidation products became significant. At the partial pressures of water used cations are considerably hydrated and good selectivity is probably associated with mechanisms similar to those in solution and presumably does not involve oxygen from the catalysts.

Surface Studies

ESCA

An AEI ES 200 with Al K_α X-radiation was used for ESCA. In some cases samples were pressed into lead shot and in others were mounted on stainless steel probes. Results for samples dehydrated for several hours at room temperature are given in Table II.

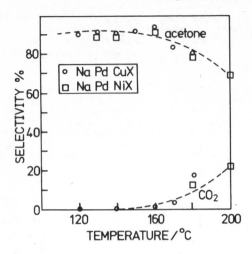

FIG. 7   Oxidation of Propene Over NaPdMX Zeolites

TABLE II   NaCuX Catalysts ESCA

Catalyst	Binding Energy /eV $\left[Cu\ 2p\frac{3}{2}\right]$
NaCuX(N)	933.0
NaCu(I)X(50)	932.9
NaCu(I)X(50)[a]	933.1
NaCuX(30)	932.0
NaCuX(40)	932.1
NaCuX(36)	932.3
NaCuX(80)	932.1, 932.4

(a) - Catalyst NaCu(I)X(50) oxidised at $400^{\circ}C$ with oxygen.   An X-ray powder photograph showed lines for CuO.

The highest binding energies, with values in the range 933.0 ± 0.1 eV, were observed on samples containing significant amounts of non-zeolitic oxygen associated with copper, for example NaCu(I)X(50), prepared from Cu(I) and then oxidised, and NaCuX(N) prepared at pH 8.

For some catalysts observations were taken at increased temperatures (Fig. 8).   In all cases there was a tendency for binding energies to decrease, presumably due to some reduction of Cu(II) (6).   Approximate relative intensities were obtained from estimates of peak areas.   The pattern of changes with temperature is shown in Fig. 9.   It is clear that migration of copper in the sub-surface layer is much more pronounced in zeolites NaCuX(30), NaCuX(40) and NaCuX(N), all of which involved at least some exchange under alkaline conditions and possessed some sites very active for oxidation of CO, than migration in NaCuX(36) and NaCuX(80) which were ion-exchanged under acidic

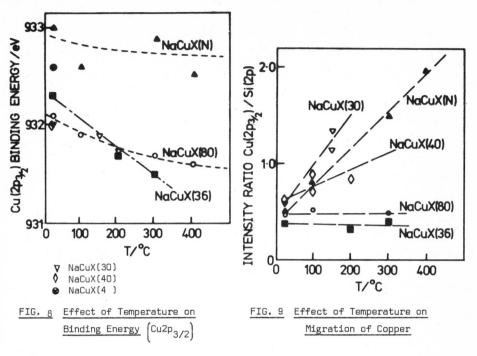

V   NaCuX(30)
◊   NaCuX(40)
⊛   NaCuX(4)

FIG. 8   Effect of Temperature on   FIG. 9   Effect of Temperature on
Binding Energy $\left[Cu2p_{3/2}\right]$ Migration of Copper

conditions and possessed little or no active sites, (Figs. 2 and 3).   It appears
that the facile migration is related to the existence of active sites formed at
higher pH presumably, by agglomerates of copper species.

SIMS

Whereas ESCA concerns atoms in a layer extending several ångstroms below
the outermost surface layer SIMS is presumed to detect only species present in
the first monolayer.   A VG Q50 quadrupole mass spectrometer was used for SIMS
with a mass-filtered argon beam (3keV) and, typically, a current density of
5×10^{-8} amp cm^{-2}.

Four samples NaCuX(30), NaCuX(36), NaCuX(80) and NaCuX(N) were examined at
room temperature using SIMS;   the latter two samples were also examined after
heating.   In all cases a wide mass range of fragments was produced many of
which contained only Si, Al and oxygen.   The major copper-containing fragments
are tabulated below (Table III).   Copper appears in the spectrum with isotopic
peaks 63 and 65.   A broad band (79-82) is observed in the room temperature
spectra of all the samples and is assigned to CuO and CuOH species.   For
NaCuX(30) the peak at mass 80 (CuOH) is predominant.   On heating NaCuX(80) and
NaCuX(N) the broad band partly resolves to give peaks with masses 79 and 81,
presumably by conversion of CuOH species to CuO species.   Copper also appears
associated with framework atoms.

The only direct evidence for bridged copper species was obtained with
NaCuX(N), which was ion-exchanged at pH 8 and was most reactive with CO, where
a peak with mass 142 (CuOCu) was observed.   Very approximate estimations of
surface copper, made by taking the ratio of peak heights for Cu(63) and Si(28),
showed that NaCuX(N) had more surface copper than NaCuX(80) although both had
similar total amounts of copper.   Similarly for the lower exchanged samples

TABLE III   NaCuX Catalysts. SIMS

Mass[a]	Assignment[b]	
63, 65	Cu	
79, 81	CuO	
80, 82	CuOH	
107,109	CuOSi	
138	CuOAlO	
139	CuOSiO	(HOCuOAlO)
142	(CuOCu)[c]	

(a) Two masses are given where observed. Generally they were in the ratio for copper isotopes.

(b) Tentative assignments; bracketed species less reliable.

(c) Observed only with NaCuX(N).

NaCuX(30) had more surface copper than NaCuX(36). This suggests that some of the active copper species may be associated with copper initially present in the outermost layer. The migration of copper in the sub-surface region observed clearly by ESCA is not detected by SIMS. In the preliminary experiments no evidence was found for accumulation of copper in the outermost layer as a result of heat treatment. The migration of copper species observed by ESCA towards the outer surfaces of the crystals may not lead, initially, to accumulation in the outermost monolayer but may do so over a sufficient period.

Discussion

The state of copper ions in NaCuX zeolites depends on method of incorporation and catalyst history. Either Cu^+ or Cu^{2+} ions may be exchanged and the exchange may be stoicheiometric or otherwise. Redox processes may lead to formation of non-zeolitic phases. The characterisation of copper species in zeolite catalysts presents some difficulty.

In the present work the conditions used for ion exchange and for catalyst pretreatment modified the properties of NaCuX zeolites. For the fresh catalysts extremes of reactivity with CO were represented by NaCuX(N), which was ion exchanged in alkaline conditions, and NaCuX(36) which was exchanged using dilute $Cu(NO_3)_2$ solution in slightly acidic conditions. NaCuX(N) reacted very rapidly and NaCuX(36) reacted very slowly with CO. Chemical analysis showed that NaCuX(N) contained excess copper ions so that the high reactivity is presumably associated with the formation of agglomerates known to be active in the CO/O_2 reaction [1].

Thermal analysis seems to support the conclusion, by Schoonheydt et al [2], that an endotherm around 600 K corresponds to loss of water from nonstoicheio-metrically exchanged species. Such endotherms were observed with NaCuX(N), with NaCuX(4) and NaCuX(5) (both very slight) and not at all with NaCuX(30) and NaCuX(40) yet all of these catalysts showed the rapid initial reaction with carbon monoxide which we attribute to copper ions associated with non-zeolitic oxygen. It may be that the endotherm was present in profiles of these last two catalysts but not observed because of the broad endotherm associated with water loss, present in all the zeolites. Clearly thermal analysis is useful in characterising zeolite catalysts but is not very sensitive.

Rapid chemisorption of CO occurs on highly active sites, presumably the agglomerated or bridged CuOCu species proposed as active sites for the CO/O_2 reaction, which are accessible to CO. This situation is found for NaCuX(N) at all temperatures studied and the rate constant shows reasonable Arrhenius behaviour with an apparent activation energy ($E_a \approx 70$ kJ mol^{-1}) close to that reported for the CO/O_2 reaction over similar catalysts.

Apparent activation energies for reduction of Cu^{2+} with CO appear to be rather lower than expected for simple migration of Cu^{2+}(3) (although data are not available for migration in presence of CO and CO_2). However, as mentioned previously, it should be borne in mind that the apparent activation energies based on the two-site model refer to a part of the reduction only and because of some variation in the degree of reduction at different temperatures are expected to be low. Further work will examine the effect of degree of reduction on activation energies.

It seems, on the basis of calculations, that mass transport can be ruled out as rate determining although in the absence of reliable values for the appropriate diffusivities it is not possible to be completely unequivocal about this. Since no reduction in rate was observed when reactions were carried out in the presence of carbon dioxide it seems that desorption is also not rate determining. Consequently the rate determining stage is associated with reduction of Cu^{2+} species or with formation of active sites. Gas-solid reactions of this kind are frequently more complex than allows for representation by the simpler chemisorption models and require consideration of the reaction interface and the possible influence of reaction products and these points will be examined in a separate publication.

ESCA measurements also divide the catalysts sharply. Those having a very rapid initial rate of reaction with CO show a substantial increase in the ratio of Cu/Si on heating. This migration in the sub-surface layer is presumably related to formation of active sites on heating. From the binding energies it does not appear that this migration is due principally to reduction to Cu^+ or Cu^0, so that it must concern Cu^{2+} species held less strongly to the zeolitic framework than stoicheiometrically exchanged Cu^{2+} ions. The smaller charge/size ratio in species of the type $(CuOCu)^{2+}$, $CuOH^+$, $Cu(OH)_2$ (which would be expected from ion-exchange under conditions favourable to cation hydrolysis) as compared with Cu^{2+} ions would suggest weaker bonding to the framework. It appears from the slightly lower binding energies associated with NaCuX(36) and NaCuX(80) (where there was virtually no evidence for non-zeolitic oxygen) compared with NaCuX(N) (which contained non-zeolitic oxygen) that coordination to the zeolite framework X compensates the charge on copper at least as much as the oxygen anions in agglomerates. However, in these experiments samples were not oxidised prior to taking spectra so that the exact state of surface reduction is uncertain. Moreover, especially for the stoicheiometric samples, spectra were rather broad and too much emphasis should not be placed on absolute values of binding energy.

Esr showed that most of the copper ions in NaCuX(36) and NaCuX(3) (33% exchanged) were present as the species giving rise to the isotropic signal (g = 2.16). Similar results were found for NaCuX(4) and NaCuX(5) (54% and 62% exchanged respectively). The isotropic signal (g = 2.16) has been assigned to weakly interacting non-linear pairs (24). It seems that weakly paired copper ions form the precursors for the active sites within the zeolite cages presumably in association with a lattice oxygen. However the only direct evidence for bridged CuOCu species comes from SIMS on NaCuX(N). This sample was most reactive with CO and also appeared to have more copper on the outermost monolayer than the other catalysts. Bridged systems within the zeolite crystals would not be readily detected by SIMS. Moreover, the appearance of a fragment corresponding to CuOCu does not necessarily mean that these atoms were bonded in the surface but rather that they were in close proximity.

In all cases water vapour reduced initial reaction rates but subsequently reaction rates were increased and, except for NaCuX(N), more Cu^+ was reduced. It appears that the activity of the most reactive sites, particularly agglomerates, is decreased, presumably by competitive adsorption of CO and H_2O. However, reaction of zeolitic copper is enhanced perhaps because of increased amounts of available Cu^{2+}. Extents of reduction of Cu^{2+} in presence of water are also related to increases in available Cu^{2+}.

The reaction with CO will be reported more fully in a separate paper. Over at least part of the reaction it is second order in Cu^{2+} ions and first order in CO which is satisfied by the basic stoicheiometry of the reduction:

$$2Cu^{2+} + O^{2-} + CO \longrightarrow 2Cu^+ + CO_2$$

Ion exchange involving alkalinity at some stage (NaCuX(N) and Series A) produced catalysts with some very active sites, and in Series A these sites existed along with extensive amounts of zeolitic copper. Reduction involves two Cu^{2+} ions and the oxygen species O^{2-} (esr and N_2O experiments show this (7)). In the dehydrated zeolites the active sites are presumably $Cu^{2+}O^{2-}Cu^{2+}$ species where the O^{2-} is not part of the zeolitic framework, and in Series A and NaCuX(N) arise from the incorporated hydrolysed cation species (e.g. by reaction (5) and perhaps (6)). Zeolitic Cu^{2+} ions can generate these sites by reactions (3) and (4) and by structural modification during heating and redox cycling. Generation of active sites by redox cycling was particularly evident with NaCuX(80) which after one redox cycle at $400^{\circ}C$ behaved more like the Series A catalysts and showed a very rapid initial rate with CO.

The slow process following reduction of the active sites involves zeolitic oxygen and overall is

$$(7)$$

This reaction is the basis of most of the reduction in catalysts prepared in acidic media and we propose that the active site involves lattice oxygen of the type removed during autoreduction (probably O_1 atoms (21)) and a non-linear pair of weakly interacting Cu^{2+} ions (g $\simeq$ 2.16). A CO_2^- species proposed as an intermediate in reduction of NaCuY(19) may be involved.

Prolonged heating in CO (7) reduces Cu^+ to Cu^0 by a very slow process. This must involve zeolitic oxygen:

$$(8)$$

An intermediate in this reduction may be the species proposed by Klier in NaCu(I)A zeolites (28):

Oxidation to reverse (8) tends to give CuO species, containing non-framework oxygen, which can be reduced under mild conditions to Cu^0. The degree of agglomeration depends on reaction conditions but it seems that isolated Cu^0 species may be formed and subsequent oxidation gives isolated CuO species in the zeolites which form new active sites (7). Reduction with hydrocarbons or higher pressures of CO forms agglomerates of Cu^0 atoms and a separate CuO phase on oxidation (7).

With reference to characterisation of the state of Cu^{2+} ions in zeolites, it seems that thermal analysis can be helpful but is not very sensitive. Chemical analysis should be sensitive for detection of non-stoicheiometry, providing the analysis of the solid is sufficiently accurate and precise and providing exchange of both H^+ and MOH^+ species (in approximately equal amounts) does not take place. In the work reported here, analysis of the solids was such that ion balances were always within a few per cent of stoicheiometry except for NaCuX(N) where excess Cu^{2+} was evident. However, reduction with CO seems to provide a very sensitive method for characterisation of Cu^{2+} in zeolites. ESCA also appears to provide a good means for characterisation, particularly via temperature dependence of intensities. Work using SIMS is in a very preliminary stage but is likely to be useful in characterising species on the outside surfaces of crystals.

With regard to selective oxidation it seems that, although it is possible to modify the reactivity of the oxygen associated with Cu^{2+} in NaCuX zeolites, the systems produced show no evidence for selective oxidation of propene at high temperature. In the NaPdMX zeolites selective oxidation, with O_2 in the gas phase, takes place at lower temperatures in the presence of water and does not involve reactive oxygen in the NaCuX catalysts.

Acknowledgement

The authors would like to thank the Surface Physics Group at UMIST for ESCA and SIMS measurements and Mr. M. Macdonald for some of the ESR measurements. Additionally we would like to thank Professor P.G. Ashmore and Dr. A.J. Parker for helpful discussion.

References

1. K.G. Ione, N.N. Bobrov, G.K. Boreskov and L.A. Vostrikova.
    Dokl. Akad. Nauk SSSR. 210 388 (1973).

2. R.A. Schoonheydt, L.J. Vandamme, P.A. Jacobs and J.B. Uytterhoeven.
    J. Catalysis 43 292 (1976).

3. R.A. Schoonheydt and F. Velghe.
    J. Chem. Soc. J.C.S. Faraday I, 72 172 (1976).

4. A. Dyer and A.L. Barri.
    J. Inorg. Nucl. Chem. 39 1061 (1977).

5. Kh. M. Minachev, G.V. Antoshin and E.S. Shpiro
    Izv. Akad. Nauk SSSR Ser. Khim. 5 1012 (1974).

6. R.G. Herman, J.H. Lunsford, H. Beyer, P.A. Jacobs and J.B. Uytterhoeven.
    J. Phys. Chem. 79 22 2388 (1975).

7. F.R. Benn, J. Dwyer, A. Esfahani, N.P. Evmerides and A.K. Szczepura.
    J. Catalysis 48 60 (1972).

8. S.Z. Roginskii, O.V. Al'tshuler, O.M. Vinogradova, V.A. Sekznev and
    I.L. Tsitovskaya.
    Dokl. Akad. Nauk Ser. Khim. SSSR 196 872 (1971).

9. F.O. Bravo, J. Dwyer and D. Zamboulis.
    Unpublished work.

10. E. White and J. Dwyer.
    Unpublished work.

11. W.J. Beek and K.K. Muttzal
    "Transport Phenomena", J. Wiley (1975).

12. O. Levenspiel.
    "Chemical Reaction Engineering", J. Wiley (1972).

13.   R.M. Barrer
       "Clay Minerals and Zeolites", Academic Press (1979),

14.   D. Zamboulis.
       M.Sc. Dissertation UMIST (1977).

15.   C. Aharoni and F.C. Tompkin.
       Adv. in Cat. 21 (1970).

16.   A.G. Ritchie.
       J.C.S. Farad. I 73 1650 (1977).

17.   J. Dwyer and A.K. Szczepura.
       Unpublished work.

18. · I.E. Maxwell and E. Drent.
       J. Cat. 41 412 (1976).

19.   C.M. Naccache and Y. Ben Taarit.
       J. Cat. 22 171 (1971).

20.   P.A. Jacobs, M. Tielen, J.P. Linart and J.B. Uytterhoeven.
       J.C.S. Farad. I 72 2793 (1976).

21.   P.A. Jacobs, W. De Wilde, R. Schoonheydt, J.B. Uytterhoeven and H. Beyer.
       J.C.S. Farad. I 72 1221 (1976).

22.   A.H. Al-Safar, J. Dwyer, A.J. Parker and P. Thornally.
       Unpublished work.

23.   I.R. Leith and H.F. Leach.
       Proc. Roy. Soc. Lond. A 330 247 (1972).

24.   C.C. Chao and J.H. Lunsford.
       J. Chem. Phys. 57 (7) 2890 (1972).

25.   J.C. Conesa and J. Soria.
       J.C.S. Farad. I 74 406 (1978).

26.   J. Dwyer and J.C. Vickerman.
       Unpublished work.

27.   T. Kuhota, F. Kuminada, H. Tominaga and T. Kunngi.
       Nippon Nagaku Kaishi 15 1621 (1972).

28.   J. Texter, D.H. Strome, R.G. Herman and K. Klier.
       J. Phys. Chem. 81 (4) 333 (1977).

Discussion on Session 5 : Catalysis

Chairman: Professor J. Haber (Krakow, Poland.)

(Throughout this section, reference numbers correspond to those in the particular paper under discussion)

Paper 20 (D.E.W. Vaughan)

Dr. Short: I have three questions:

(i) What level of coke provides just sufficient heat in the regenerator to balance the endothermicity of the cracking reactions for the average Fluid Catalytic Cracking Unit?

(ii) It was implied that addition of organo-antimony to cracker feed is giving rise to problems. I believe that Philips Petroleum are operating several FCC units using antimony addition and claiming considerable success. Could you comment further on this?

(iii) I noticed that a figure of $80,000/ton was given as the price of ZSM-5 catalyst. Is that a rental charge and, if not, what is the zeolite content of the particular catalyst quoted?

Dr. Vaughan: (i) The advent of complete CO burning and the usual vagaries of most FCC units have essentially eliminated any notion of an "average" FCC unit[38] The main elements are the balancing of the heat required in the endothermic cracking reaction plus feed preheat, by the burning of $C \rightarrow CO \rightarrow CO_2$, $S \rightarrow SO_2$ and $2H_2+O_2 \rightarrow 2H_2O$ in the regenerator. The extent of the burning of $CO \rightarrow CO_2$ is an important factor in this heat balance because of its large exothermic heat of reaction. A typical detailed analysis for a specific case has recently been published by Pierce (Catalagram, No. 57 (1978) Davison Chemical Co., Box 2117, Baltimore, Md.) and a review of this paper is strongly recommended if you are interested in the full details.

(ii) This metal passivation system seems to have been tested in several refineries for brief periods, and has undergone prolonged testing in a Phillips FCCU specifically designed for handling high metals residual feeds[43]. Injection of organometallic components to a hot regenerator (as a liquid or impregnated onto a catalyst) would be expected to give problems with the actual process of injection, and the subsequent desired uniform distribution of antimony onto the catalyst, unless the regenerator was redesigned for addition of this kind of additive. Phillips probably can accommodate this in their Borger resid FCCU, but other refiners probably cannot without major hardware modifications requiring unit shutdown. The advantage of adding the organometallic components to the feed or to the catalyst may be the formation of more viable routes (see, for example, D.L. McKay, 1977, U.S. Patent 4,025,458 and also I.A. Vasolos, 1979, U.S. Patent 4,153,534/5.)

(iii) Mobil is reported to lease rather than sell ZSM-5 at the present time (A.J. Oxenham, High Technology Surface Active Particulates (A Multiclient Catalyst Survey) v.3, 1978, Hull & Co., Greenwich, Connecticut.) The number in Table 1 is an estimate of the cost of making and selling ZSM-5 assuming:

(a) a process that yields a 30:1 SiO_2/Al_2O_3 product (A.J. Argauer and G.R. Landolt, 1974, U.S. Patent 3, 790, 471.)

(b) a 100% stoichiometric excess consumption of tetrapropyl ammonium bromide (TPA) and

(c) processing similar to that used to make synthetic mordenite.

The cost of TPA is the major factor, but the cost reductions expected from tonnage purchase of TPA were assumed to be cancelled out by the cost of pollution control in the wet end processing, and the dry end calcination to remove TPA from the product. The final catalyst is assumed to be a maximum zeolite bound fabricate (*i.e.*, pellets or pills containing 75 - 85% ZSM-5).

Paper 21 (F. Mahoney, R. Rudham and A. Stockwell)

Dr. Coughlan: In your paper you are dealing with the oxidation of hydrogen, methane and propene. I should like to point out that in the case of the last of these molecules, the possibility of complexing with Cu^{2+} cations occurs. The composition and geometry of these complexes, in which the ligands could be either framework oxygens or propene molecules or indeed both, could change as the number of Cu^{2+} cations per u.c. increases. These complexes would then be intermediates in the oxidation process and could explain the observed dependence of the activation energy on the number of Cu^{2+} cations.
(i) have you considered this possibility, and
(ii) have you examined your samples spectroscopically or visually for colour changes with increasing copper content?

Dr. Rudham: We were aware that complexes can be formed between cations and alkenes, and that the ability of cations to form dative π bonds with propene has been offered as an explanation for the activity sequence for propene oxidation on Y zeolite containing Cu^{2+} and other exchange cations (see J. Catalysis 1971, 23, 31). However, such an approach does not offer an explanation for the effects of acidity on the activity for propene oxidation presented here and in earlier work (J. Catalysis 1974, 35, 376). No spectroscopic studies have been made of our Cu/H-X oxidation catalysts.

<u>Dr. Karge</u>: You told us you have observed a significant increase in activity of your Cu-zeolite after H_2S pretreatment. Your explanation was that the H_2S was adsorbed dissociatively giving rise to new acid-OH groups. May I ask you whether you have checked this hypothesis by means of an i.r. investigation? One would expect the development of new OH bands and/or the increase of the intensity of OH bands already present before pretreatment. Furthermore, you could try to check the acidity of perhaps newly-formed OH groups using the method of pyridine adsorption and looking for i.r. bands indicative of Brønsted centres.

<u>Dr. Rudham</u>: We believe that the pronounced increase in activity for propan-2-ol dehydration is sufficient evidence for an increase in the concentration of O_zH groups in CuY after H_2S treatment, and that the activation energies indicate that the concentration of the O_zH groups is temperature independent as in HY catalysts. Comparison of the OH stretching bands in the i.r. spectra of Y/Cu(30) before and after H_2S treatment was inconclusive, since the loss of molecular water from hydrated Cu^{2+} ions profoundly modified the spectra. Furthermore, the blackening ascribed to CuS formation probably resulted in a temperature rise of the sample in the spectrometer beam. Although pyridine was a total catalyst poison, no attempt was made to characterise its mode of adsorption by i.r. spectroscopy.

There appears to be an appreciable difference in the stability of O_zH groups generated in CuY and in Na faujasites by H_2S treatment (J. Colloid & Interface Sci. 1978, <u>64</u>, 522). The catalytic activity of H_2S treated CuY was unaffected by heating at 523K in flowing nitrogen, whereas the O_zH groups generated in NaX and NaY were eliminated by evacuation at ~373K. If H_2S interaction with the unexchanged Na^+ ions occurs in our catalysts it is unlikely that the O_zH groups remain to contribute to catalysis after the system is purged of excess H_2S. We have presented evidence (Acta Physica et Chimica Szegediensis 1978, <u>26</u>, 281) to suggest that large, stable increases in propan-2-ol dehydrating activity only arise from H_2S treatment of Y zeolite containing cations which yield a sulphide which resists hydrolysis.

Paper 22 (M.S. Spencer and T.V. Whittam)

<u>Prof. van Bekkum</u>: Did you analyse the evolved gases during activation of the Nu-1 zeolite? Were, for instance, nitrogen-containing rings present?

<u>Dr. Whittam</u>: The gases evolved during the air activation of Nu-1 did not contain detectable levels of cyclic nitrogen compounds. The gases typically contained methanol and trimethylamine in the early stages and then dimethylamine, methylamine, ammonia and nitrogen.

<u>Dr. Vaughan</u>: Your descriptions of Fu-1 are similar to those given for a pair
of synthetic silica polymorphs V3 and V13, and the minerals silhydrite and
magadiite. Are you sure that Fu-1 is a zeolite, and not a platey silica or
silicate?

<u>Dr. Whittam</u>: Fu-1 always has substantial levels of Al_2O_3 associated with it,
and occurs with SiO_2/Al_2O_3 = 15 → 30. From a very large number of syntheses
the analytical and sorptive properties are very constant, suggesting the product is largely a single phase, and that Al_2O_3 is uniformly distributed. The
X-ray and electron microscope data show that Fu-1 in no way resembles magadiite
or near kenyaites which we have also synthesised from low Al-containing sodium
silicate mixes. Furthermore, magadiite and kenyaite on calcination or acid
exchange yield silica X, whereas Fu-1 remains as Fu-1, and also the silica polymorphs show negligible sorption capacities except for water. Fu-1 shows typical
zeolite sorption properties.

   Recent chemical and physical studies on Fu-1 strongly support the description of Fu-1 as a zeolite aluminosilicate material. This information will be
published in a later paper.

   It is, of course, possible that high silica zeolites are open structured
silica polymorphs with occasional replacement of Si by Al.

<u>Prof. Barrer</u>: Dr. Vaughan asked whether Fu-1 and Nu-1 could be porous crystall-
ine silicas. There is however aluminium in the analyses you showed, so that
if the crystals were silicas this aluminium must be there as an associated
impurity. It might be possible to check this by microscopic examination.
Usually zeolite crystals have characteristic forms apparent on examination,
amid which impurity can be detected, because of their different appearance. For
instance, associated amorphous materials can often be distinguished as irregu-
lar, often glass-like particles. Have you examined Fu-1 and Nu-1 in this way,
and if so what can be said about the presence of amorphous particles and the
purity of the zeolites?

<u>Dr. Whittam</u>: We have no evidence to suggest that Fu-1 and Nu-1 are porous
crystalline silicas. Considering Nu-1 first of all, X-ray data and electron
micrographs show the material is highly crystalline, with crystallite sizes in
the range 0.3 to 5 μm, and only trace amounts of glassy impurities are present.
Nu-1 can occur as at least five different crystal types by choosing appropriate
synthesis conditions. The synthesis of all these types is reproducible as are
the X-ray data.

   On the other hand, Fu-1 is a very difficult material to characterise. A
large number of samples have been synthesised, and for given conditions the

products can be obtained with very reproducible analysis both chemical and
X-ray wise, and SiO_2/Al_2O_3 can be in the range 15-30 for the most crystalline
products. X-ray diffraction data is very reproducible from sample to sample
but the peaks on a diffractometer trace are quite diffuse as would be expected
of crystals 5→10 nm thick and 50→100 nm in the other dimensions. These crystals
tend to stack perpendicularly to each other, but this can readily be changed
into random, more compact stacking. It is very difficult to focus more than
a few plates at one time on an electron microscope, and the out-of-focus plates
give an overall wispy appearance. Therefore it is not possible to say how pure
Fu-1 samples are, although we can say that recent products are more crystalline
and have higher sorptive capacities and higher catalytic activities than earlier,
less crystalline products.

Dr. Dyer: Is not the observation of sigmoidal kinetic water uptake evidence
for the presence of "bottle-top pores", *i.e.* typical silica polymorph behaviour?

Dr. Whittam: In the case of Nu-1, rate curves for water sorption and for the
larger molecules, are sigmoid, but for medium-sized molecules (*e.g.* n-hexane
and isobutane) the rate curves are rectangular. I think the most probable
explanation for the sigmoid curve for water is that at least part of the channel
system in freshly activated Nu-1 is somewhat hydrophobic, but given time, some
form of rehydration occurs rendering Nu-1 every bit as hydrophilic as low silica
zeolites such as A or X types. For Fu-1 the rate curves are in fact rectangular
but the water isotherms clearly show that capillary condensation sets in at
about P/P_0 = 0.4, presumably reflecting condensation between Fu-1 plates.

Prof. Sing: There are three possible explanations for the slow initial rate of
sorption of water vapour by the new forms of highly siliceous zeolites: (i) the
penetration of sorbate molecules into very narrow pores, (ii) the rehydration of
not easily accessible cations, (iii) the rehydroxylation of surface SiO_2. The
Type II shape of the water isotherm shown suggests that part of the solid may
be non-zeolitic *i.e.* mesoporous. If this is the case, the presence of free
SiO_2 could be responsible for the relatively slow uptake of H_2O. It would be
of interest to know whether the water isotherms are reversible, and, if not,
whether hysteresis extends over the complete range of relative pressures.

Dr. Whittam: Slow initial rates for water sorption were seen only for Nu-1.
In this case I am inclined to agree that some form of rehydration takes place,
and after an appropriate time interval the material becomes as hydrophilic as
the high alumina zeolites A and X, whereas ZSM5 zeolite remains rather hydrophobic permanently.

In the case of Fu-1 the picture is different because water uptake is rapid, *i.e.* 95% of equilibrium level within ten minutes at $P/p_0 < 0.3$. However, at $P/p_0 \sim 0.4$ it is clear from isotherms that capillary condensation occurs, and from this point isotherms become non-reversible. It seems likely that at $P/p_0 > 0.4$ the water is condensing between the very small plates of Fu-1.

Paper 24 (J. Dwyer)

Prof. Haul: You mentioned that on reduction with CO, metallic copper may be formed which, however, could not be detected by X-ray. Since you were using ESR measurements, did you look out for particle size quantum effects of the conduction electrons in small copper clusters? When reducing Ag-A zeolites we found this effect, and we were able to identify the O_2^- radical when the reduced zeolite was reacted with molecular oxygen (Hermerschmidt, Thesis, Univ. Hannover 1979).

Mr. Dwyer: Following prolonged reduction with CO at low pressure, most of the copper ions were reduced to the metal. Our ESR measurements were made after heating the reduced sample in oxygen. The final ESR signal was almost identical to the initial signal for the oxidised fresh sample of catalyst. Clusters of copper atoms tend to give some copper oxide aggregates when oxidised, which would give an ESR signal of reduced intensity. We conclude that careful reduction can give metallic copper in a very disperse, perhaps atomic state. However, we did not examine the reduced sample in any detail by ESR and therefore we have no information about quantum effects of conduction electrons which you have observed in Ag-A zeolites.

Potential Uses of Natural and Synthetic Zeolites in Industry

D. W. Breck

Linde Division

Union Carbide Corporation

Tarrytown, N.Y. 10591

U.S.A.

Introduction

The first practical use of zeolites probably occurred about 2000 years ago when natural zeolite rock was quarried for use as building stone[1] although zeolites were not recognized as a new mineral species until 200 years ago.[2] The first physical chemical property of zeolites which had an application - cation exchange - was investigated by Eichhorn about 100 years ago.[3] This led to the development of synthetic, amorphous aluminosilicates as commercial cation exchange materials (permutites) in the early 20th century[4, 5] which were primarily used in water softening.

The first application of dehydrated zeolites as molecular sieves in the separation of gas mixtures was demonstrated by Barrer in 1945 utilizing the zeolite mineral chabazite.[6] Synthetic zeolites were first introduced and utilized commercially as molecular sieve adsorbents in 1954.[7] In the space of 25 years, the synthesis, science and technology of zeolites have achieved worldwide recognition as evidenced by the appearance of about 1000 publications a year, the application of just about every modern scientific discipline, the recognition of 40 known different zeolite minerals, and the total reported synthetic types which are now well over 150.

Although only a small number of mineral and synthetic zeolites have reached the commercialization stage, the _potential_ is obviously immense. The principal zeolite types which are

commercially utilized - primarily as adsorbents, catalysts, ion
exchangers, etc. - are listed in Table 1.

TABLE 1

ZEOLITE TYPES IN COMMERCIAL APPLICATIONS

Zeolite Minerals

    Mordenite

    Chabazite

    Erionite

    Clinoptilolite

Synthetic Zeolites

A	Na, K, Ca forms
X	Na, Ca, Ba forms
Y	Na, Ca, NH_4, rare earth forms
L	K, NH_4 forms
Omega	Na, H forms
Zeolon, Mordenite	H, Na forms
ZSM-5	Various forms
F	K form
W	K form

It is evident that only twelve basic types are utilized.
Some, such as F and W are, at present, restricted to a single
application (in this case, ion exchange) as is ZSM-5 in catalysts.
Others, such as zeolite Y, are used extensively in adsorption and
catalysis (22,000 tons were used in cracking catalysts alone in
1977). Some zeolites will soon achieve commodity status such as
zeolite A as a replacement for phosphates in detergents in much
of the world.

Although several zeolite minerals are available in mineable

deposits in relatively high purity, extensive application has not yet been achieved. It is further of interest to note that the important synthetic zeolites such as zeolites A, X, Y, ZSM-5, etc. do not occur as the mineral types. The "large port" mordenite, synthesized as "Zeolon"[8], is not found as the mineral and is consequently favored in catalytic applications. Natural mordenite is limited to adsorption applications.

Structural Chemical Bases for Applications

The major factors which determine the commercial use of zeolites include (1) structural chemistry, (2) availability, and (3) cost. By structural chemistry I include all of the important structural and chemical aspects of the zeolite which determine or control the intended application. These are summarized in Table 2.

TABLE 2

STRUCTURAL CHEMICAL ASPECTS OF ZEOLITES

Framework, topology and composition

Internal voids and channels

Cation exchange properties

Physical properties - particle size, morphology

Structural stability in hostile environments

Structural defects

Framework topology refers to the geometrical array in three-dimensional space of the basic tetrahedral structural units. There are 33 known topologies (38 including the feldspathoid types).[9] Many more are theoretically possible based on known structural principles.[10]

The framework composition controls the framework charge and influences structural stability. Silica-rich zeolites such as mordenite, for example, are more stable in high-temperature and reactive environments. Zeolite "analogues" have been prepared which are essentially pure silica.[11] Framework charge influences the electrostatic fields and thus dramatically alters the interaction of the zeolite with adsorbed molecules. Zeolites showing hydrophobic character are now known.[12]

The internal void volume and accesses are determined fundamentally by the framework topology and secondarily by the presence of non-framework species (water, cations, other occlusives). In general, the internal voids consist of (1) channels - unidirectional or interconnected, or (2) cavity-like voids mutually interconnected through apertures which vary from six to twelve-membered rings of tetrahedra. Again, structures based on larger apertures (18-rings) have been postulated.[13]

Cation exchange properties such as cation selectivity and exchange capacity are controlled by the zeolite structure (composition, cation siting).

Very importantly, although zeolite structures are generally discussed in terms of idealized concepts, several types of structure defects are essential to impart the characteristics essential in many applications and catalysis in particular. These may be summarized as follows:

- Surface hydroxyl groups - Zeolite crystal surfaces must terminate in OH groups due to hydrolysis of the surface cations.

- Stacking defects - Different segments of the framework may be arrayed in more than one arrangement to produce stacking faults which obstruct the intracrystalline channels.

- Occluded ions - During synthesis, ionic and other species may be trapped in the channels and voids. Examples are OH^-, AlO_2^-, R_4N^+ and amorphous silica.

- Hydrolytic dissociation - The alkali metal form of the zeolite may act as the salt of a weak acid and undergo hydrolytic dissociation to produce free Na^+ and OH^- when suspended in water.

- Displaced cations - During thermal treatment or dehydration, poorly coordinated metal cations may be left "stranded" in unstable sites.

- Hydroxyl groups - Hydroxyl groups replace cations in the framework as the result of deammoniation or partial hydrolysis of exchanged polyvalent cations.

- Missing T atoms - Tetrahedral atoms, such as Al, can be removed by chemical or hydrothermal treatments leaving a vacancy.[14, 15]

- Interrupted framework - Some types of framework aluminosilicates have been reported which have specific, unshared oxygen atoms.

The known commercial uses of zeolites in adsorption and catalysis are summarized briefly in Table 3. For the most part, these are reviewed in detail by other participants in this symposium. Uses of zeolites in ion-exchange applications are listed in Table 4. Most of these are proposed or experimental uses at this time. Some will be reviewed briefly in this paper. A summary list of some additional possible applications is given in Table 5. These proposed applications have been suggested or reported by articles in the published literature and in patents.

The purpose of this review is to emphasize the impact that the development of zeolites has had and will have in the future on our whole technology - from the individual consumer to various large industries. All of these applications, for the most part, are based on the known structural chemistry of zeolites. However, the basis for some applications is rather obscure at the present.

TABLE 3

SUMMARY OF ZEOLITE APPLICATIONS

Adsorption

    Regenerative

        Separations based on sieving

        Separations based on selectivity

    Purification

    Bulk separations

    Non-Regenerative

        Drying

        Windows

        Refrigerants

    Cryosorption

Ion Exchange

    Regenerative Processes

        NH_4^+ Removal

        Metal separations, removal from waste water

    Non-Regenerative Processes

        Radioisotope removal and storage

        Detergent builder

        Artificial kidney dialysate regeneration

        Aquaculture - NH_4^+ removal

        Ruminant feeding of non-protein nitrogen (NPN)

        Ion exchange fertilizers

Catalysis[16]

Hydrocarbon conversion

    Alkylation

    Cracking

    Hydrocracking

    Isomerization

Hydrogenation and dehydrogenation

Hydrodealkylation

Methanation

Shape-selective reforming

Dehydration

Methanol to gasoline

Organic catalysis

Inorganic reactions

    H_2S Oxidation

    NH_3 Reduction of NO

    CO Oxidation

    $H_2O \rightarrow O_2 + H_2$

TABLE 4

ION EXCHANGE APPLICATIONS[24]

Present Applications	Advantage
Removal of Cs^+ and Sr^{++} Radioisotopes- Linde AW-500, mordenite, clinoptilolite	Stable to ionizing radiation Low solubility Dimensional stability High selectivity
Removal of NH_4^+ from waste water - Linde F, Linde W, clinoptilolite	NH_4^+ - selective over competing cations
Detergent builder Linde A, Linde X (ZB-100, ZB-300)	Remove Ca^{++} and Mg^{++} by selective exchange No environmental problem

Potential Applications	Advantage
Radioactive waste storage	Same as Cs^+, Sr^{++} removal
Aquaculture AW-500, clinoptilolite	NH_4^+ - selective
Regeneration of artificial kidney dialysate solution	NH_4^+ - selective
Feeding NPN to ruminant animals	Reduces NH_4^+ by selective exchange to non-toxic levels
Metals removal and recovery	High selectivities for various metals
Ion exchange fertilizers	Exchange with plant nutrients such as NH_4^+ and K^+ with slow release in soil

TABLE 5

SOME PROPOSED APPLICATIONS OF ZEOLITES

Adsorption

    New Adsorbents for Sieving

    Hydrophobic Adsorbents

    Gas Storage Systems

    Carriers of Chemicals

Nuclear Industry Applications

Environmental

    Weather Modification

    Solar Energy

Agricultural

    Fertilizers and Soils

    Animal Culture

Consumer Applications

    Beverage Carbonation

    Laundry Detergents

    Flame Extinguishers

    Electrical Conductors

Ceramics

New Catalysts

New Adsorbents for Sieving

Alteration of the molecular sieving character of a zeolite by means of ion exchange is a well-known means for "tailoring" basic types to perform rather unique separations. Change in the size selectivity effects of zeolite A by K^+ or Ca^{++} exchange is well known. New separations are possible by using combinations of different ions.[17] It has been shown that combinations of Zn and K exchange into zeolite A. The separation of trans-but-2-ene and cis-but-2-ene may be possible by using zeolite A with a Zn fraction of 0.6 and K of 0.4. The kinetic diameters of these molecules are 0.45 and 0.48 nm respectively (Figure 1). This

FIGURE 1

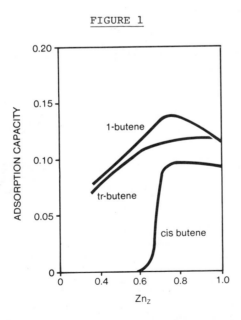

FIGURE 1. Adsorption of unsaturated hydrocarbons on Zn, KA at 50 torr, 0°C. The kinetic diameter of trans-but-2-ene is 0.45 nm and of cis-but-2-ene, 0.48 nm. (From Takaishi et al. ref. 17) Journal of the Chemical Society, Faraday Transactions

material is referred to as "4.5" A implying a zeolite molecular
sieve with an apparent pore size between 4A and 5A. It would
appear that further "fine tuning" of this type may be practical
to achieve other unique separations.

Hydrophobic Zeolites

The structure and composition of zeolites usually result in
pronounced selectivity for molecules which have permanent dipoles
due to the interaction of the zeolite electrostatic field and the
polar molecule. Thus, most zeolites exhibit a high affinity for
water. Recently, however, it has been shown that this affinity
can be reversed by removal of aluminum and, therefore, cations
from the structure.[12] Synthetic mordenite was dealuminized by
acid treatment. As the SiO_2/Al_2O_3 ratio of the zeolite was increased from 10 to about 100, the water adsorption capacity
decreased from 12 to nearly zero weight percent. Adsorption of
cyclohexane, on the other hand, decreased to a constant 5 weight
percent (Figure 2). By gradually altering the framework compo-

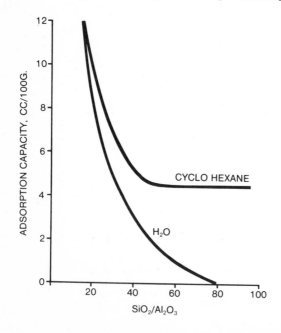

FIGURE 2.
Adsorption of cyclo-
hexane and water on
dealuminized morden-
ite. (Chen, ref. 12)
Journal of Physical
Chemistry

sition to a completely tetrahedral silica framework, the "zeolite" becomes hydrophobic.

This has been confirmed by the synthesis of an all-silica zeolite analogue, silicalite.[11] This crystalline silica polymorph encloses a three-dimensional channel system capable of adsorbing molecules 0.6 nm in diameter such as n-hexane. Comparative isotherms for water and n-hexane are shown in Figure 3. Thus, provided the economics are suitable, this material is capable of removing trace organic compounds from water in the same manner as carbon adsorbents but with the advantage of higher stability to regenerative processes.

FIGURE 3

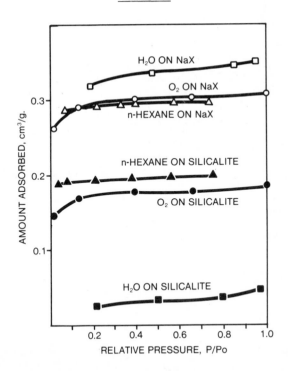

FIGURE 3. Adsorption Isotherms on Silicalite and NaX. O_2 at -183°C, H_2O and n-hexane at RT. (Flanigen et al. reference 11) NATURE

Gas Storage Systems

Under particular conditions, zeolites can encapsulate gases at high pressures and elevated temperatures. Zeolite KA adsorbs substantial quantities of CH_4, argon and krypton at $350^\circ C$ and 2000-4000 atm (Figure 4). When quenched to room temperature,

FIGURE 4

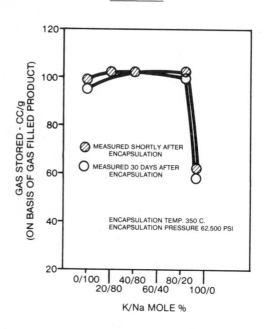

FIGURE 4. Encapsulation of krypton in zeolite A at $350^\circ C$ and 62,500 psi. The cation composition of 40% potassium exchange is optimum for maximum storage capacity and stability. (Sesny and Shaffer. reference 18)

the gases remain trapped and the zeolite-gas system remains stable for very long periods of time. To desorb, the zeolite is destroyed by chemical dissolution or it is heated to a higher temperature.[18] The application of this to a system for storing H_2 has been reported by Fraenkel.[19] Using a CsA ($Cs_z = 0.25$) at $200^\circ C$, he was able to store 85 cm^3 (STP) of H_2 per gram at 8700 psi which is equal to about 0.76 wt.% (Figure 5).   Certain alloys

FIGURE 5

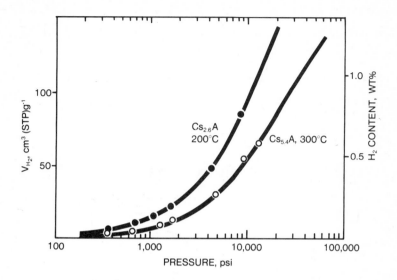

FIGURE 5.   Hydrogen encapsulation in $Cs_{2.6}$ A at 200°C, (●) and $Cs_{5.4}$A at 300°C, (o).   (Fraenkel and Shabtai.   reference 19)
Journal of the American Chemical Society

such as FeTi store ⋏ 1% as the hydride.   Encapsulation of Ar and Kr in the felspathoids synthetic cancrinite and sodalite was observed by Barrer and Vaughan (1971).[20]   At T = 504 K, nearly 40 cm^3 (STP) were encapsulated in a synthetic sodalite at 1000 atm.

Encapsulation of helium to the extent of one weight percent in a Na-Li-Cs A at 40,000 psi and 400°C has been reported.   The system was stable at least for two months at ambient temperature.[21]

Adsorption of methane on high surface area materials such as zeolites is a potential means for increasing the fuel storage capacity of vehicles using natural gas.   Adsorption of methane on zeolites up to pressures of 1000 psig has been reported.[22] Free volume in the system was determined by using a non-adsorbed

fluid - for CaA, carbon tetrachloride was employed. Up to 3.2 g.
of CH_4 was contained in a 75 cm^3 cylinder filled with CaA zeolite
pellets with a total system weight of 63 g. The zeolite NaX has
an adsorption volume of 0.5 cm^3/cm^3 of zeolite. This would correspond to about 10.5 g of CH_4 as the liquid per 100 g of zeolite
(Figure 6). Although this was shown to be preferable to storage
of CH_4 in a cylinder under pressure in terms of weight and volume,
the overall system does not compare to gasoline.

FIGURE 6

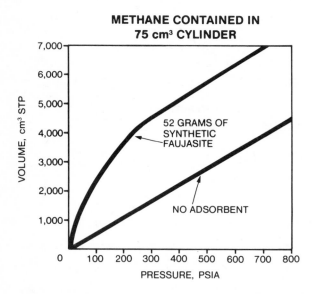

**METHANE CONTAINED IN
75 cm³ CYLINDER**

Figure 6. Methane storage at elevated pressure on zeolite NaX,
pellet form. (Munson and Clifton. reference 22) Bureau of
Mines

Carriers of Reactive Chemicals

In resin and rubber systems, zeolites can be used to isolate
a reactive chemical or catalyst until such time as it is needed
for reaction. In these compositions, referred to as Chemically
Loaded Molecular Sieves, the chemical is isolated by adsorption

within the zeolite cavities. Release, when desired, is achieved
by (a) displacement with water, or (b) heating. Release is
accomplished over a narrow temperature range. The release
temperature can be modified by changing the adsorption characteristics of the zeolite. For example, di-tertiary-butyl peroxide
(DTBP) is used as a curing agent in high-temperature polyester
resin cures. When adsorbed on zeolite X, the problems due to
flammability and vapor pressure are slight and it is widely used
as a primary curing agent in certain silicone rubber formulations.
Pot life of a resin catalyst system is prolonged indefinitely.
After use, the zeolite remains in the resin and acts only as a
filler. In other cases, the zeolite may be added to scavenge
a by-product such as HCl from a curing polyvinyl chloride.[23]

Applications in the Nuclear Industry

One of the earliest applications in ion exchange was in the
removal and purification of cesium and strontium radioisotopes.
They offer significant advantages over organic resin ion exchangers because of their stability in the presence of ionizing radiation, low solubility and dimensional stability. Several new processes have been developed and summarized (Table 6).[24] Similarly,
zeolites can be used for the long-term storage of long-lived
radioisotopes by drying the zeolite and sealing in a container.
At high temperature, the zeolite containing the radioisotopes
can be converted to a glass which has an extremely low leach-rate.

The noble gases krypton and xenon are formed by nuclear
fission in a light water reactor. The ^{85}Kr with a half-life of
10.7 years is present long enough to be present during fuel
reprocessing. To date, all ^{85}Kr in reprocessed fuel is released
by the plant off-gas streams to the atmosphere. Thus, methods
of recovery and storage are being developed. One method being
studied is the storage of ^{85}Kr by encapsulation in zeolite

TABLE 6

Cs and Sr RADIOISOTOPE RECOVERY AND PURIFICATION

USING MOLECULAR SIEVE ZEOLITE ION EXCHANGERS[24]

Service	Location	Zeolite	Bed Size	Remarks
^{137}Cs/High Level Wastes	Hanford	Linde AW-500	300 ft^3	First charge treated 3 x 10^6 gallons, then pressure drop increased (almost plugged) due to Al salt precipitation from solution. Second charge treated several million gallons without significant loss in capacity.
^{137}Cs/Rb, K, Na Purification of Product from Linde AW-500 column	Hanford	Large Port Mordenite	1 liter	Pilot plant. (Full scale facility now in operation at Hanford)
^{90}Sr, ^{137}Cs/Low level wastewater from fuel storage basin	National Reactor Testing Station	Clinoptilolite	5.3 ft.^{3} 4 beds	Non-regenerative use. Capacity: $\sim$ 12,000 gals. wastewater treated per cu. ft. of zeolite
^{137}Cs/Evaporator overheads and miscellaneous wastewater	Savannah River	Linde AW-500	9.2 ft^3	Non-regenerative use. Capacity: $\sim$ 23-76,000 gals. of overheads or 8-12,000 gals. of miscellaneous wastewater per cu. ft. Plugging by suspended solids reduced by using zeolite prefilter.
^{137}Cs/Process condensate wastewater	Hanford	Large Port Mordenite		

materials such as sodalite from which leakage has been demonstrated to be very low.[25, 26]

Carbon-14 (half life 5760 years) is released during the dissolution of spent fuel as $^{14}CO_2$. The maximum concentration in the off-gas is approximately 250 ppm. Among the methods proposed for the removal of $^{14}CO_2$ is selective adsorption on a molecular sieve (type 5A). The $^{14}CO_2$ is desorbed and recovered as $CaCO_3$ by scrubbing with a lime slurry. Total annual yield of $CaCO_3$ is 16-160 kg, since the $^{14}CO_2$ is mixed with $^{12}CO_2$ from the air that leaks into the dissolver system. The solid $CaCO_3$ is packaged for storage in metal containers.[27]

Iodine-129 (^{129}I) is produced in nuclear power reactors and then released into the process off-gas as elemental iodine or organic iodides during fuel reprocessing. Since ^{129}I has a 1.6×10^6 year half-life, permanent contamination of the environment occurs. About 330 kg/y of Iodine-129 are released from a plant handling 5 metric tons of uranium per day.

One method for separation and retention of gaseous iodine species is based upon using silver-loaded adsorbents such as a silver-exchanged zeolite, AgX. Retentions of better than 99.98% of I_2 and CH_3I were measured for a variety of conditions up to $1000^{\circ}C$.[28, 29] Regeneration of the silver zeolite loaded with I_2 is accomplished by passing H_2 through the bed at $400 - 550^{\circ}C$ to produce HI and storing the HI on Pb-exchanged zeolite. The lead exchanged PbX chemisorbs the HI at $150^{\circ}C$. The spent PbX containing the chemisorbed HI is sent to a solid waste disposal and the expensive AgX is reused.

Isotopic Enrichment

A unique application in the preparation of curium-247 was reported.[30] A lanthanide or trivalent actinide is first exchanged into zeolite X or Y and the ions fixed by dehydration.

Neutron irradiation produced americium or curium by neutron
capture. The product isotope is ejected from the exchange site
by γ recoil and thus selectively eluted from the zeolite.
Typically, from rare earths products of Am and Cm were obtained
which contained 50% of the isotope and only about one percent
of the target.

Weather Modification

There are several reports that powdered synthetic zeolites
such as zeolite X act as crystallization nuclei for water and
water vapor. For example, zeolite X powder was injected into a
50 1 chamber containing supercooled water vapor and observed to
nucleate ice crystals. It is reported that only the dehydrated
zeolite crystals are effective. Quantities of zeolite used were
small, about 10 milligrams in the 50 liter chamber.[31]

Solar Energy

Partially related in terms of environmental applications is
the proposal to use zeolites in a solid-gas adsorption cooling
system.[32] In this case, using a natural zeolite mineral chabazite
or clinoptilolite, proposals for providing hot water, space heating and cooling were studied. Zeolites typically adsorb about
30 wt.% of gases such as water or ammonia and the isotherms are
not particularly pressure-dependent. The adsorbent is saturated
at very low partial pressure. Therefore, at ambient temperature,
a large quantity of water vapor is adsorbed and when heated most
is desorbed even at high partial pressures. Therefore, the
difference in adsorbed gas between the high- and low-temperature
state depends only slightly on pressure.

Figure 7 illustrates the principle of the system which provides combined heating and cooling. The zeolite is contained in
hermetically sealed panels that are irradiated by the sun. Water
vapor is desorbed during the day and condensed in the condensing

unit, and the liquid is stored below in the storage tank until evening. Heat of condensation is rejected to the external water loop or it is used for providing domestic hot-water and space heating during winter. When there is a demand for heat, hot water is circulated from the storage tank through a coil in the air ducts of the forced air heating system. Excess heat can be rejected to the outside air-cooled coil if needed. Water from the condensate storage is circulated back into the condenser-evaporator unit where it is evaporated from the same surfaces. The vapor is adsorbed on the cool zeolite. The external water loop provides the heat of vaporization which produces chilled water for air conditioning. Chilled water can be similarly stored during the summer.

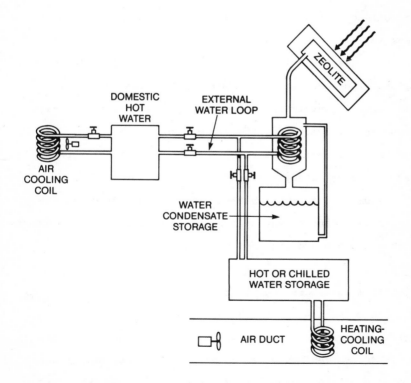

FIGURE 7. Diagram of zeolite combined heating and cooling system. (Tschernev 1978. reference 32)

In an experimental unit consisting of 9 one-foot square panels, a solar input of 20,000 BTU/day produced 9000 BTU/day of cooling. It is estimated to require about 1 ton of zeolite (a natural chabazite zeolite was used) to produce one ton of air conditioning.

Zeolites in Fertilizers and Soils

Zeolites exchanged with the nutrient ions ammonium and potassium can be added to soils as ionic-type fertilizers. The nutrient is released gradually over a long period of time. Organic cation resins introduced for this purpose were found to be uneconomical. Some experiments in our laboratory have indeed shown that NH_4^+ and potassium-exchanged zeolites can function as ionic fertilizers in plant growth. In addition to carrying the primary nutrients, K and N, micronutrients such as iron, copper, manganese and zinc can be supplied by the zeolite. For soil application, relatively impure natural zeolites may be applicable in order to keep the cost low. Additionally, the zeolite may function as a diluent in fertilizer mixtures and as an additive for improving physical properties of the soil. In Japan, it is reported that clinoptilolite rock is added to soils and greatly improves the growth and yield of farm crops (Table 7).[33]

The assumption is made that the enhancement effect is due to the adsorption and retention of ammonia nitrogen and potassium, maintenance of water content and prevention of root decay.

Animal Culture

Zeolite minerals in Japan, in particular clinoptilolite and mordenite, have been added to the diets of pigs, chickens and ruminants.[34] Significant increases in gain of body weight per unit of feed were achieved. In chickens, it was reported that feed efficiency ratios were markedly higher.

TABLE 7

EFFECTS OF CLINOPTILOLITE-TUFF AS SOIL CONDITIONERS

ON YIELDS OF SEVERAL FARM PRODUCTS[33]

Crops	Year	Amount of Zeolite Used (tons/10 ares)*	Ratio of Yield Index** (%)
Wheat	1962	1	113
		2	115
Paddy	1964	0.5	106
		1	103
Eggplant	1964	1	155
		2	119
Apple	1964	1	113
		2	128
Paddy	1965	0.5	102
		1	117
Carrot	1965	1	163
Apple	1965	1	114
		2	110

*10 ares = 0.25 acre

**Compared with control plots

When fed to pigs at levels of 5% clinoptilolite in the diet, the gain in weight of the animals was from 25 to 29% greater than that of animals receiving normal diets. Also, the presence of the zeolite in their diet contributed to the well-being of the animals. Sickness and mortality were considerably reduced. The functions of the zeolites in dietary and antibiotic behavior are not understood. Ammonium cation selectivity of the zeolite indicates that in ruminant animals the zeolite stores nitrogen in the digestive system and results in a slower release and more efficient use. This behavior has been utilized in controlling the level of NH_4 in the rumen of an animal fed urea as a non-protein nitrogen (NPN) source. In ruminant animals, the plants and other feed materials are broken down in the rumen into smaller molecules and amino acids are synthesized. Rumen bacteria can be fed inorganic nitrogen such as ammonia or urea and can convert these NPN sources to amino acids which go to form protein. If NPN is substituted for part or all of the protein in the animal's diet, major economies in the feed cost can be achieved. The ammonia concentrations in the rumen must be kept below a toxic level since urea is hydrolyzed rapidly, releasing NH_4^+.[35]

$$(NH_2)_2CO + 2 H_2O \longrightarrow 2 NH_4^+ + CO_3^=$$

By introducing an ion exchanger into the rumen prior to feeding of urea, the NH_4^+ level is partly reduced by cation exchange and then subsequently released by the Na^+ and K^+ ions which enter the rumen in the saliva. NH_4^+-selective ion exchangers such as zeolite F provide excellent performance in this application.

Zeolite minerals are used to control malodor due to animal wastes. In poultry houses, ammonia and hydrogen sulfide levels can apparently be greatly reduced by use of natural clinoptilolite.

Consumer Applications

Beverage Carbonation

The considerable adsorption capacity of zeolites for carbon dioxide is utilized in a novel method for beverage carbonation at the time of use - rather than bottling under pressure. The zeolite as powder or pellets is loaded with CO_2 and sealed in some suitable container which is impervious to atmospheric moisture. The zeolite may be incorporated in the container to which water is added at the time of consumption. Other variations include teabag and swizzle stick devices, or a cup with the zeolite in the form of a tablet secured to the interior bottom.[36]

Laundry Detergents

The principal divalent ions in water which are harmful in the laundry process are calcium and magnesium.[37] For ideal detergent performance, the Ca^{++} concentration should be below 1×10^{-5} M and the magnesium below 1×10^{-4} M. Although zeolite A (NaA) is known to be very effective as an insoluble ion exchange builder for removal of Ca^{++}, it is not as effective in the removal of magnesium, particularly at ambient temperature due to the slow rate of exchange. Zeolite X exchanges magnesium rapidly. A combination of the two zeolites results in a synergistic effect - zeolite A is more effective in removing Ca^{++} and zeolite X is more effective in the removal of Mg^{++} (Figure 8). It is estimated that the total market for zeolites in heavy-duty powder detergents may be as much as 400 million pounds in 1982.[37]

Zeolites as Flame Extinguishers

A zeolite powder may function as a dry solid flame extinguisher by one or more of the following actions:

a. Blanketing effect by the release of an adsorbed material such as H_2O vapor or CO_2.

b. Produces a cooling effect. One kilogram of zeolite 13X can release 14.4 moles of water vapor and absorbs a quantity of heat equal to the integral heat of adsorption of 200 K cal/kg.

c. Mechanically effects a reduction of heat flow from the flame to the cold combustible.

   d. Can release a chemical flame inhibitor, such as an
      organic halide, which acts as a free radical chain-
      breaking species.

   e. On burning pyrophoric metals, the zeolite may decompose
      to form a glass which results in a smothering action.

<u>FIGURE 8</u>

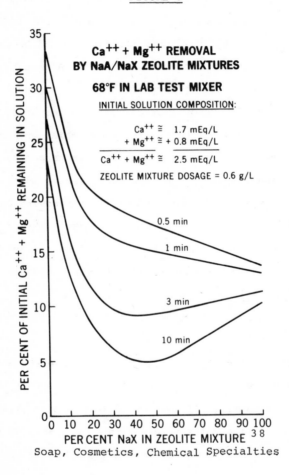

Ca^{++} + Mg^{++} REMOVAL
BY NaA/NaX ZEOLITE MIXTURES
68°F IN LAB TEST MIXER
INITIAL SOLUTION COMPOSITION:

Ca^{++} ≅ 1.7 mEq/L
+ Mg^{++} ≅ + 0.8 mEq/L
Ca^{++} + Mg^{++} ≅ 2.5 mEq/L
ZEOLITE MIXTURE DOSAGE = 0.6 g/L

PER CENT OF INITIAL Ca^{++} + Mg^{++} REMAINING IN SOLUTION

PER CENT NaX IN ZEOLITE MIXTURE [38]

Soap, Cosmetics, Chemical Specialties

Limited studies done in Union Carbide laboratories confirmed
that zeolites can act as solid extinguishers.[39] Economics, however, may be unfavorable. Some laboratory results are given in
Table 8 which compares an established dry chemical extinguisher

with various zeolites in their ability to extinguish a six-inch
diameter flame of burning fuel oil. The time required for total
extinguishment and weight of powder dispensed were measured.
The powder dispenser consisted of a simple device made from a
glass container with a venturi-type nozzle connected to a refrig-
erant 12 (CF_2Cl_2) reservoir as propellant. The flame, after
ignition, was allowed to burn one minute to reach maximum temperature.

TABLE 8

COMPARISON OF ZEOLITES AS FLAME EXTINGUISHERS[39]

Powder	Weight dispensed, g.	Time, sec.	Comments
Dry chemical	15	1.0	Good powder flow
AW-500, hydrated	20	2.5	Good powder flow
CaX powder, hydrated	10	2.5	Good powder flow
4A powder, hydrated	19	5.0	Some powder caking
Kaolin-type clay	Failed		Good powder flow
Attapulgite-type clay	Failed		Good powder flow
Clinoptilolite, hydrated	25	5.0	Good powder flow
13X with 15% HCl	11	10	Poor powder dispensing
AW-500 with 20% CH_3Br	8	1.5	Poor powder dispensing

Various hydrated zeolites appear to be roughly equivalent to
a commercial dry chemical. Chemical loaded zeolites (20 wt.%
methyl bromide on a natural chabazite zeolite, AW-500 (chabazite)
and 15 wt.% HCl on zeolite 13X) were very effective. More work
is needed to confirm the effectiveness and economics of this
application.

Zeolites as Electrical Conductors

Mobile cations located in zeolite channels impart ionic
electrical conductivity.[40] The conductivity depends on the compo-

sition, cation type and degree of hydration. Potential use of
this property in solid state batteries and conducting paper has
been described.

Zeolite-based primary cells can be made which have open
circuit voltages up to 2.0 volts and are capable of delivering
current in the microampere range from -78° to about 500°C. The
cell consists of a "catholyte" of silver, mercury or copper
exchanged zeolite X bonded to a sodium X or calcium A separator
which in turn is bonded to a magnesium, zinc or aluminum foil
anode. The cells are compacted into hard discs with soldered
lead wires. Zeolitic water enhances performance at ambient
temperatures but is not necessary at high temperatures. Experimentally, cells have shown good performance under continuous load
for several weeks.

Zeolite powders coated or incorporated in paper increase
the conductivity such that the paper is suitable for use in an
electrostatic copying process. Normally, paper has a specific
resistance of 1×10^{12} ohm cm or greater; a resistance of 1×10^7
to 1×10^8 ohm cm is desired. The zeolite powder can be coated
on the paper or incorporated (5-30 wt.%) in the paper during
manufacture. Resistivities within the desired range are readily
attained.[41]

It has been reported that certain ion-exchanged forms of
zeolites can be used in the fabrication of electrodes for fuel
cells.[42] Catalytic ions such as nickel are first exchanged into
the zeolite powder which is then coated onto a ceramic cylinder.
In other instances, the zeolite is first shaped by bonding with
a clay binder into a cylinder and then calcined at 500°C. The
nickel is then incorporated by ion exchange and then the electrode
is heated at 1200°C to stabilize the nickel ions. Actually, the
zeolite structure is destroyed and the nickel becomes unexchange-

able (which is required for actual use in a fuel cell).

Ceramics - Starting Materials

Jadeite, a pyroxene ($NaAlSi_2O_6$) is noted as a gem stone and for its physical properties such as hardness and toughness. It is a high-pressure mineral of density 3.33 to 3.35 g/cm^3. Zeolite Y is an excellent starting material, due to its chemical composition, for jadeite synthesis. Introduction of iron as Fe III produces a material with a green color. At high pressure and temperature, zeolite Y converts to jadeite. The reaction:

$$Na(AlO_2)(SiO_2)_2 \cdot 4\ H_2O \longrightarrow NaAlSi_2O_6 + 4\ H_2O$$

Zeolite Y

occurs at $500^{\circ}C$ and 2000 M Pa (20 kilobars).[43]

Zeolite X in the magnesium-exchanged form converts to the well-known ceramic material cordierite after first melting to the glass at $1600^{\circ}C$ and then being crystallized at $1000^{\circ}C$.

$$Mg_2(AlO_2)_4(SiO_2)_5 \xrightarrow[\ 1000^{\circ}C\ ]{\ 1600^{\circ}C\ } Mg_2Al_4Si_5O_{18}$$

MgX   Cordierite

Applications of zeolites as starting materials in ceramics should be advantageous since compositions can be altered through ion exchange. Conversion of MgX to cordierite would probably be accelerated by introducing appropriate nucleating agents such as dispersed TiO_2.

The intumescent behavior of zeolites, as originally observed by Cronstedt, has been utilized in the preparation of a low density porous ceramic. Polyvalent cation forms of zeolites, such

as Ca^{++}-exchanged A, contain hydroxyl groups equivalent to 3%
water. Upon fusing rapidly at high temperatures, the hydroxyl
water is released and produces a foaming action. When heated
rapidly to $1100^{\circ}C$ in a container, calcium-exchanged zeolite A
powder was converted to a product of 85 volume percent porosity
with 75 percent of the macropores in the 5 to 30 μm range (equal
to about 2.4 cm^3/g). Bulk densities were 0.2 - 0.25 g/cm^3.
Crystallographically, the product is similar to the feldspar
anorthite.[44]

New Catalyst Applications

Thermochemical Decomposition of Water

A two-step thermochemical cycle for the decomposition of
water using the strong ionizing property of zeolites has been
reported.[45] Zeolites, for example, will induce electron transfer
reactions between molecules and reduction of multivalent cations
in zeolites by adsorbed molecules. A reaction cycle as shown
below using the Cr^{3+}/Cr^{2+} couple in mordenite was devised. Upon
dehydration of the Cr III mordenite, reduction to Cr II occurs
with the evolution of oxygen. Upon rehydration, the divalent
chromium oxidizes to Cr III with the evolution of hydrogen.
Similar results were obtained with an In^{3+} exchanged mordenite.
The starting mordenite was a hydrogen form of the synthetic type,
Zeolon.

$$\{M^{n+} + mH_2O\}_{Zeolite} \xrightarrow{\Delta\ T} \{M^{(n-1)+} + H^+\}_{Zeolite} + (m-\tfrac{1}{2})H_2O + \tfrac{1}{4}O_2$$

$$\{M^{(n-1)+} + H^+\}_{Zeolite} + mH_2O \longrightarrow \{M^{n+} + mH_2O\}_{Zeolite} + \tfrac{1}{2}H_2$$

New, high-silica zeolites referred to as the ZSM-5 type have pore system sizes between the "small pore" and the "large pore" zeolites. Straight chain monomethyl-substituted paraffin hydrocarbons are adsorbed at room temperature faster than dimethyl substituted hydrocarbons. Consequently, it exhibits a molecular sieving effect among molecules such as alkyl benzenes and alkyl toluenes as well as paraffins with different branching degrees and chain lengths.

Perhaps the most recent developments in catalysis are due to the properties of the ZSM-5 zeolite.[46] A process for converting methanol to gasoline is based on the general reaction:

$$xCH_3OH \longrightarrow (CH_2)_x + xH_2O$$

Hydrocarbons produced are in the gasoline boiling range C_4 to C_{10} and the gasoline produced has RON of 90 to 100. In a single pass, 99 percent conversion is reported with 80 percent of the hydrocarbon product in the gasoline range. Of course, 56% of the methanol is converted to water. This process can be an important step in the conversion of coal to gasoline via gasification and conversion to methanol.[47, 48]

The ZSM-5 zeolite is also used as a catalyst in benzene alkylation to make ethyl benzene, in removing n-paraffins from fuel oil, dewaxing, and in isomerizing xylenes to para-xylene.

Summary

This is a broad summary of some potential applications of synthetic zeolites and zeolite minerals. Space does not permit an encyclopedic treatment. However, an attempt is made to illustrate the applicability of utilizing zeolites to solve, or improve the solution of, many technological problems of the present as well as of the future. New types of zeolites and variations are being synthesized at present and many more will be discovered in the future. At present, we try to adapt known zeolites to meet a particular application. In the future, hopefully, we will be able deliberately to synthesize a zeolite-type material to meet an application. Initially, zeolites were used in industrial applications involving separations, followed by important catalytic uses. Recently, they have been applied in solving environmental problems and today it appears that they will certainly enter the arena of consumer products.

REFERENCES

[1] F. A. Mumpton, Short Course Notes, Mineral. Soc. Am., 1977, 4, 177.

[2] A. F. Cronstedt, Akad. Handl. Stockholm, 1756, 17, 120.

[3] H. Eichhorn, Poggendorf Ann. Phys. Chem., 1858, 105, 126.

[4] R. N. Shreve, "Greensand Bibliography to 1930", U.S. Bur. Mines, Bull. 328, 1930.

[5] D. W. Breck, "Zeolite Molecular Sieves", Wiley-Interscience, New York, 1974, Chapter 1, p. 11.

[6] R. M. Barrer, J. Soc. Chem. Ind., 1945, 64, 130.

[7] R. M. Milton, Mol. Sieves, Pap. Conf., 1967, (Pub. 1968), 199.

[8] L. B. Sand, Mol. Sieves, Pap. Conf., 1967, (Pub. 1968), 71.

[9] W. H. Meier and D. H. Olson, "Atlas of Zeolite Structure Types", International Zeolite Association, Zurich, 1978.

[10] Ref. 5, Chapter 2, p. 47.

[11] E. M. Flanigen, J. M. Bennett, R. W. Grose, J. P. Cohen, R. L. Patton, R. M. Kirchner, and J. V. Smith; Nature (London), 1978, 271, 512.

[12] N. Y. Chen, J. Phys. Chem., 1976, 80, 60.

[13] R. M. Barrer and H. Villiger, Z. Kristallogr, 1969, 128, 352.

[14] G. T. Kerr, Advan. Chem. Ser., 1973, 121, 219; Acta Phys. Chem., 1978, 24, 169.

[15] D. W. Breck and G. W. Skeels, 1977, Proc. Int. Congr. Catal., 6th, 1976, 2, 645.

[16] J. A. Rabo, R. D. Bezman, M. L. Poutsma, Acta Phys. Chem., 1978, 24 (1-2), 39.

[17] T. Takaishi, Y. Yatsurugi, A. Yusa, and T. Kuratomi, J. Chem. Soc., Faraday Trans. 1, 1975, 71, 97.

[18] W. J. Sesny and L. H. Shaffer, 1967, U.S. Patent 3, 316,691.

[19] D. Fraenkel and J. Shabtai, J. Am. Chem. Soc., 1977, 99, 7074; Proc. Condens. Pap. Miami Int. Conf. Alternative Energy Sources, 1977, 557.

[20] R. M. Barrer and D. E. W. Vaughan, J. Phys. Chem. Solids, 1971, 32, 731.

[21] L. E. Mosher, Private Communication.

[22] R. A. Munson and R. A. Clifton, Jr., Bureau of Mines, Non-Metallic Minerals Program Technical Report No. 38, 1971.

[23] R. J. Neddenriep, Adhesives Age, 1966, 9, 23.

[24] J. D. Sherman, Adsorption and Ion Exchange Separations, AIChE Symposium Series, 1978, 74, No. 179, 98.

[25] C. L. Bendixen and Knecht, Proceedings of the International Symposium on the Management of Wastes from the LWR Fuel Cycle, July 11-16, 1976, 343.

[26] B. A. Foster, D. T. Pence, and B. A. Staples, Proc. AEC Air Clean. Conf., 13th, 1974 (Pub. 1975), 293.

[27] R. A. Brown, Proceedings of the International Symposium on the Management of Wastes from the LWR Fuel Cycle, July 11-16, 1976, 364.

[28] W. J. Maeck, D. T. Pence, J. H. Keller, U.S. Atom. Energy Comm. 1968, IN-1224.

[29] C. Donner and T. Tamberg, Z. Naturforsch. A, 1972, 27, 1323.

[30] D. O. Campbell, Advan. Chem. Ser., 1973, 121, 281.

[31] L. Krastanov, N. Genadiev and L. Levkov, Dokl. Bolg. Akad. Nauk, 1970, 23, 1071; ibid, 1976, 29, 1281.

[32] D. I. Tchernev, in "Natural Zeolites, Occurrence, Properties, Use", L. B. Sand and F. A. Mumpton, Ed. 1978, Pergamon, p. 479.

[33] K. Torii, in "Natural Zeolites, Occurrence, Properties, Use", L. B. Sand and F. A. Mumpton, Ed., 1978, Pergamon, p. 441.

[34] F. A. Mumpton and P. H. Fishman, J. Anim. Sci., 1977, 45, 1188.

[35] J. L. White and A. J. Ohlrogge, Can. Patent 936,186, 1974.

[36] R. L. Sampson and D. D. Whyte, U.S. Patent 3,888,998 (1975).

[37] Chem. and Eng. News, 1978, May 22, 11.

[38] J. D. Sherman, A. F. Denny, A. J. Gioffre; Soap, Cosmet. Chem. Spec., 1978, December, 33; U.S. Patent 4,094,778 (1978).

[39] F. M. O'Connor, Union Carbide Corp., Unpublished Results.

[40] D. C. Freeman, Jr., U.S. Patent 3,106,875, (1965).

[41] D. W. Breck, U.S. Patent 3,884,687, (1975).

[42] A. W. Moos, U.S. Patent 3,097,115 (1963).

[43] Ref. 5, p. 497, 494.

[44] W. H. Flank, J. E. McEvoy, J. R. Stuart; U.S. Patent 3,574,647 (1971); W. H. Flank, U.S. Patent 3,775,136 (1973).

[45] P. H. Kasai and R. J. Bishop, Jr., J. Phys. Chem., 1977, 81, 1527; U.S. Patent 3,963,830 (1976).

[46] N. Y. Chen and W. E. Garwood, J. Catalysis, 1978, 52, 453.

[47] S. L. Meisel, J. P. McCullough, C. H. Lechtaler and P. B. Weisz, Chem. Tech., 1976, 6, 86.; C. D. Chang and A. J. Silvestri, J. Catalysis, 1977, 47, 249.

[48] C. D. Chang, W. H. Lang, A. J. Silvestri, J. Catalysis, 1979, 56, 268.

Hazards of Fibrous Mineral Dusts including Zeolites

P.C. Elmes

Pneumoconiosis Research Unit, Llandough Hospital,

Penarth, Glamorgan, U.K.

Introduction

Human lungs are well defended against dust particles which arise naturally. Particles greater than 5µm in diameter are caught on the mucus lining the nose, trachea and bronchi. The mucus is moved by ciliary action and when it reaches the larynx it is swallowed. This system takes approximately twenty minutes to clear dust from the most remote parts of the bronchial tree, and it is not easy to overload it. When subjected to continuous heavy loading in people who are constantly or repeatedly exposed to dust, the system responds by increasing the rate of mucus production, causing the symptom complex known as chronic bronchitis. This is well illustrated by the effect of smoking cigarettes.

As the dust particle size falls below 5µm an increasing proportion of particles get past the airway filtration system to the alveolar part of the lung where gas exchange takes place. Here the larger particles (> 0.5 µm) are trapped but many of the smaller ones remain in suspension and are breathed out again. Trapped particles are picked up by special scavenging cells called macrophages which when fully laden carry the dust to the airways either directly or indirectly. The dust-laden macrophages are then cleared by the muco-ciliary system in the same way as the larger dust particles.

This system can be overloaded in two ways:
1. If an inert dust is inhaled in high concentration, the macrophages fail to reach the airway but collect in groups round the endings of the airways in the centre of each lung lobule. Some reach the lymph channels around the blood vessels or on the surface of the lung and then move on to the lymph nodes where the lung is attached to the heart and trachea. Entirely inert dust may remain in these sites for the rest of a lifetime. Others may dissolve gradually and disappear or, like crystalline silica, may gradually destroy the macrophages and lead to the slow development of scars which cause a stiffening of the lung and interference with its function.
2. If the particles are elongated,but still have a diameter of less than 3 to 5 µm, they may by-pass the airways filter system even though they may be 50 or even 100 µm in length.

This difference between fibrous and non-fibrous mineral dusts was not

appreciated until people began to study the problems arising from asbestos
exposure.

The Hazards of Asbestos

When asbestosis was first studied in the early 1930's it appeared to be
similar to silicosis, in that at relatively high levels of exposure workers
developed a progressive scarring of the lung leading to death from respiratory
failure within ten to fifteen years of first starting dusty work. The situation
was similar in severity to that found in workers exposed to freshly fractured
silica, but although in silicosis large solid scars appeared in the lungs, in
asbestosis the scarring was diffuse, forming a honeycomb around islands of
surviving lung. As time went by further differences became apparent:
1.   In the late 1940's an excess of cancer of the lung was noted in workers
suffering from asbestosis who survived middle age. Primary lung cancer now
affects well over 50% of those men who have lung scarring due to asbestosis and
who also smoke. Lung cancer is uncommon in non-smoking asbestos workers
although even in them it is more common than in non-smoking men who also do not
work with asbestos.
2.   In the late 1950's a rare cancer of the lining of the chest wall was recognised to be associated with past exposure to asbestos in some cases. Since
then this cancer has become increasingly frequent in industrial countries. In
Britain at present there is a background ("normal"?) incidence of this cancer
of about one case for every two to three million of the population per year.
In total, there are between two and three hundred cases per year in Britain,
and over 80% of these are associated with a past exposure to asbestos at work.
Cases associated with non-occupational exposure are rare in this country. The
tumour is called a mesothelioma; it sometimes arises in the peritoneum rather
than the chest. It does not respond well to surgery or other forms of treatment, and is therefore almost invariably fatal.

During the investigation of occupations leading to mesothelioma it became
apparent that in some parts of the asbestos industry the tumour was rare in
spite of a heavy and prolonged exposure (sufficient to cause asbestosis and
primary lung cancer) whereas in other parts of the industry it was relatively
common, even after exposure too brief or too trivial to cause any asbestosis.
Miners and millers of chrysotile asbestos in Canada and Swaziland show a low
incidence of mesothelioma, so low that it is hard to be sure there is a significant increase above the "natural" incidence. Most asbestos processors and
insulation workers in Britain had a mixed exposure to crocidolite (blue
asbestos) and chrysotile (see Plates 1a and 1d) whereas in the United States
most of the exposure was to chrysotile. The incidence of mesothelioma resulting

(a) CROCIDOLITE   (b) AMOSITE

(c) ANTHOPHYLLITE   (d) CHRYSOTILE

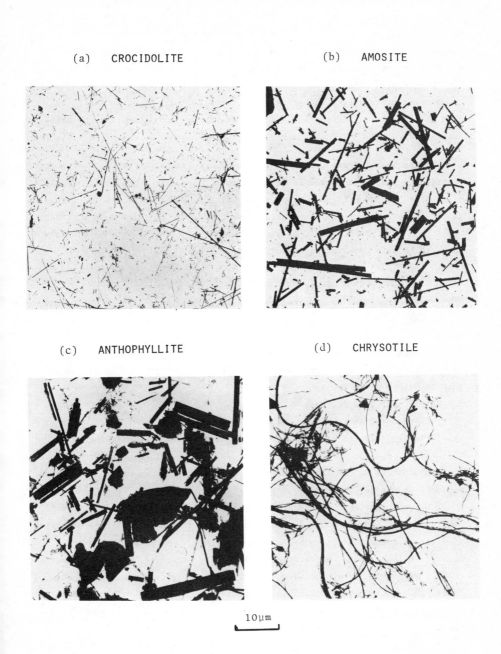

10μm

Plate 1.   A comparison of particle size distributions
            and morphologies of different fibrous minerals.

from industrial exposure is lower in the U.S.A. than in Britain, and evidence
is slowly accumulating indicating that it is crocidolite which is most likely
to cause mesothelioma. The incidence of mesothelioma is high amongst South
African crocidolite miners and in factory workers who processed a lot of this
material in Britain. Incidence is very high (up to 30%) in workers who mined
blue asbestos in Western Australia, and amongst workers who used this material
to make gas masks during the early part of the 1939–45 war, even though the
dust levels were relatively low and the exposure in many cases was for less
than a year.

Environmental cases have been reported in relation to the South African
crocidolite and are frequent in relation to the Australian material. Due to
mixed exposure and difficulties over nomenclature it is difficult in retrospect
to get a clear picture of the risk involved in using the third most important
commercial asbestos material, amosite (Plate 1b). Anthophyllite (Plate 1c)
seems to be safe but fibrous tremolite may not be.

Investigations into the Cause of Mesothelioma Production

Fundamental studies on the mechanism of mesothelioma production by these
fibrous minerals has been going on since 1961 and, briefly, the results may be
summarised as follows:

1. The mechanism of the carcinogenesis is still not known. It does not appear
to be due to a chemical carcinogen within the mineral fibre or to one adsorbed
onto it before it is inhaled.

2. The current theories are:

(a) that the asbestos adsorbs environmental carcinogens during or after
inhalation and acts as a focus for their action in the lung over the years.
This may be true for the primary lung cancers and the carcinogens from cigarette
smoke which have a mean latent interval of 35 years, but is unlikely to be an
explanation for mesothelioma;

(b) that the fibre damages the body's normal mechanism for dealing with
spontaneous malignant change in the tissue where it lodges. This is an attractive theory but difficult to prove or disprove.

3. Cell culture tests for carcinogenicity using bacterial, animal and human
cell lines do not reveal the premalignant responses from these carcinogenic
fibres that most chemical carcinogens produce. Analog cell culture responses
may prove useful but are as yet unreliable. They do not help in elucidating
the mechanism of the carcinogenesis.

4. Studies of the fibres in human tissues after death, and in their distribution after experimental exposure of animals, indicate that:

(a) the fibres at the long end of the respirable range, especially if they are curly (like chrysotile Plate 1d) tend to be trapped in the lung close to the centres of the lobules and remain there indefinitely;

(b) intermediate length fibres in the 5 - 20 µm length range are widely distributed in the lung, and tend to get coated to form "asbestos bodies" which only reach the pleura and hilar lymphnodes in small numbers;

(c) short straight fibres in the 3 - 10 µm range which are usually very thin (< 0.5 µm) migrate to the pleura and hilar lymph nodes in relatively larger numbers. Smaller fibres than this are probably cleared like non-fibrous dust;

(d) in so far as they have been tested, ceramic, glass and other synthetic and natural mineral fibres behave and migrate in the same way as asbestos.

5. Mesotheliomas can be produced experimentally with great regularity by injecting suspensions of fibres into the pleura or peritoneum. The most potent materials are fibres in the length range 7 - 15 µm and with a diameter range of 0.2 - 0.5 µm, but of course these samples always contain fibres both above and below these size ranges. Chrysotile fibres within these size ranges are not found frequently in airborne dust encountered in industry, and for experimental purposes they were produced deliberately by milling. The milled fibres were as potent in causing mesothelioma as crocidolite fibres, as were some synthetic glass and ceramic fibres and some natural minerals of the sepiolite group, provided they are in the relevant size range.

Animal studies are time-consuming and expensive, and it will be some time before all the relevant natural and synthetic mineral fibres have been tested. In so far as they are valid, the analog cell culture studies tend to confirm the importance of fibre size as opposed to chemical constitution.

Current Research

Due to the importance of this work in guiding occupational preventive medicine services, it is not ethical to wait for the outcome of the animal work, or to allow workers to be exposed to these fibrous dusts and wait for the 40 years latent period to learn the outcome. Two lines of research are being pursued:

1. The lung dust of many cases of mesothelioma is being examined by electron microscopy using EDAX to see if fibres other than asbestos can be causing the cases in Britain that were previously thought to be naturally occurring. This study should be completed this year.

2. Reports of mesotheliomas not attributable to asbestos in other countries are being investigated.

We have investigated reports from India and Turkey. There is as yet no indication that the Indian causes are due to a local environmental factor.

However, in Turkey there exists a complex situation both from a medical and a
mineralogical point of view. The upland central part of Turkey (Anatolia) is
snow covered in winter and hot and dusty in summer. It supports a sparse
agriculturally based population which has only recently had the benefit of
centralised medical services or been subject to mineral exploration and development.

Initial medical surveys in the early 1960's suggested that tuberculous
disease of the pleura was widespread, and it was not until preventive and
therapeutic programmes against tuberculosis proved ineffective that it was real-
ised that the diseases were not tuberculosis. It was recognised by Baris[1] and
others that both the malignant and non-malignant pleural diseases which were
prevalent were similar in many ways to the conditions we associate with
asbestos exposure. However, although outcrops of asbestos exist (and are in
some places worked commercially) these did not account for the pleural disease
in all the areas. Prof. Baris sought the help of the International Agency for
Research in Cancer, and after discussion with various experts, that agency is
helping Prof. Baris pursue his investigation by concentrating on the problem
in one particular village called Karain. Here the problem is particularly
severe and yet the possibility of asbestos exposure is remote. The study is by
no means complete, but it has already been established that in this village
more that 50% of the adult population has died of mesothelioma of the pleura
over a period of many years. Neighbouring and apparently similar villages are
not affected and there is no sociological, occupational or racial reason to
account for this. The communities are agricultural, and cultivate a soil
derived from a soft rock of volcanic origin. This rock is excavated for dwell-
ings and storage caves and the cut stone is used for building. The airborne
dust in these villages seems to originate from this rock and soil rather than
from the occasional outcroppings of basalt which are used uncut in building.
The soft rock in this area contains several zeolites including chabazite and
erionite. The erionite is in fibrous form (Plate 2b) but in many cases the
fibres are too coarse to form an airborne respirable dust. However, so far,
in this area, the only respirable mineral fibre found in the airborne dust
samples which is of the right size to cause mesothelioma has been erionite.
It has been found in airborne dust, street dust and in the soil of the affected villages but has not been found in the other villages. Coarser (non-respir-
able) erionite has been found in non-affected villages. No concentrated source
of fine erionite fibres has yet been found in or near Karain. The fibre is
present in low concentration (considerably less that 1%) in airborne dust and
in an even lower concentration in the stone, dust and soil. It has not so far
been possible to collect a pure enough sample to test this Turkish erionite in

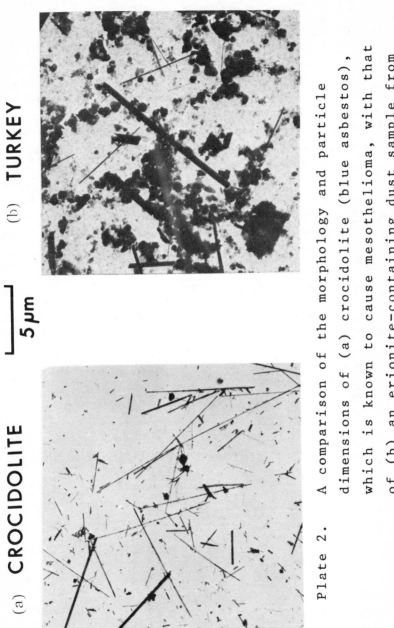

(a) CROCIDOLITE

(b) TURKEY

5 μm

Plate 2. A comparison of the morphology and particle dimensions of (a) crocidolite (blue asbestos), which is known to cause mesothelioma, with that of (b) an erionite-containing dust sample from Karain in Turkey.

the whole animal or cell culture systems mentioned earlier.

The disease in Karain affects adults of either sex in middle age, *i.e.* fifteen to twenty years younger than the occupational cases occurring in Britain. This suggests a childhood rather than an adult or occupational exposure. There are also cases of chronic pleural inflammation and thickening (fibrosing pleuritis) which can be temporarily relieved by surgery. Such cases are rare in Britain. Religious and local beliefs in Karain make the application of modern surgical techniques difficult, and autopsies almost impossible. Lung dust studies, such as we are carrying out in Britain and which would solve this problem quickly, cannot be carried out. We are left with slow and indirect techniques of epidemiology and environmental monitoring to solve a local problem far more serious than has been caused by asbestos in modern industry. However, it has relevance for modern industry in that the findings may indicate that mineral fibres other than asbestos can cause mesothelioma in man, just as they have been shown to cause mesothelioma experimentally by injection into animals.

References

1.   Baris *et al.*, "An Outbreak of Pleural Mesothelioma and Chronic Fibrosing Pleurisy in the Village of Karain/Urgup in Anatolia", *Thorax*, 1978, 33, 181.